APPLIED
COMBINATORICS

Applied Combinatorics
Second Edition

Alan Tucker

Department of Applied Mathematics and Statistics
State University of New York at Stony Brook

John Wiley & Sons

New York   Chichester   Brisbane   Toronto   Singapore

Library of Congress Cataloging in Publication Data

Tucker, Alan, 1943–
  Applied combinatorics.

  Includes index.
  1. Combinatorial analysis. 2. Graph theory. I. Title.
QA164.T83   1984   511'.6   84-7393
ISBN 0-471-86371-8

Printed in the United States of America

10 9 8 7 6 5 4 3 2

Preface

Combinatorial reasoning underlies all analysis of computer systems. It plays a similar role in discrete operations research problems and in finite probability. Two of the most basic mathematical aspects of computer science concern the speed and logical structure of a computer program. Speed involves enumeration of the number of times each step in a program can be performed. Logical structure involves flow charts, a form of graphs. Analysis of the speed and logical structure of operations research algorithms to optimize efficient manufacturing or garbage collection entails similar combinatorial mathematics. Determining the probability that one of a certain subset of equally likely outcomes occurs requires counting the size of the subset. Such combinatorial probability is the basis of many nonparametric statistical tests. Thus, enumeration and graph theory are used pervasively throughout the mathematical sciences.

This book teaches students in the mathematical sciences how to reason and model combinatorially. It seeks to develop proficiency in basic discrete math problem solving in the way that a calculus textbook develops proficiency in basic analysis problem solving.

The three principal aspects of combinatorial reasoning emphasized in this book are: the systematic analysis of different possibilities, the exploration of the logical structure of a problem (e.g., finding manageable subpieces or first solving the problem with three objects instead of n), and ingenuity. Although important uses of combinatorics in computer science, operations research, and finite probability are mentioned, these applications are often used solely for motivation. Numerical examples involving the same concepts use more interesting settings such as poker probabilities or logical games.

Theory is kept to a minimum. It is always first motivated by examples, and proofs are only given when their reasoning is needed to solve applied problems. Elsewhere, results are stated without proof, such as the form of solutions to various recurrence relations, and then applied in problem solving. Occasionally, a few theorems are stated simply to give students a flavor of what the theory in certain areas is like.

The Mathematical Association of America recently formed a Panel on a General Mathematical Sciences Program to redefine the MAA's model mathematical major to be a mathematical sciences major. This panel recommended that all majors should study combinatorial problem solving. The discrete structures course in the Association of Computing Machinery's Curriculum 78 covers set theory, Boolean algebra, propositional logic, some modern algebra, and basics of counting and graph theory. This course is far too ambitious and has become a source of controversy. The limited amount of applied set theory, modern algebra, and logic in this book would be sufficient mathematical preparation in these areas for most computer science students, when supplemented with weekly practice in induction proofs (see Appendix A.2). More importantly, basic skills in combinatorial reasoning and problem solving are, to this author, more fundamental mathematics for computer scientists.

This book is designed for use by students with a wide range of ability and maturity (sophomores through beginning graduate students). The stronger the students, the harder the exercises that can be assigned. The book can be used for a one-quarter, two-quarter, or one-semester course depending on how much material is used. It may also be used for a one-quarter course in applied graph theory or a one-semester, one-quarter course in enumerative combinatorics (starting from Chapter 5). A typical one-semester undergraduate discrete methods course should cover most of Chapters One to Three and Five to Eight, with selected topics from other chapters if time permits.

Instructors are strongly encouraged to obtain a copy of the instructor's guide accompanying this book. The guide has an extensive discussion of common student misconceptions about particular topics, hints about successful teaching styles for this course, and sample course outlines (weekly assignments, tests, etc.).

The second edition of this book puts the graph theory chapters first, as recommended by the MAA Mathematical Sciences Panel. This edition also gives more attention to computer science's use of combinatorics: a new chapter (Ten) on topics in theoretical computer science, a new section (7.3) on recursive programs, an enlarged discussion of algorithms to generate combinatorial sets, and additional programming exercises throughout the text.

Many people gave useful comments about early drafts and the first edition of this text; Jim Frauenthal and Doug West were especially helpful. The idea for this book is traceable to a combinatorics course taught by George Dantzig and George Polya at Stanford in 1969, a course for which I was the grader. Ultimately, my interest in combinatorial mathematics and in its effective teaching rests squarely on the shoulders of my father, A. W. Tucker, who has long sought to give finite mathematics a greater role in mathematics as well as in the undergraduate mathematics curriculum. Finally, special thanks go to former students of my combinatorial mathematics courses at Stony Brook. It was *they* who taught *me* how to teach this subject.

Alan Tucker
Stony Brook, New York

Contents

APPLIED
COMBINATORICS

PART ONE
GRAPH THEORY

Chapter One
Elements of Graph Theory

1.1 GRAPH MODELS

The first four chapters deal with graphs and their applications. A **graph** $G = (V,E)$ consists of a finite set V of **vertices** and a set E of **edges** joining different pairs of distinct vertices. Figure 1.1 shows a depiction of a graph with $V = \{a,b,c,d\}$ and $E = \{(a,b),(a,c),(a,d),(b,d),(c,d)\}$. We represent vertices with points and edges with lines joining the prescribed pairs of vertices. Sometimes the edges are ordered pairs of vertices, called **directed edges.** We write $(b\vec{,}c)$ to denote a directed edge from b to c. The two ends of an undirected edge can be written in either order, (b,c) or (c,b). We say that vertex a is **adjacent** to vertex b when there is an edge from a to b.

Graphs have proven to be an extremely useful tool for analyzing situations involving a set of elements in which various pairs of elements are related by some property. The most obvious examples of graphs are sets with physical links, such as electrical networks, where electrical components (transistors) are the vertices and connecting wires are the edges; or telephone communication systems, where telephones and switching centers are the vertices and telephone lines are the edges. Road maps, oil pipelines, and subway systems are other examples.

Another natural form of graphs is sets with logical or hierarchical sequencing, such as computer flow charts, where the instructions are the vertices and the logical flow from one instruction to possible successor instruction(s) defines the edges; or an organizational chart, where the people are the vertices

3

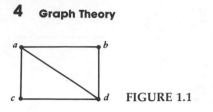

FIGURE 1.1

and if person A is the immediate superior of person B then there is an edge (A,B). Computer data structures, evolutionary trees in biology, and the scheduling of tasks in a complex project are other examples.

Any of the preceding examples can generate enough graph-related questions for a whole book. Some of these questions can be settled quickly by inspection of the associated graph; some require general mathematical theory; and some require very application-specific analysis where the graph models are only of limited conceptual help. In the chapters on graph theory, we will survey instances of all three types of questions.

The emphasis will be on problem solving, with problems about general graphs and applied graph models. Our problem solving will rely on logical reasoning and clever insights. Observe that we will not have any numbers to work with, only some vertices and edges. At first, this may seem to be highly nonmathematical. It is certainly very different from the mathematics that students learn in high school or in calculus courses. However, computer science and operations research contain as much graph theory as they do standard numerical mathematics.

Let us consider some problems in which graphs provide a natural mathematical model. A **path** is a sequence of vertices $(x_1, x_2, \ldots, x_n)$ such that x_i is adjacent to x_{i+1}. A **circuit** is a path where $x_n = x_1$.

Example 1

Suppose the graph in Figure 1.2 represents a network of telephone lines (or electrical transmission lines). We are interested in the network's vulnerability to accidental disruption. We want to identify those lines and switching centers that must stay in service to avoid disconnecting the network.

There is no line whose disruption (removal) will disconnect the graph (network), but there is one vertex, vertex d, whose removal disconnects the graph. Suppose we want to find a minimal set of edges needed to link together the six vertices—that is, form paths between all pairs of vertices. There are several possible minimal sets. One is $\{(a,b),(b,c),(c,d),(d,e),(d,f)\}$. In Chap-

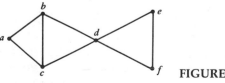

FIGURE 1.2

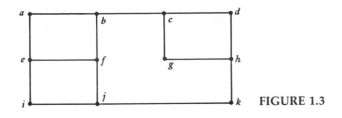

FIGURE 1.3

ter 3 we discuss such minimal connecting sets of edges further. One interesting general result about these sets is that if the graph G has n vertices, then a minimal connecting set for G (if any exists) always has $n - 1$ edges. ∎

Example 2

Consider the section of a city's street map represented by the graph in Figure 1.3. We want to position police at corners so that they can keep every block under surveillance, that is, every block should have police at (at least) one end. How many police are needed? How many police are needed to keep every corner under surveillance (being at most one block from every corner)?

The map has 14 blocks (edges). Corners b, c, e, f, h, and j each are at the end of 3 blocks, and corners a, d, g, i, and k each at the end of 2 blocks. Since 4 corners can be incident to at most $4 \cdot 3 = 12$ blocks, the fewest number of police we can hope for is five. Of these five police, at least four would have to be positioned at 3-block corners in order to cover all 14 blocks. Trying different sets of five out of the six 3-block corners, we find that corners b, c, e, h, and j cover all blocks. The reader is challenged to find another set of 5 covering corners.

Next we turn to the problem of keeping all corners under surveillance, that is, have police on, or one block away from, every corner. There are 11 corners and 6 of these corners can cover 4 corners (including themselves). Thus three is the theoretical minimum. This minimum can be achieved, but only by including a corner that covers 3 corners. We leave it to the reader to find this minimum corner-covering set.

In general, minimum surveillance is a very difficult problem that requires exhaustive systematic enumeration of the possibilities (such enumeration is discussed in detail in Chapter 3). ∎

Example 3

Suppose that we have five people A, B, C, D, E and five jobs a, b, c, d, e, and various people are qualified for various jobs. The problem is to find a feasible one-to-one matching of people to jobs, or show that no such matching can exist. We can represent this situation by a graph with vertices for each person and for each job, and edges joining people with jobs for which they are

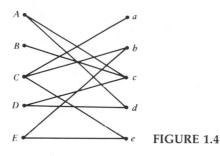

FIGURE 1.4

qualified. Does there exist a feasible matching of people to jobs for the graph in Figure 1.4.

The answer is no. The reason can be found by considering people A, B, and D. These three people as a set are collectively qualified for only two jobs, c and d, hence there is no feasible matching possible for these three people, much less all five people. An algorithm for finding a feasible matching, if any exists, will be presented in Chapter 4. Such matching graphs with edges between two sets of vertices are called **bipartite**. ∎

Example 4

Consider the problem of identifying an unknown control word X in a compiler. Suppose the possible control words are *ADD*, *DO*, *FILL*, *GET*, *RE-PLACE*, *STORE*, and *WAIT*. In a computer, words are actually numbers. We can test X in a three-way branch (less than, or equal to, or greater than) against any control word. Devise an efficient scheme of three-way branch tests to identify X. (By efficient, we mean a scheme that minimizes the maximal number of tests that might be needed to identify X.)

Figure 1.5 displays an efficient scheme in the form of a graph. A left downward edge from a vertex represents a "less than" answer to a three-way branch and a right downward edge from a vertex represents a "greater than" answer. We start by testing X against *GET*. Depending on the answer, either we know: (a) $X = GET$, or (b) $X < GET$ and now we test X against *DO*, or (c) $X > GET$ and we test X against *STORE*. After the second test (if $X \neq GET$), we will have identified X.

A graph of the form in Figure 1.5 is called a tree. Chapter 3 is devoted to trees. A **tree** has a root vertex and a unique path from the root to each other vertex in the tree (in Figure 1.5, *GET* is the root).

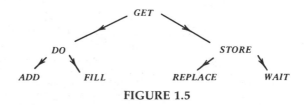

FIGURE 1.5

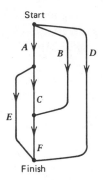

Start

A   *B*   *D*

C

E

F

Finish   **FIGURE 1.6**

The simple testing scheme of successively checking the unknown X against each of the eight possible words could require eight tests, as opposed to three tests with a tree. When the set of possible matches becomes large, as in a spelling checking program (with 50,000 or more recognizable words), the efficiencies of tree-based searching are essential—the number of tests needed with trees is proportional to the logarithm (base 2) of the number of words. ■

Example 5

Consider a curtain-making company's project of starting a new production and sales operation in a foreign country. The key tasks might be: A—market studies to assess the demand for various types of curtains; B—hiring and training plant employees; C—renting or building the plant; D—hiring and training sales personnel; E—renting or building showrooms; and F—initial production (start-up) of curtains. We can model the project with a graph having an edge for each task and vertices to link edges according to task precedences. Suppose for this project that the appropriate task graph is the graph in Figure 1.6.

To each task, one can assign a number indicating the number of months required to complete the task. Actually, the standard procedure is to use three numbers: a reasonable estimate, a best possible time, and a worst possible time. See Table 1.1. Using these numbers, one can calculate the maximum duration of the project under various assumptions about how many tasks require worst-case times, and so forth. One also tries to identify "bottleneck"

TABLE 1.1

Duration	Task					
	A	*B*	*C*	*D*	*E*	*F*
Estimate	2	5	4	3	4	4
Best case	2	2	1	1	2	4
Worst case	3	8	12	4	15	5

tasks, for which the estimated or worst-case time adversely affects the project duration, that is, the initiation of many tasks depends on the completion of a "bottleneck" task (once identified, extra manpower can be allocated to "bottlenecks" to speed their completion).

If only one task takes the worst-case time (and other tasks are completed in the estimated times), what is the longest time that might be required to finish the project—the length of the longest path from *Start* to *Finish*.

By inspection, we see that the longest project duration occurs when task C takes its worst-case time of 12 months. Then the task sequence A–C–F will require $2 + 12 + 4 = 18$ months.

This type of graph-based project analysis is called *PERT* (*Program Evaluation and Review Technique*). ∎

Example 6

Consider the following problem in measurement theory.

One has a collection S of items and a person is asked to state a preference between each pair X_i, $X_j \in S$. We write $X_i < X_j$ if X_j is preferred to X_i and $X_i \sim X_j$ if the person is indifferent between them. We can attempt to represent this preference relation by assigning numerical values $f(X)$ to each item X in S such that $f(X_i) < f(X_j)$ if and only if $X_i < X_j$, and $f(X_i) = f(X_j)$ when $X_i \sim X_j$.

There are two reasons why this numerical scheme may fail. First, we might have $X_1 < X_2$, $X_2 < X_3$, but $X_3 < X_1$. However, it is reasonable to require a person to be consistent ($<$ should be a transitive relation). The other, more serious, difficulty is that some of the items could form a sequence of closely related choices, for example, $\frac{1}{2}$ of an apple pie, $\frac{49}{100}$ ths of an apple pie, . . . to $\frac{1}{100}$ th of an apple pie. If X_k represents the choice of $k/100$ths of pie, $1 \leqslant k \leqslant 50$, then it is likely that $X_{50} \sim X_{49}$ (the person does not care about, or cannot distinguish, a decrease of $\frac{1}{100}$ th of the pie) and in general $X_k \sim X_{k-1}$. Then $f(X_k) = f(X_{k-1})$, for $1 \leqslant k \leqslant 50$, and hence $f(X_1) = f(X_{50})$. However, clearly $X_1 < X_{50}$. Half a pie is preferred to no pie. The numerical model forced indifference to be transitive when it really was not.

Suppose $f(X_k)$ is not a single point but rather an interval I_k of unit length (on the real line). So $f(X_i) < f(X_j)$ now means that interval I_i is to the left of interval I_j, and $X_i \sim X_j$ means that intervals I_i and I_j overlap. This new version accurately models the fact that indifference means close rather than exactly equal. The existence of such a function $f(X)$ can be shown to depend solely on which pairs are indifferent. Thus, given a set of items, we are led to define an **indifference graph**, a graph with vertices for the items, edges between indifferent pairs of items, and such that there exists a unit-length interval I_k associated with each vertex x_k such that I_i and I_j overlap if and only if x_i and x_j are adjacent. For obvious reasons, indifference graphs are also called **unit-interval graphs**.

Consider the graph in Figure 1.7a. Is it an indifference graph, that is, does the required set of unit-length intervals exist?

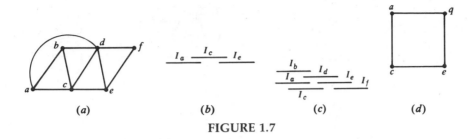

FIGURE 1.7

The only way to model the adjacency relation among vertices a, c, and e in Figure 1.7a with overlapping unit intervals is as shown in Figure 1.7b (or inverting the positions of I_a and I_e), for I_a and I_e cannot overlap and so must each overlap a different end of I_c. By similar reasoning we can extend this unit-interval model of a, c, and e to a model of the whole graph as shown in Figure 1.7c. Thus the graph in Figure 1.7a is an indifference graph. ∎

Interval models for graphs arise in many settings. The next example involves an interval model in ecology. In Section 2.3 an interval model arises in VLSI (*Very Large Scale Integrated*) circuitry design. Other applications are given in the exercises at the end of this section.

Example 7

A competition graph, used in ecology, has a vertex for each of a given set of species and an edge joining each pair of species that feed on a common prey (i.e., the two species compete for food). See the competition graph in Figure 1.8a. Attempts have been made to model competition graphs on the real line in the spirit of indifference graphs. That is, we try to associate an interval (ecological niche), not necessarily of unit length, with each vertex (species) such that adjacent vertices (competing species) correspond to overlapping intervals. Is such a model possible for the graph in Figure 1.8a?

The answer is yes. See the model in Figure 1.8b. Note that a unit interval model is not possible; that is, Figure 1.8a is not an indifference graph. ∎

Example 8

Suppose psychological studies of a group of people determine which members of the group can influence the thinking of others in the group. We can

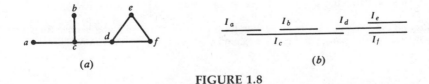

FIGURE 1.8

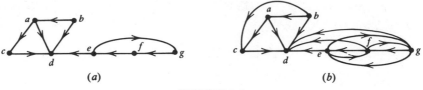

(a)   (b)

FIGURE 1.9

make a graph with a vertex for each person and a directed edge (p_1, p_2) whenever person p_1 influences p_2. Let the graph in Figure 1.9a represent a set of such influences. Now let us ask for a minimal subset of people who can spread an idea through to the whole group, either directly or by influencing someone who will influence someone else and so forth. In graph-theoretic terms, we want a minimal subset of vertices with directed paths to all other vertices (a *directed path* from p_1 to p_k is an edge sequence (p_1, p_2), (p_2, p_3) . . . (p_{k-1}, p_k)). Such a subset is called a **vertex basis**.

To aid us, we can build a directed-path "graph"* for the original graph with the same vertex set and a directed edge (p_i, p_j) if there is a directed path from p_i to p_j in the original graph. Figure 1.9b shows the directed path graph for the graph in Figure 1.9a. Now our original problem can be restated as, find a minimal subset of vertices in the new graph with edges directed to all other vertices. This is just a directed-graph version of the corner covering problem in Example 2. Observe that any vertex in Figure 1.9b with no incoming edges must be in this minimal subset (since no other vertices have edges to it); vertex b is such a vertex. Since b has edges to a, c, and d, then e, f, and g are all that remain to be "influenced." Either e, f, or g "influence" these three vertices. Then b, e, or b, f, or b, g are the desired minimal subsets of vertices. ∎

EXERCISES

SUMMARY OF EXERCISES   The first 11 exercises involve simple graph models. Exercises 12–41 present examples and extensions of the models presented in the Examples in this section. Exercises 42–48 involve other new graph models.

1.   Suppose Interstate highways join the six towns A, B, C, D, E, F as follows: I-77 goes from B thru A to E; I-82 goes from C thru D then thru B to F; I-85 goes from D thru A to F; I-90 goes from C thru E to F; and I-91 goes from D to E.

---

* A directed path "graph" is not really a graph by our definition of a graph since there may be two edges between a pair of vertices, for example, (e, f) and (f, e). This "graph" is a multigraph (see Section 2.1).

(a)   Draw a graph of the network with vertices for towns and edges for segments of Interstates linking neighboring towns.

(b)   What is the minimum number of edges whose removal prevents travel between some pair of towns?

(c)   Is it possible to take a trip starting from town C that goes to every town without using any Interstate highway for more than one edge (the trip need not return to C)?

2.   (a)   Suppose four teams, the Aces, the Birds, the Cats, and the Dogs, play each other once. The Aces beat all three opponents except the Birds. The Birds lost to all opponents except the Birds. The Dogs beat the Cats. Represent the results of these games with a directed graph.

(b)   A ranking is a listing of teams such that the ith team on the list beat the $(i + 1)$st team. Find all rankings for part (a).

3.   (a)   A schedule is to be made with five football teams. Each team is to play two other teams. Explain how to make a graph model of this problem.

(b)   Show that except for interchanging names of teams, there is only one possible graph in part (a).

4.   In the circuit below, current will flow from A to B only when both switches are closed. Design a circuit such that current will flow from A to B when:

(a)   At least one of two switches are closed.

(b)   At least two of three switches are closed, if each "switch" really is a multiple switch that closes switches on two parallel wires.

5.   (a)   Describe how one might represent interlocking (overlapping) Boards of Directors of large corporations with a graph model.

(b)   Suppose Figure 1.3 is the graph of such interlocking boards. What is the maximal number of boards that could meet at the same time?

6.   Suppose there are six people—John, Mary, Rose, Steve, Ted, and Wendy—who pass rumors among themselves. Each day John talks with Mary and Wendy; Mary talks with John, Rose, and Steve; Rose talks with Mary, Steve, and Ted; Steve talks with Mary, Rose, Ted, and Wendy; Ted talks with Rose, Steve, and Wendy; and Wendy talks with John, Steve, and Ted. Whatever people hear one day they pass on to others the next day.

(a)   Model this rumor-passing situation with a graph.

(b)   How many days does it take to pass a rumor from John to Steve? Who will tell it to Steve?

(c)   Is there any way that if two people stopped talking to each other, then it must take three days to pass a rumor from one person to all the others?

7. (a) Give a direction to each edge in Figure 1.3 so that there are directed routes from any vertex to any other vertex.
    (b) Do part a so as to minimize the length of the longest directed path between any pair of vertices. Explain why a smaller minimum is not possible.

8. (a) In Figure 1.9b, what is the minimum number of edges whose directions must be reversed to yield directed routes from any vertex to any other vertex?
    (b) In Figure 1.9a, is it possible by reversing some edge directions to produce a graph with directed paths from any vertex to any other vertex?

9. The flow chart below displays the logical flow possibilities in a program. One possible execution sequence is *ABDFGBCEFGH*.

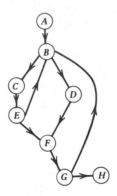

    (a) How many different execution sequences are possible?
    (b) Find all execution sequences with 15 steps.

10. (a) Draw a graph modeling the schedule of matches in a tennis tournament with 10 entrants.
    (b) How many rounds in a tournament are needed with 50 entrants?

11. Suppose eight varieties of chipmunk evolved from a common ancestral strain through an evolutionary process in which at various stages one variety split into two varieties.
    (a) Explain how one might model this evolutionary process with a graph.
    (b) What is the total number of splits that must have occurred?

12. In Example 1, find a minimal connecting set of edges containing neither (*a,b*) nor (*b,d*).

13. (a) For the graph below, find all pairs of vertices whose removal disconnects the graph of remaining vertices.

(b)   Find all pairs of edges whose removal would disconnect the graph.

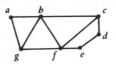

(c)   What is the minimal number of vertex deletions required to dis-connect the graph formed by the 12 edges of a cube?

14.   To measure the probability that a communication network will survive a natural disaster intact, we can look at all pairs of edges of the network and count the fraction of the pairs whose removal does not disconnect the network. Do this for the graph in Figure 1.2.

15.   Repeat Example 2 for minimal block and corner surveillance when the network in Figure 1.3 is altered by adding edges (f,g), (g,j) and deleting (b,f).

16.   (a)   In Example 2, give a argument to show that there is only one way to position five police to cover all 14 blocks.
       (b)   In Example 2, find all sets of three corners that have all 11 corners under surveillance.

17.   Repeat Example 2 for minimal block and corner surveillance when the network is formed by a regular array of north-south and east-west streets of size:
       (a)   3 by 3   (b)   4 by 4   (c)   5 by 5

18.   (a)   A queen dominates any square on an 8 by 8 chessboard in the same row, column, or diagonal as the queen. How many queens suffice to dominate all squares?
       (b)   Repeat this problem for bishops, which dominate only diagonals.

19.   Give another reason why Figure 1.4 has no matching by considering the appropriate subset of jobs (showing they cannot all be filled).

20.   Find matchings or explain why none exist for the following graphs:

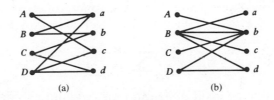

                             (a)   (b)

21.   Find all matchings in Exercise 20(a).

22.   We generalize the idea of matching in Example 3 to arbitrary graphs by defining a matching to be a pairing off of adjacent vertices in a graph.

For example, one possible matching in Figure 1.1 is *a-b, c-d, e-f*. Which of the following graphs have a matching? If none exists, explain why.

(a)   Figure 1.2   (b)   Figure 1.3   (c)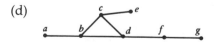

23.   (a)   Suppose a dictionary in a computer has a "start" from which one can branch to any letter, and at any letter one can go to the preceding and succeeding letters. Model this data structure with a graph.
    (b)   Suppose additionally that one can return to "start" from letters *c* or *k* or *t*. Now what is the longest path between any two letters?

24.   Build an efficient three-way branching scheme as in Example 4 to identify one of the 26 letters of the alphabet.

25.   (a)   Repeat Example 4 with a two-way branch (less than, greater than–equal to).
    (b)   In general, how many more rounds of testing will a two-way scheme for *n* words require over a three-way scheme (you do *not* need to justify your answer rigorously)?

26.   Find the longest time that might be required to complete the project in Example 5 under the following assumptions:
    (a)   All tasks take worst-case times.
    (b)   One task is best case (remaining tasks are estimated times).
    (c)   At most one task is best case and at most one is estimated time.

27.   Suppose you have the extra resources in Example 5 to reduce the estimated time for tasks by a total of three months. How should you allocate this reduction among tasks so as to reduce as much as possible the completion time if all jobs take their estimated times?

28.   (a)   In the spirit of Example 5, model the process of planning and committing a burglary (and successfully escaping).
    (b)   With the help of a legal friend, model the "routes" through the criminal justice system of a person arrested for burglary.

29.   Which of the following graphs are indifference graphs? If they are, give a model of overlapping unit intervals; if not, explain why.
    (a)   Figure 1.1   (b)   Figure 1.2   (c)   Figure 1.3

    (d)

30.   (a)   Show that a circuit of length four (without chords) cannot be an indifference graph.
    (b)   Find a graph *G* not containing a circuit of length four or more (without chords), such that *G* is not an indifference graph but if any vertex is removed from *G* the resulting graph is always an indifference graph.

(c)   Repeat part b replacing "indifference graph" with "interval graph." Now the intervals may have different lengths.

31.   Show that the graph below is an interval graph but not a unit-interval graph.

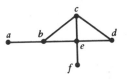

32.   Consider the following two-dimensional model of indifference. We seek to assign each item in a set to a point in x,y-space such that we are indifferent between items X_i and X_j if and only if the distance between their two points is at most 1. A graph is a *two-dimensional indifference graph* if its vertices can be assigned coordinates such that adjacent vertices are within 1 of each other.

(a)   Instead of overlapping unit intervals, what overlapping two-dimensional figures now represent adjacent vertices?

(b)   Which of the following graphs are two-dimensional indifference graph? Explain.

(i)   Figure 1.7d

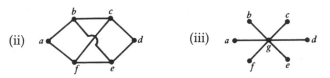

(ii)   (iii)

33.   Show that adjacency in the graph below can be modeled by overlapping rectangles in the plane.

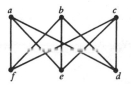

34.   A food web has a vertex for each of a set of species and an edge (A,B) if A feeds on B. Suppose the graph in Figure 1.9b is a food web. Draw the competition graph implied by Figure 1.9b (see Example 7). Is this graph an interval graph?

35.   One of the first experiments to confirm the linear structure of a gene DNA involved testing a special collection of viruses each of which had a "blemished" segment in certain gene DNA. A laboratory test could tell whether or not the blemished segments in two of these viruses overlapped. Give an interval graph model of the problem of determining whether this gene DNA is linear (a long chain with no rings, etc.).

36. Archeologists sometimes make the assumption that types of jewelry are in style for a certain period of time and then never used again, and that a type of jewelry appears in every grave during the jewelry's period of popularity. By examining graves, one can then determine which pairs of jewelry were in use at the same times. Suppose six types of jewelry were found in six graves as shown by the jewelry-grave incidence matrix (entry (i,j) is 1 if grave i has type j jewelry).

Graves			Jewelry			
	1	2	3	4	5	6
1	1	0	1	0	1	0
2	0	1	0	1	0	0
3	0	0	0	1	0	1
4	1	0	1	0	0	0
5	1	0	1	0	0	1
6	0	0	1	0	0	0

(a) Use this matrix to build a graph model representing which jewelry types overlap in time and state the assumption of one period of popularity in terms of an interval-graph type of model.

(b) Does the assumption apply to this set of jewelry? (*Hint:* If true, the 1's in each column will be consecutive when the graves are indexed in chronological order.)

37. One of the six students A, B, C, D, E, F stole a book from the library last Monday night. Each student is questioned about whom they saw at the library, and all tell the truth except the thief (it is possible one student saw another without in turn being observed, but we assume that one must have seen the other if both were there at the same time). A saw B and E; B saw A and F; C saw D and F; D saw A and F; E saw B and C; and F saw C and E. Who is the thief? (*Hint:* See Exercise 30(a).)

38. The matrix at the right indicates correlations of student performances on five different tests. We say two tests are correlated if the correlation number is at least .2.

	A	B	C	D	E
A	1	.2	.1	.0	.3
B	.2	1	.3	.2	.1
C	.1	.3	1	.3	.1
D	.0	.2	.3	1	.4
E	.3	.1	.1	.4	1

(a) Draw a graph representing correlated tests.

(b) Can we order the tests so that each test is correlated with a consecutive set of tests (including itself)? First show that such a model is equivalent to being an indifference graph, and then look for a unit-interval model.

(c) Show that this graph can be modeled by overlapping unit-length arcs on a circle, that is, a "circular" indifference graph.

39. Suppose we represent preference rather than indifference with a graph. That is, if item X_j is preferred to X_i, we make an edge (X_i, X_j). Such a graph should be transitive (assuming preference is consistent); that is, (X_i, X_j) and (X_j, X_k) imply the existence of an edge (X_i, X_k). Suppose we are given an unoriented graph in which an edge means that there is a discernible difference between two items (without knowing which way the preference goes). We seek to orient the edges into a transitive preference graph. Can such an orientation be given to the graphs in:
    (a) Figure 1.2   (b) Figure 1.4   (c) Figure 1.8a.

40. Suppose a certain type of CB radio can be tuned to receive frequencies ranging from R to R' but can only transmit from T to T', where $R < T < T' < R'$. We make a "CB graph" for a set of CB radios with an edge between two vertices if they represent CB radios that can mutually receive each other.
    (a) Show that any interval graph can be a CB graph.
    (b) Find a CB graph that is not an interval graph.
    (c) Modified CB radios can receive over two disjoint frequency ranges and transmit over the same two ranges. Show that the graph in Figure 1.4 is a modified CB radio graph. Is the graph in Figure 1.3 a modified CB radio graph?

41. (a) Let the vertex labels in Figure 1.5 be treated as people. Repeat the search of Example 8 for a minimal subset of dominating people for the graph in Figure 1.5.
    (b) Show that the minimal subset of dominators is unique if there is no sequence of directed edges that forms a circuit in the graph.

42. A game for two players starts with an empty pile. Players take turns putting 1 or 2 or 3 pennies in the pile. The winner is the player who brings the value of the pile up to 16¢.
    (a) Make a directed graph modeling this game.
    (b) Show that the second player has winning strategy by finding a set of four "good" pile values, including 16¢, such that the second player can always move to one of the "good" piles (when the second player moves to one of the good piles, the first player cannot next move to a good pile).

43. The parsing of a sentence can be represented by a directed graph, with a vertex S (for the whole sentence) having edges to vertices Su (subject) and P (predicate), then Su and P being decomposed into pieces, and so on. Consider the abstract grammar with decomposition rules $S \rightarrow AB$, $S \rightarrow BA$, $A \rightarrow ABA$, $B \rightarrow BAS$, and $B \rightarrow S$. For example, $BAABA$ can be "parsed" as:

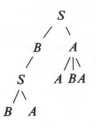

```
          S
        /   \
       B     A
      /     /|\
     S     A BA
    /\
   B  A
```

Find a parsing graph for each of the following (or explain why no parsing exists):
(a) *BABABABA* (b) *BBABAABA*

44. A signed directed graph has a plus (+) or a minus (−) associated with each edge (i,j) indicating whether person i has a positive or a negative influence on person j. If i expresses an opinion in favor of an idea and there is an edge (i,j) with a "+," then a unit of time later j will express an opinion in favor of the idea. If there are also edges (j,k) with "+" and (j,m) with "−," then two units after i first spoke, k will speak in favor and m against the idea. If a person is influenced by several people who just expressed an opinion in the previous unit of time, the person expresses a positive, negative, or no opinion, depending on whether the algebraic sum of influences is positive, negative, or zero, respectively. If person a expresses a positive opinion at time 0 in the above graph, who will express opinions (of what type) at time 4?

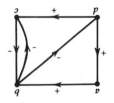

45. A Markov chain is a probability model for a many-round experiment. It may be modeled in turn with a vertex for each state in the Markov chain (e.g., amount of poker winnings), and a directed edge (i,j) with associated probability p_{ij} that if now in state i there is probability p_{ij} of being in state j in the next round.

		Next State				
		1	2	3	4	5
	1	0	.3	0	.5	.2
	2	.5	0	0	.5	0
Current	3	.4	0	0	.4	.2
State	4	1	0	0	0	0
	5	0	0	.1	0	.9

(a) Given the "transition matrix" above, draw the associated Markov chain graph.

(b) Show that if now in state 1 it is possible to be in state 1 in any subsequent round except the second (next) round.

46. Consider the $2^3 = 8$ messages that can be sent as 3-digit binary sequences. A *"snake in the box" code* for transmitting the numbers 1 thru 6 assigns each number a 3-digit binary sequence (not normally its binary representation) such that if one digit in the sequence for number i is (accidentally) switched, the resulting sequence is either the sequence for number $i - 1$ or for $i + 1$ or for none of the numbers (if $i = 1$, then $i - 1 = 6$ and vice versa). Find a "snake in the box" code with the help of a graph with a vertex for each 3-digit sequence and an edge between sequences that differ in one digit.

47. The graph below (due R. Graham) shows the precedence relations for a set of tasks: $F/7$ denotes a task named F that takes 7 units of time to perform. According to the precedence graph, F cannot be performed until B and D are completed. Two or three processors simultaneously work on these tasks. When a processor is ready for a new task and more than one task is ready to be done, the processor selects the task with the earlier alphabetical letter.

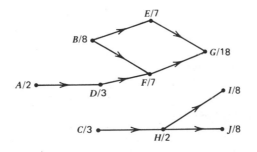

(a) If two processors are available, how many units of time will elapse before all tasks are completed?

(b) If three processors are available, show that it will take 5 units of time longer for all tasks to be completed!

48. Ask a friend to pose three subject areas in which you must find problems having graph models.

1.2   ISOMORPHISM

In this section we present some basic concepts of graph theory. We motivate this discussion with the question: When are two graphs equivalent? For example, are the two 5-vertex graphs in Figure 1.10 equivalent?

What do we mean by "equivalent"? The same graph can be drawn in several different looking ways. Clearly then graphs that are equivalent may not be drawn to look the same. If we consider just the underlying set of vertices and edges, we see that "equivalent" should mean that we can find a one-to-one correspondence between the vertices in the two graphs such that a pair of vertices are adjacent in one graph if and only if the corresponding pair of vertices are adjacent in the other graph. Such a one-to-one correspondence of vertices is called an **isomorphism**, and the two graphs are then **isomorphic**. Our original question is thus, are the two graphs in Figure 1.10 isomorphic?

To be isomorphic, two graphs must have the same number of vertices and the same number of edges. The two graphs in Figure 1.10 pass this initial test. Both graphs have one vertex, e and 5, respectively, at the end of just one edge. Then any isomorphism of these two graphs must match e with 5. Also the vertices at the other ends of the edge from e and 5 must be matched; that is, d matched with 4. The remaining three vertices in each graph are mutually adjacent (forming a triangle) and also are all adjacent to d or 4, respectively. Thus the matching a with 1, b with 2, and c with 3 (or any other matching of these two subsets of three vertices) will preserve the required adjacencies. The correspondence $a - 1, b - 2, c - 3, d - 4, d - 5$ is then an isomorphism, and the two graphs are isomorphic. To visualize how they can be made to look the same, think of moving vertex 4 and 5 in the right graph upward and to the right (past edge $(1,3)$), so that 1, 2, 3, 4 form a quadrilateral with crossing diagonals.

The **degree** $\deg(x)$ of a vertex is the number of edges incident to the vertex. Degrees are clearly preserved under isomorphism, that is, two matched vertices must have the same degree. Then in Figure 1.10, e has to be matched with 5 and d matched with 4 because they are the unique vertices of degree 1 and 4 in their respective graphs.

Another basic concept is a **subgraph**, a graph formed by a subset of vertices and edges of a larger graph. If two graphs are isomorphic, then corresponding subgraphs must be isomorphic. In Figure 1.10, removal of vertices e and 5 (and their incident edges) leaves two isomorphic subgraphs consisting of four mutually adjacent vertices. Once this subgraph isomorphism is noted, isomorphism of the whole graphs is easily demonstrated.

The configuration of n mutually adjacent vertices is called a **complete graph on n vertices**, denoted K_n. A complete graph on two vertices, K_2, is just an edge. Complete subgraphs are in a sense the building blocks of all larger

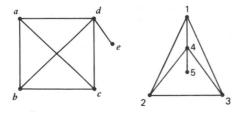

FIGURE 1.10

graphs. For example, both graphs in Figure 1.10 consist of a K_4 and a K_2 joined at a common vertex.

Before examining other pairs of graphs for isomorphism, let us mention the practical importance of determining whether two graphs are isomorphic. Researchers working with organic compounds build up large dictionaries of compounds that they have previously analyzed. When a new compound is found, they want to know if it is already in the dictionary. Large dictionaries can have many compounds with the same molecular formula but which differ in their structure as graphs (and possibly differ in other ways). Then one must test the new compound to see if its graph-theoretic structure is the same as the structure of one of the known compounds with the same formula (and same in other ways), that is, whether the new compound is graph-theoretically isomorphic to one of a set of known compounds. A similar problem arises in designing efficient integrated circuitry for an electrical network. If the design problem has already been solved for an isomorphic network (or if a subpiece of the new network is isomorphic to a previously designed network), then valuable savings in time and money are possible.

Example 1

Are the two graphs in Figure 1.11 isomorphic?

Both graphs have 8 vertices and 10 edges. Let us examine the degrees of the different vertices. We see that b, d, f, h and 3, 4, 7, 8 have degree 2, while the other vertices have degree 3. Then the two graphs have the same number of vertices of degree 2 and the same number of degree 3. The respective subgraphs of the 4 vertices of degree 2 (and the edges between these vertices) in each graph must be isomorphic. However, there are no edges between any pair of b, d, f, h, while the other subgraph of degree-2 vertices has two edges: $(4,3)$ and $(8,7)$. So the subgraphs of degree-2 vertices are not isomorphic, and hence the two full graphs are not isomorphic.

There is also a difference in the structure of the subgraphs formed by the 4 vertices of degree 3 in each graph. Further, only the right graph possesses a circuit passing through each vertex exactly once. ■

The vertices of degree 2 in the left graph in Figure 1.11 form a subgraph of mutually nonadjacent vertices. Such a subgraph is called a set of **isolated vertices**.

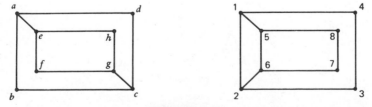

FIGURE 1.11

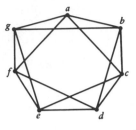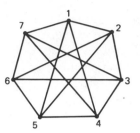

FIGURE 1.12

Example 2

Are the two graphs in Figure 1.12 isomorphic?

The two graphs both have 7 vertices and 14 edges. Every vertex in both graphs has degree 4. Further, both graphs exhibit all the symmetries of a regular 7-gon. With no distinctions possible even among vertices within the same graph, our only option is to try to construct an isomorphism. Starting with vertex a in the left graph, we can match a to any vertex in the right graph by symmetry. Let us use the match $a - 1$.

The set of neighbors of a (vertices adjacent to a) must be matched with the set of neighbors of 1. Let us look at the subgraphs formed by these neighbors of a and 1. See Figure 1.13. Both subgraphs are paths: one is f to g to b to c, and the other is 7 to 4 to 5 to 2. The matching must make these path subgraphs isomorphic. Thus f and c must be matched with 7 and 2 (matching ends of the two paths). By the horizontal symmetry of the graphs, it makes no difference which way f and c are matched—say $f - 7$ and $c - 2$. Then to complete the isomorphism of neighbors of a and 1, we must match g with 4 and b with 5. Now there remain only two unmatched vertices in each graph: d,e and 3,6. Vertex g is adjacent to e but not d, and its matched vertex 4 is adjacent to 3 but not to 6. Thus we must match e with 3 and d with 6.

In sum, allowing for symmetries to match a with 1 and f with 7, we conclude that if the graphs are isomorphic, the isomorphism must be $a - 1$, $b - 5$, $c - 2$, $d - 6$, $e - 3$, $f - 7$, $g - 4$. Checking edges, we see that the graphs are indeed isomorphic with this matching (if this matching were found not to be an isomorphism, then the two graphs would not be isomorphic). ■

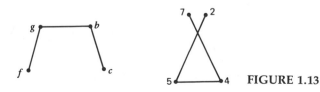

FIGURE 1.13

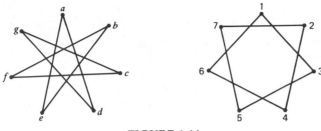

FIGURE 1.14

Given a graph $G = (V, E)$, its **complement** is a graph $\bar{G} = (V, \bar{E})$ with the same set of vertices but now with edges between exactly those pairs of vertices not linked in G. Two graphs G_1 and G_2 will be isomorphic if and only if $\bar{G}_1$ and $\bar{G}_2$ are isomorphic. The isomorphism problem in Example 2 is easy to answer using complements. Figure 1.14 shows the complements of the two graphs in Figure 1.12. Clearly, both these complementary graphs are just a (twisted) circuit of length 7 and hence are isomorphic.

In general, if a graph has more pairs of vertices joined by edges than pairs not joined by edges, then its complement will have fewer edges and thus will probably be simpler to analyze.

Example 3

Are the two directed graphs in Figure 1.15 isomorphic?

Each graph has 8 vertices and 12 edges, and each vertex has degree 3. If we break the degree of a vertex into two parts, the **in-degree** (number of edges pointed in towards the vertex) and **out-degree** (number of edges pointed out), we see that each graph has 4 vertices of in-degree 2 and out-degree 1, and each graph has 4 vertices of in-degree 1 and out-degree 2. We could try to build an isomorphism as in the previous example by starting with a match (by a symmetry argument) between a and 1 and then matching their neighbors (with edge directions also matched), and so forth.

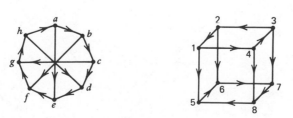

FIGURE 1.15

However, there is a basic difference in the directed path structure of the two graphs. We will exploit this difference to prove nonisomorphism. In the left graph we can draw a directed path from any given vertex to any other vertex by going clockwise around the circle of vertices. Actually, the outer edges form a directed circuit through all the vertices in the left graph. But in the right graph, all edges between the vertex subsets $V_1 = \{1,2,3,4\}$ and $V_2 = \{5,6,7,8\}$ are directed from V_1 to V_2, and thus there can be no directed paths from any vertex in V_2 to any vertex in V_1 (nor is there a directed circuit through all the vertices). Thus the two graphs are not isomorphic. ∎

EXERCISES

1. List all nonisomorphic undirected graphs with four vertices.
2. List all nonisomorphic directed graphs with three vertices.
3. Are the subgraphs of vertices of degree 3 in Figure 1.11 isomorphic?
4. If directions are ignored, are the two graphs in Figure 1.15 isomorphic?
5. Which of the following pairs of graphs are isomorphic? Explain carefully.

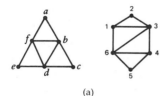

(a)

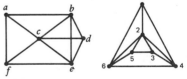

(b)

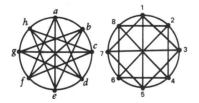

(c)

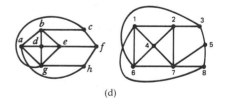

(d)

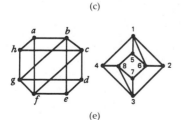

(e)

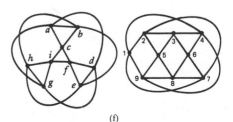

(f)

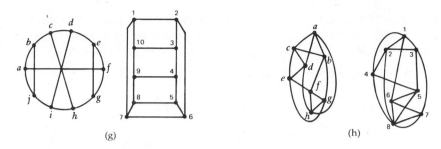

(g)   (h)

6. Which pairs of graphs in this set are isomorphic?

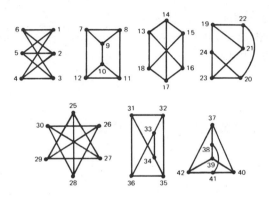

7. Which pairs of graphs in this set are isomorphic?

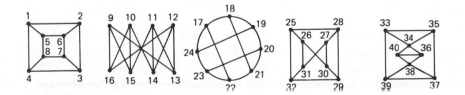

8. Suppose each edge in the graphs in Figure 1.12 is directed from the smaller (numerically or alphabetically) end vertex to the larger end vertex. Are the two resulting directed graphs isomorphic?

9. Are the following pairs of directed graphs isomorphic?

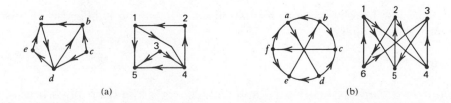

(a)   (b)

10. What is the minimum number of vertices that must be deleted from the two graphs in the following figures to get isomorphic subgraphs?
     (a) Figure 1.11   (b) Figure 1.15   (c) Exercise 5(h)

11. What are the sizes of the largest complete subgraphs in the two graphs in Exercise 5(h)?

12. What are the sizes of the largest sets of isolated vertices in the two graphs in Exercise 5(g)?

13. Build six-vertex graphs with the following degrees of vertices, if possible. If not possible, explain why not.
     (a) Three vertices of degree 3 and three vertices of degree 1.
     (b) Vertices of degrees 1, 2, 2, 3, 4, 5.
     (c) Vertices of degrees 2, 2, 4, 4, 4, 4.

14. A graph is called *regular* if all vertices have the same degree. Find all nonisomorphic regular graphs of degree 3 with:
     (a) four vertices   (b) five vertices   (c) six vertices
     (d) eight vertices

1.3   A SIMPLE COUNTING FORMULA

There is very little in the way of general assertions that can be made about graphs. The one general theorem is a straightforward counting formula.

Theorem

In any graph, the sum of the degrees of all vertices is equal to twice the number of edges.

The sum of the degrees of all vertices counts all instances of some edge being incident at some vertex. But each edge is incident with two vertices, and so the total number of such edge-vertex incidences is simply twice the number of edges. The theorem is now proved. For a set of integers to sum to an even integer, there must be an even number of odd integers in the sum. Thus we obtain:

Corollary

In any graph, the number of vertices of odd degree is even.

Let us now look at some immediate uses and a quite subtle use of this theorem and corollary.

Example 1

Suppose we want to construct a graph with 20 edges and have all vertices of degree 4. How many vertices must the graph have?

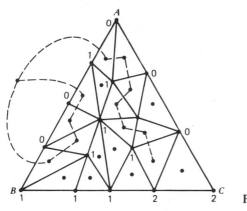

FIGURE 1.16

Let **v** denote the number of vertices. Then the sum of the degrees of the vertices will be **4v**, and by the theorem we have **4v** = 2 · 20 = 40, or **v** = 10. ∎

Example 2

Is it possible to have a group of seven people such that each person knows exactly three other people in the group?

If we try to model this problem using a graph with a vertex for each person and an edge between each pair of people who know each other, then we would have a graph with seven vertices all of degree 3. But this is impossible by the corollary, and so no such set of seven people can exist. ∎

Example 3 (*Optional*)

Consider a triangle with corners *A*, *B*, *C*, labeled 0, 1, 2, respectively. Suppose we triangulate the figure (divide it up into smaller triangles). See the example in Figure 1.16. Now make any labeling with 0, 1, 2 of the new vertices in the triangulation such that vertices on the line from *A* to *B* get a 0 or 1, vertices on the line from *A* to *C* get a 0 or 2, and vertices on the line from *B* to *C* get a 1 or 2. Show that (at least) one small triangle must have all three different labels at its three corners. (This result is known as Sperner's lemma.)

We model the triangulation with a "triangle" graph having a vertex for each small triangle, and also a vertex for the surrounding outside region, and an edge between two vertices corresponding to two neighboring triangles *whose common two corners are labeled 0 and 1*. The triangle graph is represented with dashed edges in Figure 1.16. A triangle with all three different labels at its three corners will correspond to a vertex of degree 1 in the triangle graph. All other triangles will correspond to vertices of degree 0 or 2. The outside region will correspond to a vertex of odd degree: since *A* is 0 and *B* is 1, the line from *A* to *B* will alternate between 0 and 1 labels an odd number of times. According to the corollary, there will be an even number of vertices of odd degree in the triangle graph. Since the odd-degree vertices in this graph correspond to the outside region and triangles with all three labels, there must be an odd number of—at least one—such triangles. ∎

EXERCISES

1. How many vertices will the following graphs have if they contain:
   (a) 12 edges and all vertices of degree 2.
   (b) 15 edges, 3 vertices of degree 4, and the other vertices of degree 3.
   (c) 20 edges and all vertices of the same degree.

2. For each of the following questions, describe a graph model and then answer the question.
   (a) Must the number of people at a party who do *not* know an odd number of other people be even?
   (b) Must the number of people ever born who had (have) an odd number of brothers and sisters be even?
   (c) Must the number of families in Alaska with an odd number of children be even?
   (d) For each vertex x in the graph below, let $s(x)$ denote the number of vertices adjacent to at least one of x's neighbors. Must the number of vertices with $s(x)$ odd be even? Is this true in general?

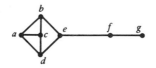

3. What is the largest possible number of vertices in a graph with 19 edges and all vertices of degree at least 3?

4. If a graph G has n vertices, all of which but one have odd degree, how many vertices of odd degree are there in $\overline{G}$, the complement of G?

5. Suppose all vertices of a graph G have degree p, where p is an odd number. Show that the number of edges in G is a multiple of p.

6. There used to be 26 football teams in the National Football League (NFL) with 13 teams in each of two conferences (each conference was divided into divisions, but that is irrelevant here). An NFL guideline said that each team's 14-game schedule should include exactly 11 games against teams in its own conference and 3 games against teams in the other conference. By considering the right part of a graph model of this scheduling problem, show that this guideline could not be satisfied!

7. Interchange the 0's and 1's for the labels of the five internal vertices in Figure 1.16 (vertices not on the boundary of the original large triangle). Now repeat the construction of the triangle graph and find a triangle with all three labels.

8.   Verify that only a triangle with all three labels will correspond to a vertex of degree 1 in the associated triangle graph in Example 3 (the surrounding region might also be of degree 1).

9.   Prove a directed version of Theorem 1: The sum of the in-degrees of vertices in a directed graph equals the sum of the out-degrees of vertices.

1.4   PLANAR GRAPHS

The most natural examples of graphs are street maps and pipeline systems. The graphs that arise from such physical networks usually have the property that they can be depicted on a piece of paper without edges crossing (different edges meet only at vertices). We say that a graph is **planar** if it can be drawn on a plane without edges crossing. We use the term **plane graph** to refer to a planar depiction of a planar graph. In this section, we discuss ways to determine whether a graph is planar and look at some basic properties of planar graphs.

Note that if a graph G has been drawn with edges crossing, this does not mean the graph is nonplanar. There may be another way to draw the graph without edges crossing. For example, the two graphs in Figure 1.10 are isomorphic, that is, the same graph, yet one has crossing edges and the other does not.

The most important use of planar graphs is in designing and building electrical circuitry. A printed circuit board or integrated semiconductor chip is a planar network, and so minimizing the amount of nonplanarity is a key design criterion. When a large planar graph such as a city's street network is key-punched for computer analysis, it is a common error-checking technique to test first whether the graph as key-punched is indeed planar (most key-punching errors would make the graph nonplanar).

Example 1

One of the oldest problems in graph theory concerns map coloring. The question is how many colors are needed to color countries on some map so that any pair of countries with a common border are given different colors. A map of countries is a planar graph with edges as borders and vertices where borders meet. See Figure 1.17a. However, a closely related planar graph called a **dual graph** of the map graph is more useful. The dual graph is obtained by making a vertex for each country and an edge between vertices corresponding to two countries with a common border. See Figure 1.17b. (Normally a vertex is also included for the unbounded region surrounding the map.)

The question now is how many colors are needed to "color" the vertices such that adjacent vertices have different colors. In Figure 1.17b vertices A, B, C, D form a complete subgraph and so each requires a different color, four

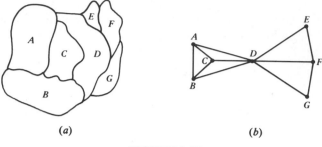

(a)   (b)

FIGURE 1.17

colors in all. With four colors, we can also properly color the remaining vertices.   ■

One of the most famous unsolved problems in all of mathematics during the last century was the conjecture that all planar graphs can be properly colored with only four colors. In 1976 the conjecture was proven true by Appel and Haken using an immense computer-generated, case-by-case, exhaustive analysis (there were 1955 classes of graphical configurations to be considered). We will take a closer look at graph coloring in the next chapter.

Let us now try to find a systematic way to draw a graph on the plane without edges crossing. As with isomorphisms between two graphs, we want to be able to conclude that a graph is not planar if our construction fails. The basic idea we will use is that if we have drawn a circuit, then any pair of adjacent vertices that are not on the circuit must be both inside or both outside the circle. A second less intuitive fact is that the roles of being inside and outside a circle are symmetric: If one considers two maps of the earth, the first with the North Pole at the center of the map, the second with the South Pole at the center, then the Northern Hemisphere is inside the circle of the equator in the first map and is outside the equator in the second map.

Example 2

Show that $K_{3,3}$, the graph in Figure 1.20a (top of page 32), is nonplanar.

$K_{3,3}$ consists of two sets of three vertices with each vertex in one set adjacent to all vertices in the other set. We form a circuit containing all six vertices in $K_{3,3}$, and then try to add the remaining edges (not in the circuit). There are several choices for a 6-vertex circuit. Suppose we use the circuit $(1,4,2,5,3,6,1)$ shown in Figure 1.18. Without loss of generality in trying to draw $K_{3,3}$ as a plane graph, we can assume this circuit is drawn as a circle.

Next the edges $(1,5)$, $(2,6)$, and $(3,4)$ must be added. First draw $(1,5)$. By the inside-outside symmetry of a circle discussed above, we can assume that $(1,5)$ is drawn inside the circuit (as in Figure 1.18). Then $(2,6)$ must be drawn outside the circuit (as in Figure 1.18): if we start to make a path from 2 inside

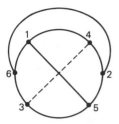

FIGURE 1.18

the circuit, the subcircuit (1,4,2,5,1) would block this path fragment from reaching 6. Finally, we must draw (3,4): a path from 3 going outside the circuit is blocked by the subcircuit (2,5,3,6,2), and a path from 3 inside the circuit is blocked by the subcircuit (1,5,3,6,1). So $K_{3,3}$ has no possible planar depiction. ∎

Using a mixture of theory and careful, case-by-case analysis, it is possible to prove that all nonplanar graphs contain two basic nonplanar configurations. A pair of graphs represents the same **configuration** if one graph can be obtained from the other by inserting or erasing vertices of degree 2; that is, an edge and a path of vertices of degree 2 may be interchanged in a configuration. Figure 1.19 shows a $K_{3,3}$ configuration. Note that such edge-path interchanges do not affect planarity.

Theorem 1

A graph is planar if and only if it does not contain a K_5 or $K_{3,3}$ configuration.

The graph K_5, the complete graph on five vertices, is shown in Figure 1.20b. The graph $K_{3,3}$ was discussed in Example 2. It is left as an exercise to show that the method used on $K_{3,3}$ in Example 2, can also be used to show that K_5 is nonplanar.

To determine whether or not a graph is planar, we can use the following combination of Theorem 1 and the previous construction: start trying to construct a planar drawing, and if difficulties are being encountered, then look for a K_5 or $K_{3,3}$ configuration.

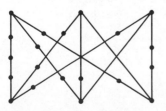

FIGURE 1.19

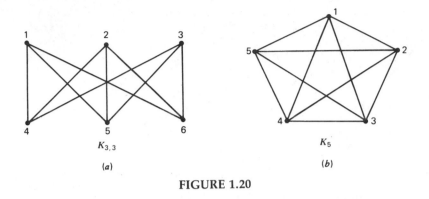

$K_{3,3}$

(a)

K_5

(b)

FIGURE 1.20

Example 3

Is the graph in Figure 1.21*a* planar?

Let us look for as long a circuit as possible. We can find a circuit with all eight vertices: *a-f-c-h-d-g-b-e-a*. Now try to add the other four edges (*a,h*), (*b,f*), (*c,g*), (*d,e*). By inside-outside symmetry, we can start by drawing (*a,h*) inside. See Figure 1.21*b*. Then (*b,f*) and (*c,g*) must go outside. Then (*d,e*) must and can legally go inside. So the graph is planar. ∎

Example 4

Is the graph in Figure 1.22*a* planar?

Let us look for as long a circuit as possible. We find the circuit *a-b-c-d-e-f-g-h-a* with all eight vertices. See Figure 1.22*b*. By inside-outside symmetry, we can start by drawing (*a,c*) inside the circuit. Then (*b,e*) and (*b,g*) must go outside. Then (*a,e*) must go inside. Now *d* and *f* are on opposite sides of the triangle (*a,b,e*), and hence *d* and *f* cannot be joined by a line as required.

To confirm the nonplanarity, we must be able to find a K_5 or $K_{3,3}$ configuration. First we can erase the degree-2 vertex *h* and join *a* and *g* directly. For K_5, we require five vertices of degree 4. This graph has exactly five vertices of degree at least 4: *a, b, d, e, g*. But we cannot find disjoint paths from *d* to *a, b, e,*

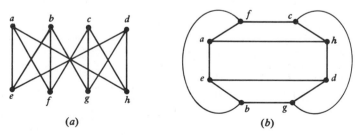

(a)

(b)

FIGURE 1.21

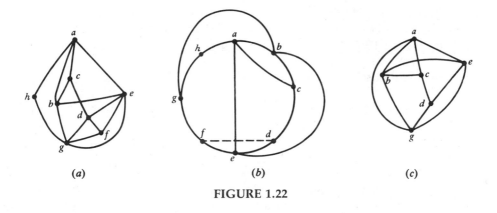

(a) (b) (c)

FIGURE 1.22

and g as required in a K_5 configuration: one of the four edges at d—namely (d,f)—is useless since from f one can only go on to e or g (which are already joined directly to d). So K_5 appears impossible. Further, vertex f appears superfluous (it is so similar to d, see Exercise 4 for details). Let us delete f.

Let us try to find a $K_{3,3}$ configuration. We have six vertices, but only c and d have degree 3. See Figure 1.22c. $K_{3,3}$ has all six vertices of degree 3. An obvious approach is to delete two edges among a, b, e, g (to make all vertices of degree 3). Instead we note that the neighbors of any vertex of degree 3 in Figure 1.22c must be either the upper or lower set of three vertices in $K_{3,3}$. In this way, the match between Figure 1.22c and $K_{3,3}$ in Figure 7.21 is easily found. For example, setting the vertex c of degree 3 equal to vertex 1 in Figure 7.21 and c's neighbors equal to vertices 4, 5, 6, we obtain the isomorphism $a - 4, b - 5, c - 1, d - 6, e - 2, g - 3$ (edges (a,b) and (e,g) must be deleted).

∎

There are many different plane graphs possible for a planar graph. For example, see the two plane graphs for the graph in Figure 1.23. However, any two planar depictions of a given graph will always divide the plane into the same number of regions. For simplicity, assume that G is a **connected** planar graph (connected means "in one piece," i.e., with paths between every pair of vertices). If **v** and **e** denote the number of vertices and edges, respectively,

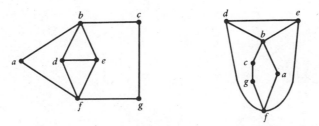

FIGURE 1.23

in G, then a planar depiction of G will always form $\mathbf{r} = \mathbf{e} - \mathbf{v} + 2$ regions, including the unbounded region. This remarkable formula for $\mathbf{r}$ was discovered by Euler in 1752.

Theorem 2 (Euler's Formula)

If G is a connected planar graph, then any plane graph for G has $\mathbf{r} = \mathbf{e} - \mathbf{v} + 2$ regions.

Proof

Let us draw a plane graph of G edge-by-edge. Let G_n denote the graph obtained after n edges have been added, and let $\mathbf{v}_n$, $\mathbf{e}_n$, and $\mathbf{r}_n$ denote the number of vertices, edges, and regions in G_n, respectively. Initially we have G_1, one edge, its two end vertices, and the one (unbounded) region. Then $\mathbf{e}_1 = 1$, $\mathbf{v}_1 = 2$, $\mathbf{r}_1 = 1$, and so Euler's formula is valid for G_1: $1 = 1 - 2 + 2$. We obtain G_2 from G_1 by adding an edge at one of the vertices in G_1. In general, G_n is obtained from G_{n-1} by adding an nth edge at one of the vertices of G_{n-1}. The new edge might link two vertices already in G_{n-1}. If it does not, the other end vertex of the nth edge is incorporated in G_n.

   We will now use the method of induction (see Appendix A.2) to complete the proof. We have shown that the theorem is true for G_1. Then we assume that it is true for G_{n-1}, for any $n \geqslant 2$, and prove it is true for G_n. Let (x,y) be the nth edge that is added to G_{n-1} to get G_n. There are two cases to consider. If x and y are both in G_{n-1}, then they are on the boundary of a common region K of G_{n-1}, possibly the unbounded region (if x and y were not on a common region, edge (x,y) could not be drawn in a planar fashion, as required). See Figure 1.24. Since edge (x,y) splits K into two regions, we conclude that $\mathbf{e}_n = \mathbf{e}_{n-1} + 1$, $\mathbf{v}_n = \mathbf{v}_{n-1}$, $\mathbf{r}_n = \mathbf{r}_{n-1} + 1$. So each side of Euler's formula grows by 1. Hence if the formula was true for G_{n-1}, it will also be true for G_n.

   In the second case, one of the vertices x,y is not in G_{n-1}, say it is y. Then adding (x,y) implies that y is also added, but no new regions are formed (i.e., no existing regions are split). Thus $\mathbf{e}_n = \mathbf{e}_{n-1} + 1$, $\mathbf{v}_n = \mathbf{v}_{n-1} + 1$, $\mathbf{r}_n = \mathbf{r}_{n-1}$, and again the validity of Euler's formula for G_{n-1} implies its validity for G_n. By induction, the formula is true for all G_n's and hence for the full graph G. ∎

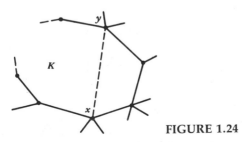

FIGURE 1.24

Example 5

How many regions would there be in a plane graph with 10 vertices each of degree 3?

By the theorem in Section 1.3, the sum of the degrees $10 \cdot 3$ equals $2e$, and so $e = 15$. By Euler's formula, the number of regions r is

$$r = e - v + 2 = 15 - 10 + 2 = 7$$

∎

Theorem 2 has the following corollary that can often be used to show quickly that a graph is nonplanar.

Corollary

If G is a connected planar graph with $e > 1$, then $e \leq 3v - 6$.

Proof

Let us define the degree of a region to be the number of edges on its boundary. If an edge occurs twice along a boundary, as does (x, y) in region K in Figure 1.25, the edge is counted twice in region K's degree. Observe that each region in a plane graph must have degree ≥ 3, for if < 3 then either two edges would connect the same pair of vertices or one edge would loop back to have both ends at the same vertex (but parallel edges and loops are not allowed in graphs).

Since each region has degree ≥ 3, the sum of the degrees of all regions will be at least $3r$. But this sum of degrees of all regions must equal $2e$, since this sum counts each edge twice, that is, each side of an edge is part of some boundary (this is the same argument as used to show that the sum of the vertices' degrees equals $2e$). Thus $2e =$ (sum of regions' degrees) $\geq 3r$, or $(\frac{2}{3})e \geq r$. Combining this inequality with Euler's formula (Theorem 2), we have

$$(\tfrac{2}{3})e \geq r = e - v + 2 \qquad \text{or} \quad 0 \geq (\tfrac{1}{3})e - v + 2$$

Solving for e, we obtain $e \leq 3v - 6$. ∎

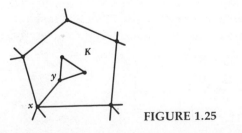

FIGURE 1.25

Example 6

Use the corollary to prove that K_5, the complete graph on five vertices, is nonplanar.

The graph K_5 has $\mathbf{v} = 5$ and $\mathbf{e} = 10$. Then $3\mathbf{v} - 6 = 3 \cdot 5 - 6 = 9$. But the corollary says that $\mathbf{e} \leqslant 3\mathbf{v} - 6$ in a connected planar graph, and so K_5 cannot be planar. ∎

The corollary *should not be misinterpreted to mean* that if $\mathbf{e} \leqslant 3\mathbf{v} - 6$, then a connected graph is planar. Many nonplanar graphs also satisfy this inequality.

Our two theorems and the corollary have laid the foundation for a mathematical theory of planar graphs. In the process, we have acquired some practical aids for showing that a graph is nonplanar. There are two ways to try to extend the corollary for use in showing that other graphs are nonplanar. One way is to make the inequality "stronger," that is, get a smaller upper bound on $\mathbf{e}$. Recall that the key step in proving the corollary was the observation that every region has degree at least 3. This led to the inequality $2\mathbf{e} \geqslant 3\mathbf{r}$. Suppose that a certain connected graph G (with at least 3 edges) is known to contain no circuits of length 3 or 4. Then every boundary of a region must have at least five edges; that is, every region has degree $\geqslant 5$. Summing the degrees of all regions, we now obtain the inequality $2\mathbf{e} \geqslant 5\mathbf{r}$. Assuming G is planar, we then have

$$(\tfrac{2}{5})\mathbf{e} \geqslant \mathbf{r} = \mathbf{e} - \mathbf{v} + 2$$

and hence

$$\mathbf{e} \leqslant (5/3)\mathbf{v} - \tfrac{10}{3}$$

If G does not satisfy this stronger inequality (even though it satisfies $\mathbf{e} \leqslant 3\mathbf{v} - 6$), then G cannot be planar.

The other way to extend the use of the corollary is to make an alteration in a graph that does not destroy planarity (assuming the graph was originally planar) and show that the altered graph does not satisfy the corollary's inequality. The easiest alteration is to contract an edge e so as to merge the two end vertices of e into a single vertex. This contraction decreases $\mathbf{e}$ by 1 but $3\mathbf{v} - 6$ by 3 (one less edge and vertex). Physically, such a contraction would deform a plane graph for the original graph into a plane graph for the altered graph. We must, however, be careful not to contract an edge that will result in a "nongraph," having two edges between the same pair of vertices (see the end of Example 7 below).

Example 7

Show that the graph G in Figure 1.26 is nonplanar.

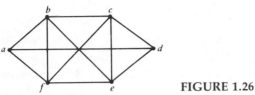

FIGURE 1.26

G has $\mathbf{v} = 6$ and $\mathbf{e} = 11 \leqslant 3\mathbf{v} - 6 = 18 - 6 = 12$. So this equality does not prove nonplanarity. Now contract edge (a,d) combining vertices a and d. The altered graph G' has $\mathbf{v}' = 5$ and now $\mathbf{e}' = 10 \nleqslant 3\mathbf{v}' - 6 = 9$. Thus G' cannot be planar, and so G also cannot be planar. Note that if we had chosen to contract edge (a,b), the resulting combined vertex would have had two edges going to f—a situation not permitted in graphs. ∎

EXERCISES

SUMMARY OF EXERCISES The first eight exercises involve determining whether various graphs are planar and drawing planar graphs in different ways. Exercises 9–12 discuss variations of the concept of planarity. Exercises 13–25 build on Euler's formula and the Corollary $\mathbf{e} \leqslant 3\mathbf{v} - 6$.

1. Draw a dual graph of the planar graph (map graph) in:
   (a) Figure 1.21*b*   (b)   Figure 1.22*c* with edge (b,e) deleted

2. Show that K_5 is nonplanar by the method in Example 2.

3. Which of the following graphs are planar? Find K_5 or $K_{3,3}$ configurations in the nonplanar graphs.

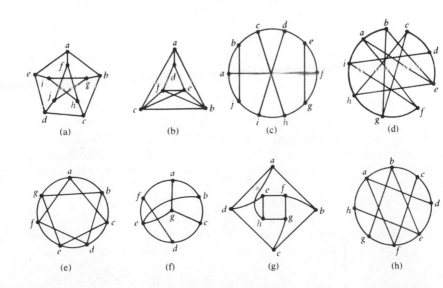

4.   Show that in Figure 1.21a, vertex f cannot be part of a $K_{3,3}$ configuration. Consider three subcases using f: (i) the edge (d,f) is in the $K_{3,3}$ configuration, (ii) (d,f) is not in the configuration but vertex d is in it, and (iii) vertex d is not in the configuration.

5.   Make a graph with one vertex for each square on a 4×4 chessboard and an edge between two vertices that correspond to squares that can be the start and end of a knight's move (go two squares in one direction and then one square to either side). Is this graph planar?

6.   A **line graph** $L(G)$ of a graph G has a vertex of $L(G)$ for each edge of G and an edge of $L(G)$ joining each pair of vertices corresponding to two edges in G with a common end vertex.
     (a)   Show that $L(K_5)$ and $L(K_{3,3})$ are nonplanar.
     (b)   Find a planar graph whose line graph is nonplanar.

7.   Show that any planar graph can be drawn on the surface of a sphere without crossing edges and vice versa.

8.   Re-draw the graph in Figure 1.21b so that the infinite region is bounded by the circuit a-f-c-h-a.

9.   Show that the following graphs can be drawn on the surface of a doughnut (torus) without crossing edges:
     (a) $K_{3,3}$ (b) K_5 (c) K_6 (d) K_7

10.   The *crossing number* $c(G)$ of a graph G is the minimum number of pairs of crossing edges in a depiction of G. For example, if G is planar then $c(G) = 0$. Determine $c(G)$ for the following graphs:
      (a) $K_{3,3}$
      (b) K_5
      (c)   Graph in Figure 1.22a.
      (d)   A $2n$-gon with diagonals joining opposite pairs of corners.

11.   A graph G is *critical planar* if G is nonplanar but any subgraph obtained by removing a vertex is planar.
      (a)   Which of the following graphs are critical planar: (i) $K_{3,3}$, (ii) K_5, (iii) graph in Figure 1.22a.
      (b)   Show that critical planar graphs must be connected and cannot have a vertex whose removal disconnects the graph.

12.   The construction of a dual $D(G)$ can be applied to any planar depiction of a planar graph G: draw a vertex of $D(G)$ in the middle of each region in the depiction of G and draw an edge e^* of $D(G)$ perpendicular to each edge e of G; e^* connects the vertices of $D(G)$ representing the regions on either side of e.
      (a)   A dual need not be a graph. It might have two edges between the same pair of vertices or a self-loop edge (from a vertex to itself). Find two planar graphs with duals that are not graphs because they contain these two forbidden situations.

(b)   Show that the duals of the two different planar depictions of the graph in Figure 1.23 are isomorphic.

(c)   Show for any planar depiction of a graph G that the vertices of G correspond to regions in $D(G)$ and that $D(D(G)) = G$.

(d)   Find a planar graph that is isomorphic to its own dual.

13.   Suppose a planar graph is not connected but instead consists of several components, that is, disjoint connected subgraphs.

(a)   Find the appropriate modification of Euler's formula for a planar graph with **c** components.

(b)   Show that the corollary is valid for unconnected planar graphs.

14.   Mimic the argument in the corollary to prove that $\mathbf{e} \leqslant 3\mathbf{r} - 6$ in a planar graph with each vertex of degree $\geqslant 3$.

15.   If a planar graph with n vertices all of degree 4 has 10 regions, determine n.

16.   (a)   Prove that every connected planar graph has a vertex of degree at most 5. (*Hint:*   Assume that every vertex has degree at least 6 and obtain a contradiction.)

(b)   Show that part (a) immediately generalizes to any (unconnected) planar graph.

17.   (a)   Prove that every connected planar graph with less than 12 vertices has a vertex of degree at most 4. (*Hint:*   Assume every vertex has degree at least 5 to obtain a lower bound on **e** (together with the upper bound on **e** in the corollary) that implies $\mathbf{v} \geqslant 12$.)

18.   Prove that if a graph G has at least 11 vertices, then either G or its complement $\bar{G}$ must be nonplanar.

19.   If G is a connected planar graph with all circuits of length at least k, show that the inequality $\mathbf{e} \leqslant 3\mathbf{v} - 6$ can be strengthened to

$$\mathbf{e} \leqslant \frac{k}{k-2}(\mathbf{v} - 2)$$

20.   Suppose l lines are drawn through a circle and these lines form p points of intersection (involving exactly two lines at each intersection). How many regions **r** are formed inside the circle by these lines? (*Hint:* Assume that the lines end at the edge of the circle (at $2n$ distinct points).)

21.   (a)   Show that every circuit in $K_{3,3}$ has at least four edges.

(b)   Use part a and the result of Exercise 19 to show that $K_{3,3}$ is nonplanar.

22.   (a)   Show that every circuit in the graph in Exercise 3(a) has at least five edges.

(b) Use part a and the result of Exercise 19 to show that this graph is nonplanar.

23. Draw the graph G' in Example 7.

24. Use the edge reduction technique in Example 7 (possibly repeated use) to show that each of the following graphs is nonplanar. Remember not to combine vertices with a common neighbor.
    (a) Exercise 3(a)   (b) Exercise 3(b)   (c) Exercise 3(c)

25. A *Platonic graph* is a planar graph in which all vertices have the same degree d_1 and all regions have the same number of bounding edges d_2, where $d_1 \geq 3$ and $d_2 \geq 3$. A Platonic graph is the "skeleton" of a Platonic solid, for example, an octahedron.
    (a) If G is a Platonic graph with vertex and face degrees d_1 and d_2, respectively, then show that $\mathbf{e} = \frac{1}{2}d_1\mathbf{v}$ and $\mathbf{r} = (d_1/d_2)\mathbf{v}$.
    (b) Using part a and Euler's formula, show that $\mathbf{v}(2d_1 + 2d_2 - d_1d_2) = 4d_2$.
    (c) Since $\mathbf{v}$ and $4d_2$ are positive integers, we conclude from part b that $2d_1 + 2d_2 - d_1d_2 > 0$. Use this inequality to prove that $(d_1 - 2)(d_2 - 2) < 4$.
    (d) From part c, find the five possible pairs of (integral) values of d_1, d_2 and draw the respective Platonic graphs.

1.5 SUMMARY AND REFERENCES

This chapter introduced graphs, their applications, and some of their basic structures. The initial section on graph models was meant to emphasize the diverse uses of graphs. This text takes a user-oriented approach to graph theory. Readers interested in a more formal graph theory text presenting the subject as an interesting area of pure mathematics should see the books by Bondy and Murty [1], Harary [3], or Wilson [4]. The basic structure of graphs was explored in Section 1.2 under the guise of determining what makes two graphs different. Section 1.3 presented a simple proposition about graphs that has unexpected consequences. The final section introduced the important class of planar graphs. It surveyed ad hoc and formal theories for determining whether a graph is planar.

The history of graph theory begins with the work of L. Euler in 1736 on Euler circuits (discussed in Section 2.1). Euler's formula for planar graphs, originally stated in terms of polyhedra, was proved in 1752. Bits of graph theory appeared in papers about topology and geometric games, but it was around 1850 that formal studies of graphs began to appear. One was A. Cayley's 1857 paper counting the number of trees (discussed in Chapter 3). Another was G. Kirchhoff's 1847 paper presenting an algebra of circuits and introducing graphs in the study of electrical circuits. This same paper contains Kirchhoff's famous current and voltage laws (Kirchhoff was 21 when he wrote this historic paper). The Four Color Problem was first mentioned in correspondence in 1852 in a letter from W. Hamilton to A. De Morgan. The term graph was first used by J. Sylvester in 1877. The first book on graph theory, by D. Konig, did not appear until 1936. An excellent sourcebook on the history of graph theory is *Graph Theory 1736–1936* by Biggs, Lloyd, and Wilson [2].

See the General References (at the end of the book) for a list of other introductory texts on graph theory.

1. M. J. Bondy and U. Murty, *Graph Theory with Applications,* American Elsevier, New York, 1976.
2. N. Biggs, E. Lloyd, and R. Wilson, *Graph Theory 1736–1936,* Clarendon Press, Oxford, 1976.
3. F. Harary, *Graph Theory,* Addison-Wesley, New York, 1969.
4. R. Wilson, *Introduction to Graph Theory,* Academic Press, New York, 1972.

SUPPLEMENT I REPRESENTING GRAPHS INSIDE A COMPUTER

This supplement is meant as a review. Students who are unfamiliar with data structures would benefit by consulting a data structures text, such as A. Aho, J. Hopcroft, J. Ullman, Data Structures and Algorithms, Addison-Wesley, Reading, Mass., 1983.

Problems that involve large graphs or that must be solved repeatedly need to be programmed for computer solution. Unlike numerical computations, such as plotting the solutions of differential equations, graph problems require no arithmetic operations. Instead, the essence of these problems is systematic search and use of the right data structures for representing graphs. In this appendix, we present the most common ways of representing graphs in a computer.

The simplest but least efficient way to represent a graph is with an *adjacency matrix*. This matrix has a row and a column for each vertex. Entry $(i, j) = 1$ if vertex x_i is adjacent to x_j and $= 0$ otherwise. Figure S.2 shows the adjacency matrix for the graph in Figure S.1. In a directed graph, entry $(i, j) = 1$ if and

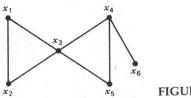

FIGURE S.1

	x_1	x_2	x_3	x_4	x_5	x_6
x_1	0	1	1	0	0	0
x_2	1	0	1	0	0	0
x_3	1	1	0	1	1	0
x_4	0	0	1	0	1	1
x_5	0	0	1	1	0	0
x_6	0	0	0	1	0	0

FIGURE S.2

only if there is an edge (x_i, x_j). Because the adjacency matrix often has 0's in most entries, it is an inefficient way to store a graph.

A better approach is to store only those adjacencies that do occur. A list of the edges (the pairs of adjacent vertices) is compact, but it requires extensive searching to determine the set of vertices adjacent to a given vertex. A compromise between an adjacency matrix and an edge list is a set of adjacency lists. For each vertex we list the set of vertices adjacent to it. These lists are stored one after another in one long adjacency array, along with an associated index list that tells where the adjacency list of each vertex begins. Figure S.3 shows the adjacency array and associated index list for the graph in Figure S.1.

For example, according to entry 3 in the index list, the adjacency list of vertex x_3 starts at entry 5 in the adjacency array, and x_3's adjacency list continues to entry 8 in the adjacency array (since the adjacency list for x_4 begins at entry 9). We need an extra "phantom" entry at the end of the index list to tell us where the end of the adjacency list of the last vertex is, that is,

Adjacency Array		Index List	
1	2	1	1
2	3	2	3
3	1	3	5
4	3	4	9
5	1	5	12
6	2	6	14
7	4	7	15
8	5		
9	3		
10	5		
11	6		
12	3		
13	4		
14	4		

FIGURE S.3

	Data List	Pointer List	Index List	
1	2	19	1	1
2	—	—	2	3
3	1	16	3	4
4	1	5	4	6
5	2	13	5	10
6	3	17	6	11
7	—	—	7	0
8	6	0	8	0
9	—	—		
10	3	14	Unused	8
11	4	0	Entries	
12	—	—	1	2
13	4	20	2	18
14	4	0	3	7
15	—	—	4	9
16	3	0	5	12
17	5	8	6	15
18	—	—	7	21
19	3	0	8	22
20	5	0		
21	—	—		
22	—	—		

FIGURE S.4

where the next adjacency list would begin if there were another vertex; thus we put 15 in entry 7 of the index list.

Finally, if edges and vertices can be added or deleted in the middle of a graph analysis program, we need to use linked lists in the adjacency array in Figure S.3. A **linked list** is really a pair of lists. The main list has the entry values of interest to us, but these values are not stored in consecutive order. The second list contains "pointers" that tell where the next value in the list is stored. These pointers allow the list to jump around to skip deleted entries and to incorporate added entries. For example, if a portion of a linked list has the sequence of entries →6→14→3→, then to delete entry 14 from the list, we simply change pointer entry 6 from 14 to 3, and now the list is →6→3→. Conversely, we can later insert entry 23 after entry 6 (and before entry 3) by changing 6's pointer from 3 to 23 and setting pointer entry 23 to 3. We mark the end of a list by setting the pointer of the last entry in the list equal to 0. As before, we put the linked adjacency lists for all vertices into one long adjacency array and we use an index list to indicate the start of each vertex's adjacency list; if a vertex is deleted, its entry in the index list is set to 0. We also need an associated "housekeeping" list of unused entries in the array of linked lists: the number of a deleted entry in the linked list is added to the end of this unused entries list; when a new entry is to be inserted into the linked list, we use the entry whose number is at the end of the unused entries list (to keep track at the end of the unused entries list, we maintain an updated end-of-list-pointer; this pointer appears in the square box in Figure S.4). Figure S.4 shows a linked list representation of the graph in Figure S.1 that might have evolved if that graph were the product of several deletions and insertions.

Some graph problems require doubly linked lists having two associated pointer lists, one for the next entry in the list and one for the preceding entry in the list.

SUPPLEMENT II SUPPLEMENTARY EXERCISES

SUMMARY OF EXERCISES   Graph theory is a field famous for its interesting problems. Exercises 21–29 involve programming exercises and most require use of the data structures for representing graphs introduced in Supplement I. The rest of the exercises range over a variety of questions in graph theory. Several exercises introduce new graph concepts, such as strong connectedness (Exercise 11) and cut-set (Exercise 16). Exercise 31 is a very famous problem in Ramsey Theory (see Appendix A.4). For more problems in graph theory, see any of the graph theory texts listed in the References (at the end of this text).

1. Suppose that there are seven committees with each pair of committees having a common member and each person is on two committees. How many people are there?

2. (a) Show that the complement of K_n, a complete graph on n vertices, is a set of n isolated vertices.
    (b) Show that K_n has $\frac{1}{2}n(n-1)$ edges.

3. A graph is *regular* if all vertices have the same degree. If a graph with n vertices is regular of degree 3 and has 18 edges, determine n.

4. Show that at least two vertices have the same degree in any graph with at least two vertices. (*Hint:* Be careful about vertices of degree 0.)

5. Show that if a graph is not connected, then its complement must be connected.

6. Show that an undirected graph with all vertices of degree ≥ 2 must contain a circuit (edges cannot be repeated in a circuit).

7. Show that removal of some vertex x disconnects the connected undirected graph G if and only if there are two vertices a and b in G such that all paths in G from a to b pass through x.

8. Show that if an undirected graph G contains exactly two vertices of odd degree, then they must be in the same component of G.

9. If a graph has 50 edges, what is the least number of vertices it can have?

10. If every vertex in a directed graph G has positive out-degree (at least one outwardly directed edge), then:
    (a) Must G contain a directed circuit?
    (b) Must every vertex of G be on a directed circuit?

11. A directed graph is called *strongly connected* if there is a path from x to y for any two vertices x, y in G. Prove that G is strongly connected if and only if G's vertices cannot be partitioned into two sets V_1, V_2 such that there are no edges from a vertex in V_1 to a vertex in V_2.

12. Direct the edges in the following graphs to make the graphs strongly connected (see Exercise 11). If not possible, explain why. (An application to this problem is making streets in a city one-way.)
    (a) Figure 1.3   (b) Figure 1.8a   (c) Figure 1.21b

13. Prove that the edges of a connected undirected graph G can be directed to create a strongly connected graph (see Exercise 11) if and only if there is no *bridge* in G, an edge whose removal disconnects G.

14. Suppose there are three farms each with a child, a goat, and a rabbit. Arbitrarily name the three sets of inhabitants $C_a, C_b, C_c, G_a, G_b, G_c$, and R_a, R_b, R_c. Use the following information and a graph model to determine the three groups of a child, goat, and rabbit on each farm:
    (i) The child on the farm with goat G_a and the child on the farm with rabbit R_c are competing for the attention of the (female) child C_b on the third farm.
    (ii) Goat G_c and rabbit R_b are not on the same farm.
    (iii) The boy on the farm with rabbit R_a is not C_a.

15. A *simple* path, or circuit, is a path (circuit) in which a vertex is visited at most once, whereas a general path (circuit) may repeat vertices but not edges.
    (a) Show that a subset of the vertices on a general path from x to y can be used to make a simple path from x to y.
    (b) Prove or give a counterexample: if x and y lie on a general circuit, then they must lie on a simple circuit.
    (c) Show that the edges in a general circuit can be partitioned into a collection of simple circuits.
    (d) Show that if C is an odd-length general circuit, then a subset of C's edges form an odd-length simple circuit.

16. A *cut-set* S is a set of edges in a connected undirected graph G whose removal disconnects G, but no proper subset of S can disconnect G.
    (a) Find a cut-set of minimal size in Figure 1.26.
    (b) Show that every cut-set has an even number of edges in common with any circuit (remember that 0 is an even number).

17. Show that if an n-vertex graph has more than $\frac{1}{2}(n-1)(n-2)$ edges, then it must be connected. (*Hint:* The most edges possible in a discon-

nected graph will occur when there are two components, each complete subgraphs.)

18. Show that an n-vertex graph cannot be a bipartite graph (see Example 3 of Section 1.1) if it has more than $\frac{1}{4}n^2$ edges.

19. (a) Show that in the square A^2 of a graph's adjacency matrix A (see Supplement I), the value of entry (i, j), for $i \neq j$, is the number of different 2-edge paths between x_i to x_j.
    (b) Extend part a to k-edge paths.

20. Modify the construction of linked lists for directed graphs (we want to have lists of incoming and outgoing edges for each vertex).

21. Given an undirected graph in linked list form, write a program to produce the graph's complement in linked list form.

22. Write a program to find all triangles in a given graph (in linked list form).

23. Given a graph in linked list form, write a program to find a path (if one exists) between any specified pair of vertices.

24. Use Exercise 19 to write a program to find, for a given pair of vertices x, y in a given graph G (in adjacency matrix form) and a given k, whether there is a path of k edges between x and y in G.

25. (a) Use Exercise 19 to write a program to find, for every pair of vertices x, y in a given graph (in adjacency matrix form), the length of the shortest path between x and y.
    (b) Repeat part a with graph given in linked list form (do not just convert to adjacency matrix form).

26. Given a directed graph (in any form you want), write a program to determine whether it is strongly connected, or else find the two vertex sets V_1, V_2 described in Exercise 11.

27. Given an undirected graph G (in any form you want), write a program to direct the edges to make G strongly connected (assuming G has no bridge, see Exercise 13).

28. Write a computer program to read in two graphs of up to six vertices and test whether they are isomorphic.

29. (a) If every vertex in a graph G has degree $\geq d$, then show that G must contain a simple circuit (with no repeated vertices) of length at least $d + 1$.
    (b) Write a program to find such a circuit for a given graph and given d.

30. (a) Find a graph that is isomorphic to its own complement.
    (b) Show that any self-complementary graph (as in part (a)) must have either $4k$ or $4k + 1$ vertices, for some integer k.

31. Let each edge of a complete graph on six vertices be painted red or white. Show that there must always be either a red triangle of three edges or a white triangle of three edges.

32. Draw planar graphs with the following types of vertices, if possible:
    (a) Six vertices of degree 3.
    (b) i vertices of degree $i, i = 1, 2, 3, 4, 5$.
    (c) Two vertices of degree 3 and four vertices of degree 5.

33. Suppose that each path in a certain 7-vertex planar graph contains an even number of edges (0 edges or 2 edges or 4 edges, etc.). Draw the graph.

34. Show that a directed graph has no directed circuits if and only if its vertices can be indexed $x_1, x_2, \ldots, x_n$ so that all edges are of the form (x_i, x_j), $i < j$.

35. (a) A *clique* is a complete subgraph not contained as a subset of a larger complete subgraph. Find all cliques in the following interval graphs:
        (i) Figure 1.7a (ii) Figure 1.8a
    (b) Show that the cliques in an interval graph correspond to points of maximal overlap in an interval model and that when cliques are indexed according to the order of the points in an interval model, each vertex will be in a consecutive set of cliques.
    (c) Show that if the cliques of a graph G can be indexed so that each vertex is in a consecutive set of cliques, then G is an interval graph.

36. A *line graph* $L(G)$ of a graph G has a vertex of $L(G)$ for each edge in G and an edge between two vertices in $L(G)$ corresponding to two edges of G with a common end vertex.
    (a) Draw a line graph of the graph in Figure 1.26.
    (b) Show that each vertex in $L(K_n)$ has degree $2(n - 2)$.
    (c) Write a program to build $L(G)$ given G (in any form you want).
    (d) Find all graphs that are isomorphic to their own line graph.
    (e) Show that if a graph H is the line graph of some graph, then the edges of H can be partitioned into a collection of complete subgraphs such that each vertex of H is in exactly two such complete subgraphs.

37. (a) An *independent set* I of vertices in a graph is a set of mutually nonadjacent vertices. Find the largest independent set in Figure 1.3.
    (b) If I is an independent set in G, show that I is a complete subgraph in $\bar{G}$, the complement of G.
    (c) A *dominating set* is a set of vertices that collectively are adjacent to all other vertices in a graph. Draw a graph whose smallest dominating set is not an independent set.

38. An *automorphism* of a graph is an isomorphism (1 − 1 mapping preserving adjacency) of the vertices of a graph with themselves.
    (a) Find all automorphisms of the graph in Figure 1.1.
    (b) Show that the set of automorphisms of a graph forms a group.

39. At a complicated intersection with special signals for left turns, and so forth, each particular signal has an associated arc on the signal timing cycle when the signal is green. A general approach to constructing a timing cycle is first to make a graph with a vertex for each signal and an edge between two vertices (signals) that control noncrossing streams of traffic. Then we try to build a family of arcs on a timing cycle such that if two arcs overlap then they must represent adjacent vertices. Build the graph and arc model for this intersection.

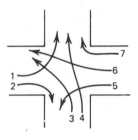

40. (a) Show that there is no way to pair off the 14 vertices in this graph with 7 edges.
    (b) Generalize part (a) to the problem of trying to use 31 dominoes to cover the 62 squares of an 8 × 8 chessboard with two opposite corner squares removed.

41. (a) A round-robin tournament can be represented by a complete directed graph with vertices for competitors and an edge $(\vec{a,b})$ if a beats b. Each competitor plays every other competitor once. A vertex's (competitor's) score will be its out-degree (number of victories). Show that if vertex x has a maximum score among vertices in a round-robin tournament, then for any other vertex y, either there is an edge $(\vec{x,y})$ or for some w there are edges $(\vec{x,w})$ and $(\vec{w,y})$.
    (b) Suppose the round-robin tournament graph has no directed circuits. We define a ranking of vertices (competitors) as follows. A vertex with no outward has a rank of 0. In general, a vertex has rank k if it has an edge directed to a rank $k − 1$ vertex and all other

edges directed to lower ranks ($\leq k - 1$). Show that a directed complete graph with no directed circuits always has such a ranking and that each vertex will have a different rank.

42. Suppose circuits C_1 and C_2 have common edges (but $C_1 \neq C_2$). Show that the edges in $(C_1 \cup C_2) - (C_1 \cap C_2)$ form a circuit (or collection of circuits).

43. If the graph G has $2n$ vertices and no triangles, then show that G cannot have more than n^2 edges.

44. A set of n different medicines is going to be tested. Each test uses three of the medicines. Each pair of medicines must be used together in exactly one test.
    (a) Restate this testing procedure in graph-theoretic terms (let each test be a triangle).
    (b) Show that the procedure is impossible for $n = 4$ or 5. Is it possible for $n = 6$?

Chapter Two
Covering Circuits and Graph Coloring

2.1 EULER CIRCUITS

In this chapter we examine two graph-theoretic concepts that have many important applications. One is covering circuits, circuits that traverse all the edges in a graph once and circuits that visit all the vertices once. The other is graph coloring, a concept introduced in Example 1 of Section 1.4.

An **Euler circuit** is a circuit that traverses all the edges in a graph once and visits each vertex at least once. In some applications of Euler circuits, such as a highway network linking a group of cities, it is natural to permit several edges to join the same pair of vertices. Such generalized graphs are called **multigraphs** (edges of the form (a,a) from a vertex to itself are also allowed in multigraphs).

Example 1

The old Prussian city of Konigsberg (now Kaliningrad) was located on the banks of the Pregel River. Part of the city was on two islands that were joined to the banks and each other by seven bridges, as shown in Figure 2.1a. The townspeople liked to take walks, or "Spaziergangen," about the town across the bridges. Several people were apparently bothered by the fact that no one could figure out a walk that crossed each bridge just once, for they brought this problem to the attention of the famous mathematician Euler. He solved

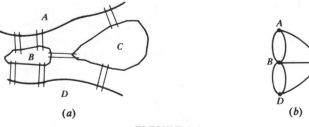

(*a*)   (*b*)

FIGURE 2.1

the Spaziergangen problem, thereby giving birth to graph theory and immortalizing the Seven Bridges of Konigsberg in mathematics texts.

We can model this walk problem with a multigraph having a vertex for each body of land and an edge for each bridge. See Figure 2.1*b*. The desired type of walk corresponds to what we now call an Euler circuit. Readers should convince themselves that no Euler circuit exists for this multigraph. ∎

A multigraph possessing an Euler circuit will have to have an even degree at each vertex, since each time the circuit passes through a vertex it uses two edges. A second obvious property is that the multigraph must be connected. Euler showed that these two properties are also sufficient to guarantee the existence of an Euler circuit. We will prove this theorem for undirected multigraphs. Let us first try to build an Euler circuit by ad hoc methods in a graph that is connected and has even-degree vertices. Then we extend our construction to a general proof of the Euler circuit theorem.

Example 2

Build an Euler circuit for the graph in Figure 2.2*a* that is connected and has even-degree vertices.

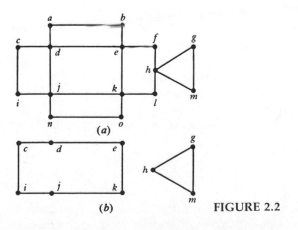

(*a*)

(*b*)   **FIGURE 2.2**

Let us start by blindly tracing out a path from vertex a. Suppose we go a-d-j-n-o-k-l-h-f-e-b-a. Now we are back to a. Note that the even degree property means that we can always leave any vertex we enter except a (there are no vertices of degree one or that are reduced to degree one after being visited several times). Thus any path we trace from a must eventually come back to a, forming a circuit.

Next consider the graph of remaining edges, shown in Figure 2.2b. It is no longer connected, but all vertices still have even degree (removing the circuit reduced each degree by an even amount). Each connected piece of the remaining graph has an Euler circuit: d-c-i-j-k-e-d and h-g-m-h. These two circuits can be inserted into the original circuit at vertices d and h, respectively, yielding the Euler circuit a-d-c-i-j-k-e-d-j-n-o-k-l-h-g-m-h-f-e-b-a. ■

Theorem

An undirected multigraph has an Euler circuit if and only if it is connected and has all vertices of even degree.

Proof

The proof generalizes the construction in Example 2. As noted above, when an Euler circuit exists, the multigraph must be connected and have all vertices of even degree.

Suppose a multigraph G satisfies these two conditions. Then pick any vertex a and trace out a path. As in Example 2, the even degree condition means that we are never forced to stop at some other vertex, and so eventually the path must terminate at a (possibly it passes through a several times before finally stopping at a). Let C be the circuit thus generated and let G' be the multigraph consisting of the remaining edges of $G - C$. As in Example 2, G' may not be connected but it must have all vertices of even degree.

Since the original graph was connected, C and G' must have a common vertex (or else there is no path from vertices in C to vertices in G'). Let a' be such a common vertex. Now build a circuit C^* tracing through G' from a' just as C was traced in G from a. Incorporate C^* into the circuit C at a', as done in Example 2, to obtain a new larger circuit C'. Repeat this process by tracing a circuit in the graph G'' of still remaining edges and incorporate the circuit into C' to obtain C''. Continue until there are no remaining edges. ■

Let us now consider an application of Euler circuits to an urban systems analysis problem.

Example 3

Suppose the graph of solid edges in Figure 2.3 represents a collection of blocks to be swept by a street sweeper in a certain district of some city from 10

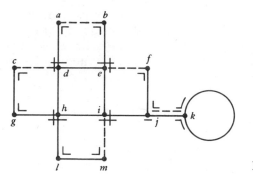

FIGURE 2.3

A.M. to 11 A.M. (when parking on these blocks is forbidden). The looping edge at *k* represents a circle. We want a tour that sweeps each solid edge once. That is, we want an Euler circuit of the solid edges. Unfortunately, the graph of edges to be swept in such applications rarely has all vertices of even degree. Frequently the graph is also not connected. In such cases we must traverse extra edges, called *deadheading edges,* to obtain the desired tour. This creates a new problem, minimizing the number of deadheading edges.

The desired tour will be an Euler circuit of the multigraph of sweeping and deadheading edges; a multigraph because some edges may be repeated in deadheading. By the theorem, this multigraph must be connected and have all vertices of even degree. Thus to minimize deadheading, the deadheading edges should be a minimal set of extra edges that when added to the original graph make all vertices of even degree and the new multigraph connected. Suppose the dashed edges in Figure 2.3 are such a minimal set of extra edges.

There is one more problem to be faced once we have a multigraph possessing an Euler circuit. We want to build an Euler circuit that minimizes turns at corners, especially U-turns. Even right-hand turns can tie up traffic in busy cities. We can successively look at each corner (vertex) and pair off the edges at the corner so as to minimize disruption on the tour's passes through that corner. For the multigraph in Figure 2.3, we would go straight through corners *d, e, h, i* on visits to these corners; we would go straight through *j* once and turn once (going between (*f,j*) and (*k,j*)); we must turn both times at *k*; and at all other corners (of degree 2) we would have forced turns. The short lines near each vertex in Figure 2.3 indicate these edge pairings. Of course, such edge pairings are very unlikely to produce a single Euler circuit. In Figure 2.3, we get two circuits from these pairings: *a-b-e-i-m-l-h-d-a* and *c-d-e-f-j-k-k-j-i-h-g-c*. Now we break the two circuits at a common vertex and fuse them together. For example, at *d*, change the edge pairings to

The result is an Euler circuit with optimal pairings at all but one corner.

In practice, a large street network with 200 edges typically forms only three or four circuits when minimal-disruption edge pairing is performed at each corner, and so only a few pairings need to be changed. (In practice, it is also necessary to use directed graphs since street sweepers cannot move against the traffic on their side (curb) of the street.)　　　　　■

It is interesting to note that the street sweeping problem gives rise to a nice alternative proof of the Euler circuit theorem:

Alternative Proof of Theorem

The even-degree condition means that we can pair off (and link together) the edges at each vertex. These linked edges form a set of circuits. The connectedness condition means that these circuits have common vertices where they can be joined to form a single circuit—an Euler circuit.　　　　　■

This proof is not as intuitive or pictorial as our original path-tracing method, but it is both simpler and more applicable.

In closing we mention a basic corollary of the Euler circuit theorem. Its proof is left as an exercise. An **Euler path** is a path containing all edges and touching all vertices.

Corollary

A multigraph has an Euler path, but not Euler circuit, if and only if it is connected and has exactly two vertices of odd degree.

EXERCISES

SUMMARY OF EXERCISES　The first five exercises involve trying to build Euler circuits. Exercises 6–13 present extensions and other questions related to the Euler Circuit Theorem. The remaining questions involve some modeling and further theory, and the last two exercises ask for computer programs.

1. (a) Build an Euler circuit for the right graph in Figure 1.12.
   (b) Build an Euler path for the graph in Figure 1.22a with edge (e,g) removed.

2. (a) For which values of n does K_n, the complete graph on n vertices, have an Euler circuit?
   (b) Are there any K_n that have Euler paths but not Euler circuits?

3. Suppose a new artificial island were placed in the Pregel River inbetween islands B and C in Figure 2.1a. If the bridge from B to C is removed and several bridges built connecting the new island to all other bodies of land (any number of bridges to other bodies of land allowed),

it is possible to have a Spaziergangen that is an Euler circuit. Demonstrate.

4. What is the minimum number of times one must raise one's pencil in order to draw the graph in Figure 1.3?

5. Try to find a minimal set of edges in the graph below whose removal produces an Euler circuit. (*Hint:* Tricky.)

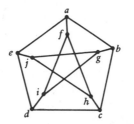

6. Prove the corollary. (*Hint:* Add an edge between the two vertices of odd degree.)

7. Prove the directed version of the theorem: A directed multigraph has a directed Euler circuit if and only if the multigraph is connected (when directions are ignored) and the in-degree equals the out-degree at each vertex.
    (a) Model your proof after the argument in the proof of the theorem.
    (b) Model your proof after the argument in the alternate proof of the theorem.
    (c) Build a directed Euler circuit for the graph below.

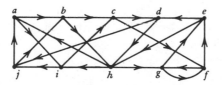

8. State and prove a directed multigraph version of the corollary.

9. A directed graph is called *strongly connected* if there is a directed path from any given vertex to any other vertex. Show that if a directed graph possesses a directed Euler circuit, then it must be strongly connected.

10. (a) Can a graph with an Euler circuit have a bridge (an edge whose removal disconnects the graph)? Prove or give a counterexample.
    (b) Give an example of a 10-edge graph with an Euler path that has a bridge.

11. Why would a mathematician say that a 1-vertex graph (with no edges) has an Euler circuit?

12. Suppose that in the definition of an Euler circuit, we drop the seemingly superfluous requirement that the Euler circuit visit every vertex and require only that the circuit include every edge. Show that now the

theorem is false. Draw a graph that illustrates why the theorem is now false.

13. (a) Prove that if a connected graph has a $2k$ vertices of odd degree, then there are k disjoint paths that contain all the edges.

(b) State and prove the directed graph counterpart of part (a).

14. Is it possible for a knight to move around an 8 by 8 chessboard so that it makes every possible move exactly once (consider a move between two squares connected by a knight to be completed when the move is made in either direction)?

15. The matrix below marks with a 1 each pair of the set of racers A, B, C, D, E, F who are to have a drag race together. It is most efficient if a racer can run in two races in a row (but not three in a row). Is it possible to design a sequence of races such that one of the racers in each race (except the last race) also runs in the following race (but not three in a row)? If possible, give the sequences of races (pairs of racers); if not, explain why not.

	A	B	C	D	E	F
A	—	1	0	1	1	0
B	1	—	1	1	0	1
C	0	1	—	0	1	0
D	1	1	0	—	1	1
E	1	0	1	1	—	1
F	0	1	0	1	1	—

16. The *line graph* $L(G)$ of a graph G has a vertex for each edge of G and two of these vertices are adjacent if and only if the corresponding edges in G have a common end vertex.

(a) Show that $L(G)$ has an Euler circuit if G has an Euler circuit.

(b) Find a graph G which has no Euler circuit but for which $L(G)$ has an Euler circuit.

17. Show that the dual of a planar Eulerian graph is bipartite; dual is defined in Example 1 of Section 1.4, bipartite in Example 3 of Section 1.1.

18. Show that in Example 3 the minimum set of deadheading edges in any sweeping problem will be a collection of edges forming paths between pairs of different odd-degree vertices.

19. Consider the following algorithm due to Fleury for building an Euler circuit, when one exists, in a single pass through a graph (without later adding side circuits as in the proof of the theorem). Starting at a chosen vertex a, build a path and erase edges after they are used (also erase vertices when they become isolated points). The one rule to follow in the path-building is that one never chooses an edge whose erasure will disconnect the resulting graph of remaining edges.

    (a)   Apply this algorithm to build Euler circuits for the graph in: (i) Figure 2.2*a* and (ii) Figure 2.3 (including deadheading edges).

    (b)   Prove that this algorithm works.

    (c)   Does this algorithm work for Euler paths? Explain.

20.   Suppose we are given an undirected connected graph representing a network of two-way streets.

    (a)   Show that there always exists a tour of the network in which a person drives along each side of every street once.

    (b)   Show that the tour in part a can be generated by the following rule: At any intersection do not leave by the street first used to reach this intersection unless all other streets from the intersection have been used.

21.   A set of 16 binary digits (0 or 1) are equally spaced about the edge of a disk. We want to choose the digits so that they form a circular sequence in which every subsequence of length four is different. Model this problem with a graph with eight vertices, one for each different subsequence of three digits. Make a directed edge for each subsequence of four digits whose origin is the vertex with the first three digits of the edge's subsequence and whose terminus is the vertex with the last three digits of the edge's subsequence.

    (a)   Build this graph.

    (b)   Show how an Euler circuit (which exists for this graph) will correspond to the desired 16-digit circular sequence.

    (c)   Find such a 16-digit circular sequence with this graph model.

    (d)   Repeat the problem for ternary (0, 1, or 2) sequences: build a 27-digit circular ternary sequence in which each three-digit subsequence is different. Draw the associated graph model.

22.   Explain how to alter the various representations of graphs given in Supplement I of Chapter 1 for multigraphs.

23.   Write computer programs for finding an Euler circuit, when one exists, in a multigraph using:

    (a)   The method in the proof of the theorem.

    (b)   Repeat part a for a directed graph.

    (c)   The method in the alternate proof of the theorem.

    (d)   Modify the program in parts b or c to build the *k* paths in Exercise 13(a).

24.   Write a program to implement the algorithm in Exercise 19.

2.2 HAMILTON CIRCUITS

In this section we explore **Hamilton** circuits and paths, simple circuits and paths that visit each vertex in a graph once. Here, **simple** means that the circuit or path passes through a vertex at most one time. Hamilton circuits arise in operations research problems involving the route of, say, a traveling salesperson who must visit a set of locations. In these applications, we want to find a minimal-cost Hamilton circuit (this problem is discussed in Section 3.3).

Finding Hamilton circuits by inspection, when they exist, is usually not too hard in moderate-sized graphs; but proving that no Hamilton circuit exists in a given graph can be very difficult. Such a proof involves the same types of reasoning needed to show that two graphs are not isomorphic. At the end of this section we present some special conditions on a graph that guarantee the existence of a Hamilton circuit. We also give a necessary condition about planar graphs with a Hamilton circuit that can be very useful.

We will concentrate on the problem of showing that a Hamilton circuit does not exist, for the nonexistence problem requires the type of systematic logical analysis that is the essence of most applied graph theory. To prove nonexistence, we must begin building parts of a Hamilton circuit and show that the construction must always fail, that is, we cannot get to all vertices without visiting some vertex twice. The following examples will demonstrate how such contradictions can be obtained. First let us state three basic rules that must be followed in building Hamilton circuits. The idea underlying these rules is that a Hamilton circuit must contain exactly two edges incident at each vertex.

Rule 1.   If a vertex x has degree 2, both edges incident at x must be part of any Hamilton circuit.

Rule 2.   No proper subcircuit, that is, a circuit not containing all vertices, can be formed when building a Hamilton circuit.

Rule 3.   Once the Hamilton circuit we are building has passed through a vertex x, all other (unused) edges incident at x can be deleted (since they cannot be used later in the circuit).

Note that in deleting edges as permitted by Rule 3, we may reduce the degree of other vertices to 2, so that Rule 1 can be applied.

Example 1

Show that the graph in Figure 2.4 has no Hamilton circuit.

We can apply Rule 1 at vertices a, b, d, and e. To indicate that the two edges at each of these vertices must be used, we draw a little line segment from one edge to the other, as shown in Figure 2.4. There are two types of contradic-

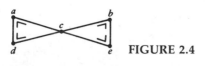

FIGURE 2.4

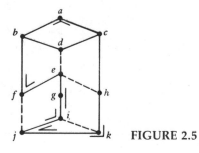

FIGURE 2.5

tions that have now arisen. First, using Rule 1 for vertices a and d forces the Hamilton circuit to use edges (a,c), (a,d), (d,c) forming a triangle; this violates Rule 2. Second, using Rule 1 for all four vertices of degree 2 forces the Hamilton circuit to contain more than two edges incident at c. Either of these difficulties proves that this graph has no Hamilton circuit.   ■

Example 2

Show that the graph in Figure 2.5 has no Hamilton circuit.

We can apply Rule 1 at vertices a and g, so that the subpaths b-a-c and e-g-i must be part of any Hamilton circuit. Next consider vertex i. We already know that (g,i) must be part of the circuit. Since the graph is symmetric with respect to edges (i,j) and (i,k), it does not matter which of these two edges we choose as the other edge incident to i on the circuit. Suppose we pick (i,j) (if we obtain a contradiction using (i,j), then we would also obtain a contradiction with (i,k)). Now by Rule 3 we can delete the other edge at i (i,k). See Figure 2.5.

Deleting (i,k) reduces the degree of k to 2, and so Rule 1 says we must use both remaining edges incident at k, (j,k) and (h,k). Adding the edge (j,k) to the circuit now means that vertex j has two edges on the circuit, and j's other edge (f,j) must be deleted. This deletion reduces the degree of f to 2. See Figure 2.5. So we must use the two remaining edges at f, (b,f) and (f,e). Adding (f,e) gives e two edges on the circuit, and so e's other edges, (e,d) and (e,h), must be deleted.

Now we run into some contradictions. Looking at d which now has degree 2, we must use edges (b,d) and (c,d), but we then have the subcircuit a-b-d-c-a. We also have placed three edges incident to b in the Hamilton circuit. Either of these inconsistencies proves that this graph has no Hamilton circuit.

Note that this graph does have a Hamilton path, for example, a-b-f-e-g-i-j-k-h-c-d.   ■

Example 3

Show that the graph in Figure 2.6 has no Hamilton circuit.

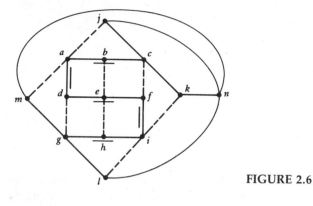

FIGURE 2.6

Note that this graph has vertical and horizontal symmetry (although vertex *n* is off to one side, its adjacencies are symmetrical). It sometimes takes a bit of trial-and-error experimenting to find a good vertex at which to start trying to build a Hamilton circuit when there are no vertices of degree 2. We seek a vertex with the property that once two edges are chosen at the vertex, then the use of Rules 1 and 3 will force the successive deletion and inclusion of many edges. Vertex *e* is such a vertex. We can either use two edges incident at *e* from opposite sides (180° apart) or use two edges incident at *e* that form a 90° angle. We examine both cases.

Case I

Suppose we use two edges incident at *e* from opposite sides. By symmetry, they can either be edges from *d* and *f* or from *b* and *h*. Suppose we choose (*d,e*) and (*e,f*). Then by Rule 3, we can delete edges (*e,b*) and (*e,h*). Then at *b* and at *h* we must use both remaining edges (*a,b*), (*b,c*) and (*g,h*), (*h,i*). Now at *d* we can either use edge (*d,a*) or (*d,g*). The two cases are symmetrical with respect to the edges chosen for the circuit thus far. Thus without loss of generality, we can choose edge (*d,a*).

At *f*, we cannot use (*f,c*) or else subcircuit *a-b-c-f-e-d-a* results. So we must use (*f,i*). See Figure 2.6. Having used two edges at vertices *a* and *i*, the other edges at these vertices can be deleted by Rule 3. See Figure 2.6. We obtain several inconsistencies. Vertices *j* and *k* currently have degree 2, but adding their two remaining edges to the circuit creates three edges on the circuit incident to *c*, (*b,c*), (*j,c*), and (*k,c*) (a similar difficulty also occurs at *g*). We conclude that there is no Hamilton circuit in Case I.

Case II

Suppose we use two edges incident at *e* that form a 90° angle. By symmetry it does not matter which of the four 90° angle pairs of edges we choose. Suppose we choose (*b,e*) and (*d,e*). Then by Rule 3 we can delete edges (*e,f*) and (*e,h*). Then at *f* and *h* we must use the two remaining edges (*c,f*), (*f,i*) and

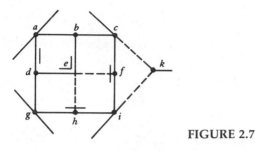

FIGURE 2.7

(g,h), (h,i). See Figure 2.7. If at d we choose (d,a) and at b choose (b,a), we get the subcircuit a-b-e-d-a. If at d we choose (d,g) and at b (b,c), we get the subcircuit b-c-f-i-h-g-d-e-b. So either at d and b we must choose (d,a) and (b,c), or must choose (d,g) and (b,a). By symmetry either choice is equivalent. Let us choose (d,a) and (b,c). Now having used two edges at c and i, we can delete the other edges at these vertices. However this leaves only one edge at k. This impossibility proves that there is no Hamilton circuit in Case II.

Having obtained contradictions in both cases at e, we can conclude that the graph has no Hamilton circuit. ∎

Often it is possible to use symmetry or other arguments to find a unique way to begin building a Hamilton circuit. For example, in Figure 2.6 only two of the vertices b, d, f, and h will use their edge into e, and so by symmetry we could assume that, say, d is a vertex not using its edge to e. That is, edges (d,a) and (d,g) are on the Hamilton circuit. Then at m we could use either edge (m,a) or (m,g) along with (m,n) (we cannot use (m,a) and (m,g) or else the subcircuit a-d-g-m-a results). By symmetry we could assume we choose (m,a). However now at g or at n we have different (nonsymmetric) choices and will have to consider several cases.

We present a few of the theoretical results about the existence of Hamilton circuits and paths. (See [4] for proofs of these theorems.)

Theorem 1

A graph cannot have a Hamilton circuit if the deletion of some set of k vertices results in more than k connected components in the remaining graph.

Theorem 2

A connected graph with n vertices, $n > 2$, has a Hamilton circuit if the degree of each vertex is at least $n/2$.

Theorem 3

A complete directed graph on n vertices (with an edge in one direction or the other between every pair of vertices) must have a directed Hamilton path.

Theorem 4

Let G be a connected graph with n vertices, and let the vertices be indexed x_1, $x_2, \ldots, x_n$ so that $\deg(x_i) \leq \deg(x_{i+1})$. If for each $k \leq n/2$, either $\deg(x_k) > k$ or $\deg(x_{n-k}) \geq n - k$, then G has a Hamilton circuit.

Theorem 5

Suppose a planar graph G has a Hamilton circuit $\mathcal{H}$. Let G be drawn with any planar depiction, and let r_i denote the number of regions inside the Hamilton circuit bounded by i edges in this depiction. Let r_i' be the number of regions outside the circuit bounded by i edges. Then the numbers r_i and r_i' satisfy the equation

$$\sum_i (i - 2)(r_i - r_i') = 0 \tag{*}$$

Just as the inequality $\mathbf{e} \leq 3\mathbf{v} - 6$ for planar graphs in the corollary in Section 1.4 could be used to prove that some graphs are not planar, Theorem 5 can be used to show that some planar graphs cannot have Hamilton circuits.

Example 4

Show that the planar graph in Figure 2.8 has no Hamilton circuit.

We have indicated the number of bounding edges inside each of the regions in the planar depiction of the graph in Figure 2.8. There are three regions with four edges and six regions with six edges. Thus no matter where a Hamilton circuit is drawn (if it existed), we know that $r_4 + r_4' = 3$ and $r_6 + r_6' = 6$. Observe that for this graph, (*) reduces to

$$2(r_4 - r_4') + 4(r_6 - r_6') = 0$$

We cannot have $r_6 - r_6' = 0$, that is, $r_6 = r_6' = 3$, for then (*) would require $r_4 - r_4' = 0$ or $r_4 = r_4'$—which is impossible since $r_4 + r_4' = 3$. If $r_6 - r_6' \neq 0$, then $|r_6 - r_6'| \geq 2$ and so $|4(r_6 - r_6')| \geq 8$. Now it is impossible to satisfy (*) since even if $r_4 = 3$, $r_4' = 0$ (or $r_4 = 0$, $r_4' = 3$), $|2(r_4 - r_4')| \leq 6$. Thus it is impossible for (*) to be valid for this graph, and so no Hamilton circuit can exist. ∎

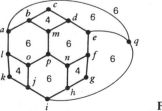

FIGURE 2.8

EXERCISES

SUMMARY OF EXERCISES   The first seven exercises involve the existence or nonexistence of Hamilton paths and circuits. Exercises 8 and 9 introduce other potential aids for proving nonexistence. Exercises 10–16 involve applications of Hamilton circuits. The last five exercises involve theory.

1.   (a)   Draw a graph with a Hamilton circuit but no Euler circuit.
      (b)   Draw a graph with an Euler circuit but no Hamilton circuit.

2.   Find a Hamilton circuit in each of the following graphs.

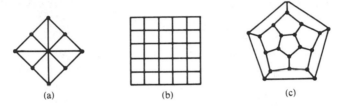

3.   Find another subcircuit in Example 2 formed by undeleted edges.

4.   Find a Hamilton path in each of the following graphs and prove that no Hamilton circuit exists:

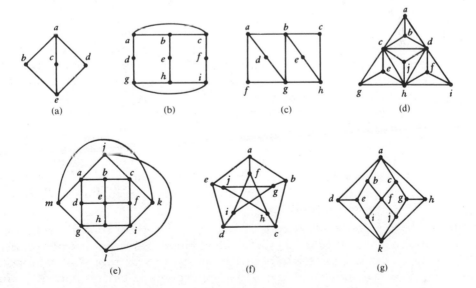

5.   Find a Hamilton path in the Figure 2.8 and show that no Hamilton circuit exists.

6.   (a)   Prove that the following graphs have no Hamilton circuits:

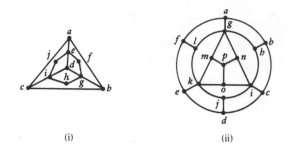

(i)   (ii)

(b) Prove that neither graph in part a has a Hamilton path.

7. Use Theorem 5 to show that the following planar graphs have no Hamilton circuit:
    (a) Exercise 4(a)   (b) Exercise 4(b)   (c) Exercise 4(g)

8. Recall from Example 3 in Section 1.1 that a graph is *bipartite* if the vertices can be divided into two sets, for convenience call them blue vertices and red vertices, such that every edge connects a blue and a red vertex.
    (a) Show that if a connected bipartite graph has a Hamilton circuit, then the numbers of red and blue vertices must be equal; if it has a Hamilton path, then the numbers of reds and blues can differ by at most one.
    (b) Use part a to show that the following graphs have no Hamilton circuit: (i) Exercise 4(g); (ii) Exercise 6(a-ii).

9. Suppose a set I of k vertices in a graph G is chosen so that no pair of vertices in I are adjacent. Then for each x in I, $\deg(x) - 2$ of the edges incident to x will not be used in a Hamilton circuit. Summing over all vertices in I, we have $e' = \Sigma_{x \in I}(\deg(x) - 2) = \Sigma_{x \in I}\deg(x) - 2k$ edges that cannot be used in a Hamilton circuit.
    (a) Let v and e be the numbers of vertices and edges in G, respectively. Show that if $e - e' < v$, then G can have no Hamilton circuit.
    (b) Why is part a valid only when I is a set of nonadjacent vertices?
    (c) With a suitably chosen set I, use part a to show that the following graphs have no Hamilton circuits: (i) Figure 2.6; (ii) Exercise 4(g); (iii) Exercise 6(a-ii).

10. A *Gray code* is a binary code in which each number from 1 through n, where $n = 2^r$, is coded as some r-digit binary sequence (not necessarily the binary representation of that number) with the property that for all consecutive numbers i, $i + 1$, the corresponding binary sequences are the same in all but one position.
    (a) Show that a Gray code is equivalent to a Hamilton path in a certain graph whose vertices represent all r-digit binary sequences. Describe this graph.

(b)   For $n = 2^3$, use a graph model to find a Gray code.

(c)   Repeat part b for $n = 2^4$.

11.   Show without citing any Theorems stated in this section that any six-vertex undirected graph with all vertices of degree 3 has a Hamilton circuit.

12.   (a)   Describe how to construct a circuit including all squares of an $n \times n$ chessboard, n even, using a rook. Using a king.

(b)   Repeat part a for n odd.

13.   Find a path of knight's moves visiting all squares exactly once on an 8×8 chessboard.

14.   Suppose a classroom has 25 students seated in desks in a square 5×5 array. The teacher wants to alter the seating by having every student move to an adjacent seat (just ahead, just behind, on the left, or on the right). Show by a parity argument that such a move is impossible.

15.   Consider 27 little cubes arranged in a 3 by 3 by 3 array (as in Rubik's Cube). Form an associated graph with 27 vertices, one for each little cube, and with two vertices adjacent if they have touching common faces. Does this graph have a Hamilton circuit starting at the vertex corresponding to the middle inside cube and ending at one of the vertices corresponding to a corner cube?

16.   (a)   How many different Hamilton circuits are there in K_n, a complete graph on n vertices?

(b)   Show that K_n, n odd, can have its edges partitioned into $\frac{1}{2}(n - 1)$ disjoint Hamilton circuits.

(c)   If 15 professors dine together at a circular table during a conference, and if each night each professor sits next to a pair of different professors, how many days can the conference last?

17.   (a)   If a graph G has an Euler circuit, show that $L(G)$, the line graph of G (see Exercise 16 of Section 8.1), has a Hamilton circuit.

(b)   If G has a Hamilton circuit, show that $L(G)$ has a Hamilton circuit.

(c)   Show that the converses of parts a and b are false by finding counterexamples.

18.   Prove Theorem 1.

19.   (a)   Prove Theorem 3. (*Hint:* Use induction.)

(b)   Show that in a round-robin tournament it is always possible to rank the contestants so that the person ranked ith beat the person marked $(i + 1)$st. (*Hint:* Use part a.)

(c)   Show that if G is not complete graph, then it is possible to direct the edges of G so that there is no directed Hamilton path.

20.   Show that Theorem 2 is false if the requirement of degree $\geq \frac{1}{2}n$ is relaxed to just $\geq \frac{1}{2}(n - 1)$.

21. (a) Prove that an undirected graph with n vertices and at least $\binom{n-1}{2} + 2$ edges must have a Hamilton circuit.

    (b) Show that part a is false if there are only $\binom{n-1}{2} + 1$ edges.

2.3 GRAPH COLORING

In Example 1 of Section 1.4 we introduced the problem of map coloring—coloring the countries of a map so that two countries with a common border are assigned different colors. The problem of showing that any map can be 4-colored tantalized mathematicians for 100 years until a computer-assisted proof was obtained by Appel and Haken in 1976. More recently, graph coloring has been applied to a variety of problems in computer science, operations research, and design of experiments. Recall that coloring countries in a map is equivalent to coloring vertices, with adjacent vertices getting different colors, in the dual graph obtained by making a vertex for each country and an edge between vertices (countries) with a common border; see Example 1 in Section 1.4. In general, a **coloring** of a graph G assigns colors to the vertices of G so that adjacent vertices are given different colors.

In this section we show how to determine the minimal number of colors required to color a given graph. This minimal number of colors is called the **chromatic number** of a graph. We also give some applications of graph coloring. In the next section we will present some theorems about graph coloring.

For graphs with 15 or less vertices, it is usually not difficult to guess a graph's chromatic number. To verify rigorously that the chromatic number of a graph is a number k, we must also show that the graph cannot be properly colored with $k - 1$ colors. Proving that a graph cannot be $k - 1$-colored is similar to proving that a graph has no Hamilton circuit or cannot be isomorphic to another particular graph. In this case, the goal is to show that any $k - 1$-coloring we might construct for the graph must force two adjacent vertices to have the same color.

Example 1

Find the chromatic number of the graph in Figure 2.9.

Looking at the inner square with crossing diagonals, we see that vertices a, c, e, g are mutually adjacent, that is, form a complete subgraph. They each require a different color in a proper coloring, four colors in all. Once four colors are available, it is easy to properly color the remaining vertices b, d, f, h. Each of them is only adjacent to two other vertices, and so at most two out of the four colors need ever be avoided with these vertices. Let us use the numbers 1, 2, 3, 4 as the "names" of our colors. Then one possible 4-coloring of the graph is shown in Figure 2.9.

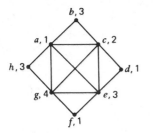

FIGURE 2.9

In this problem it is immediate that the graph cannot be 3-colored, since some adjacent pair of vertices in the complete subgraph formed by a, c, e, g would have the same color in a 3-coloring. So the chromatic number of this graph is 4. ∎

Example 1 points up two important rules. First, a complete subgraph on k vertices requires k colors (cannot be $k - 1$-colored). Second, when building a k-coloring of some graph, we can ignore all vertices of degree $<k$ (and their incident edges), since once the other vertices are colored, there will always be at least one color available (not used by any adjacent vertex) to properly color each such vertex.

Example 2

Find the chromatic number of the graph in Figure 2.10.

We note that a graph of this form is called a **wheel**. The largest complete subgraph in this graph is a triangle. Let us try to build a 3-coloring of this graph with "colors" 1, 2, and 3. We will start by coloring the vertices of a triangle. Suppose we choose the triangle a, b, f: let a be 1, b be 2, and f be 3 (the order of colors is arbitrary). See Figure 2.10. Since c is adjacent to vertices b and f of colors 2 and 3, respectively, c is forced to be color 1. Similarly, d is forced to be 2, and then e is forced to be 1. However, now the adjacent vertices a and e both have color 1. Thus the graph cannot be 3-colored. On the other hand, using a fourth color for e yields a proper coloring. So the chromatic number of this graph is 4. ∎

Observe that in Example 2 if vertex e were missing and d were adjacent to a instead, then three colors would work. In general, wheel graphs with an even

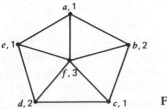

FIGURE 2.10

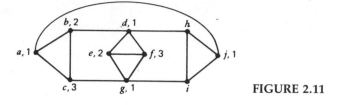

FIGURE 2.11

number of "spokes" can be 3-colored, while wheels with an odd number of spokes require four colors.

The key to the impossibility of finding a 3-coloring in Example 2 is the sequence c, d, e of forced vertex colors. In general, when attempting to build a k-coloring graph, it is desirable to start by k-coloring a complete subgraph of k vertices and then successively finding vertices adjacent to $k - 1$ different colors, thereby forcing the color choice of such vertices.

Example 3

Find the chromatic number of the graph in Figure 2.11.

The largest complete subgraph is again a triangle, and so we want to try building a 3-coloring. The only triangles whose coloring will force the color of another vertex are d, e, f and e, f, g. Suppose we color d 1, e 2, and f 3. Then g is forced to be 1. Now we are in trouble, since no more uncolored vertices are adjacent to two colors.

Observe that b and c are both adjacent to a vertex of color 1 and are adjacent to each other. Thus one of b and c must be color 2 and the other color 3. By the symmetry of the graph, we can assume b is 2 and c 3. Then b and c force the color of a to be 1. Similarly h and i will be 2 and 3, collectively, forcing j to be 1. But the adjacent pair a, j are both 1. Thus the graph cannot be 3-colored. Making j a fourth color yields a proper 4-coloring, and so the graph's chromatic number is 4. ∎

Let us now look at some applications of graph coloring in computer science and operations research.

Example 4

In an optimizing compiler, the computation in loops of a program is speeded up by temporarily storing the values of frequently used variables in index registers in the central processor (instead of storing the values in the slower regular memory). A problem that arises in this optimization is, how many index registers are needed to provide temporary storage for certain variables in a loop. The diagram in Figure 2.12a shows the periods in an 8-step loop during which four variables need temporary storage. Variable A requires two periods of storage (not both necessarily in the same index register). Variable D

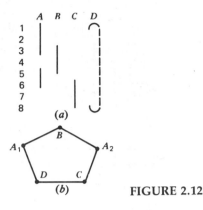

FIGURE 2.12

requires storage for the (continuous) period of step 8 and step 1 (during the next iteration of the loop). How many different index registers are needed?

It is easy to check that although at most two variables require storage at any one time, two index registers are not sufficient. The difficulty is most easily seen with a graph model. Let us build a graph with a vertex for each period of storage for some variable and an edge joining each pair of vertices corresponding to overlapping periods. See the graph in Figure 2.12b. Then coloring this graph corresponds to assigning index registers to storage periods. It is easy to show that the odd-circuit graph in Figure 2.12b cannot be 2-colored but can be 3-colored. ∎

Example 5

In a simplified (one-dimensional) form of Very Large Scale Integrated circuit (VLSI) design, one has a row of gates $G_1, G_2, \ldots, G_n$ (gates are building blocks for logical circuits) and specified connections between various pairs of gates. These connections are laid in a set of parallel tracks on top of the gates. Two connections must go in different tracks if they would overlap, even if just at a common endpoint. In Figure 2.13a five connections between five gates are realized with three tracks: track one has connections (1,3) and (4,5); track two (1,2) and (3,4); and track three (2,5).

What is the minimum number of tracks needed for the following six connections among a row of five gates: (1,3), (1,3)—two copies of this connection—(2,3), (2,4), (2,5), (4,5)?

The set of connections can be treated as a family of intervals on a line. Two intervals (connections) can have the same track if they do not overlap. We form the associated interval graph: vertex-interval, and edge-overlap between intervals. See Figure 2.13b. (Interval graphs were introduced in Section 1.1.) Now let colors stand for tracks. A minimal coloring of the interval graph will correspond to a minimal track assignment. Since the interval graph contains a complete graph on five vertices (excluding vertex (4,5)), five colors (tracks) are needed.

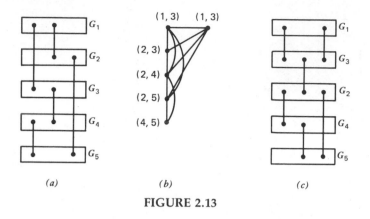

FIGURE 2.13

Frequently the order of the gates is under the designer's control. Is it possible to order the gates differently so as to reduce the number of tracks in the previous design problem?

Changing the gate order produces a different set of connection intervals, and so may change the chromatic number of the associated interval graph. By inspection, we observe that with the gate order 1, 3, 2, 4, 5, only three gates are needed. See Figure 2.13c. (No simple procedure is known for finding the best gate order.) ∎

Example 6

We now consider a complex optimization problem in which graph coloring plays only a secondary role. There is a set of sites S_i that must be serviced (visited) k_i times each week ($1 \le k_i \le 6$). We need a minimal, or near-minimal, set of day-long truck tours for a week such that each site is visited on k_i of the tours. In addition, we require that these tours can be partitioned among the six days of the week (Sunday is excluded) in a manner so that no site is visited twice on one day. Even if each site is visited just once a week, this is an extremely difficult problem which cannot be solved exactly.

When a problem of this type (involving garbage collection) was analyzed for the New York City Department of Environmental Protection, an algorithm was used that started with an inefficient set of tours and successively tried to improve the set of tours (this method gave only near-optimal tours). Suppose we had a simplified situation where the week contained just three work days, and our algorithm had generated the set of tours shown in Figure 2.14a. Can these six tours be so partitioned?

Given a set of tours, we form an associated tour graph with one vertex for each tour and with two vertices adjacent if they correspond to two tours which visit a common site. In the general six-day problem, partitioning the tours among the six days is equivalent to 6-coloring the tour graph. In the tour graph in Figure 2.14b, we desire just a 3-coloring, which indeed exists. If the optimizing algorithm were next to combine tours A and F in Figure 2.14a (to

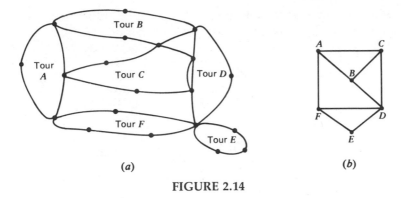

FIGURE 2.14

get a smaller set of tours), such a move would have to be blocked (and other optimizing moves tried instead) because the resulting tour graph would have a complete graph on the four vertices A, B, C, D; this would require four colors (days). In the original six-day case, to insure that the final set of tours has a tour graph that can be 6-colored, we would check each improvement tried by the algorithm to be sure that the resulting tour graph can be 6-colored. ■

EXERCISES

SUMMARY OF EXERCISES   Exercises 1–7 involve finding minimal vertex colorings and associated problems. Exercises 8–10 require minimal coloring of maps and a geometric array. Exercises 11–17 are color modeling problems.

1. Find the chromatic number of each of the following graphs. Give a careful argument to show that fewer colors will not suffice.

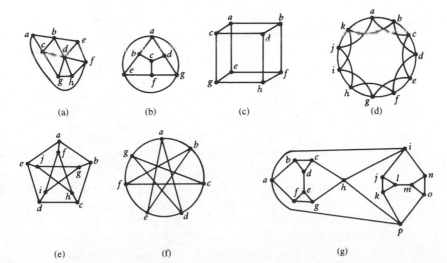

2. Find a minimal edge coloring of the following graphs (color edges so that edges with a common end vertex receive different colors).

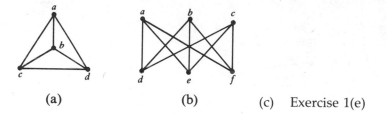

    (a)   (b)   (c)   Exercise 1(e)

3. A graph G is *color critical* if the removal of any vertex of G decreases the chromatic number.
    (a)   Which graphs in Exercise 1 are color critical?
    (b)   Find the largest color critical subgraph of each graph in Exercise 1 that is not color critical.

4. Find all sets of three or more vertices that could have the same color in a proper coloring of the graph in:
    (a)   Exercise 1(a)   (b)   Exercise 1(b)

5. A coloring partitions a graph G into sets of mutually nonadjacent vertices. In the complement $\bar{G}$ of G, this partition becomes a partition of $\bar{G}$ into sets of mutually adjacent vertices, that is, complete subgraphs. Find such a minimal set of complete subgraphs partitioning the vertices in the graph in:
    (a)   Exercise 1(a)   (b)   Exercise 1(b)   (c)   Exercise 1(d)
    (d)   Exercise 1(g)

6. An *equitable coloring* is a minimal coloring in which the numbers of vertices of each color differ by at most one. Which of the following graphs have minimal colorings that are equitable?
    (a)   Exercise 1(a)   (b)   Exercise 1(c)

7. The *chromatic polynomial* $P_k(G)$ of a graph G is a polynomial in k that gives for each positive integer k, the number $P_k(G)$ of different k-colorings of G. Determine $P_k(G)$ for the following graphs:

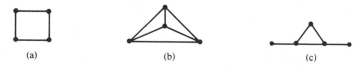

    (a)   (b)   (c)

    (d)   If we know $P_k(G)$, how can we use it to determine the chromatic number of G?

8. Can the 50 states in a map of the United States be properly 3-colored? (Note that states meeting only at a corner, such as Colorado and Arizona, are not considered adjacent.)

9. Suppose a map is made by drawing n mutually intersecting circles. Show that the regions in this map can be properly 2-colored.

(a)   Solve by an inductive argument.
(b)   Solve by showing that the dual graph is bipartite.
(c)   Solve by assigning colors based on the number of circles that contain a region.

10.   How many colors are needed to color the 15 billiard balls in this triangular array with touching balls different colors?

11.   A state legislature has a set of 23 committees. Each committee is supposed to meet one hour each week. What is wanted is a weekly schedule of committee meeting times that uses as few different hours in the week as possible (so as to maximize the time available for other legislature activities). The one constraint is that no legislator should be scheduled to be in two different committee meetings at the same time. Model and restate this scheduling problem as a graph coloring problem.

12.   A banquet center has eight different special rooms. Each banquet requires some subset of these eight rooms. Suppose that there are 12 evening banquets that we wish to schedule in a given week (seven days). Two banquets that are scheduled on the same evening must use different special rooms. Model and restate this scheduling problem as a graph coloring problem.

13.   (a)   A set of solar experiments are to be made at observatories. Each experiment begins on a given day of the year and ends on a given day (each experiment is repeated for several years). An observatory can only perform one experiment at a time. The problem is, what is the minimum number of observatories required to perform a given set of experiments annually? Model this scheduling problem as a graph coloring problem.
      (b)   Suppose experiment A runs from Sept. 2 to Jan. 3, experiment B from Oct. 15 to March 10, experiment C from Nov. 20 to Feb. 17, experiment D from Jan. 23 to May 30, experiment E from April 4 to July 28, experiment F from April 30 to July 28, and experiment G from June 24 to Sept. 30. Draw the associated graph and find a minimal coloring (show that fewer colors will not suffice).

14.   A convention of mathematicians will have rooms available in six hotels. There are n mathematicians and because of personality conflicts, various pairs of mathematicians must be put in different hotels. The organizers wonder whether six hotels will suffice to separate all conflicts. Model this conflict problems with a graph and restate the problem in terms of vertex coloring.

15. Which of the following pairs of tours in Figure 2.14a can be combined without violating the 3-colorability requirement of the tour graph?
    (a) Tours *D* and *E*.   (b) Tours *C* and *D*.

16. In a round-robin tournament where each pair of *n* contestants plays each other, a major problem is scheduling the play over a minimal number of days (each contestant plays at most one match a day).
    (a) Restate this problem as an edge coloring problem (see Exercise 2).
    (b) Solve this problem for *n* = 6.

17. Consider a graph representing games played between a set of football teams with a directed edge from vertex *A* to vertex *B* if team A beats team B. Suppose that it is known that this graph has *no directed circuits* (i.e., no situation such as A beats B, B beats C, and C beats A). We can define a set of levels in this football graph as follows: A vertex with no outward edges (the team beat no other team) is at level 0; if all a vertex's outward edges go to level-0 vertices (no-win teams), then the vertex is at level 1; and in general, a vertex is at level *k* if the greatest level of a team it beat is level *k* − 1. Show that the level number of each vertex is a proper "coloring" of the vertices.

2.4 COLORING THEOREMS

We now present five representative coloring theorems. We will prove only the last one, the 5-color theorem for planar graphs. For proofs of Theorems 1–4 see [4]; Theorems 2 and 4 are also proved in the exercises. Let $\chi(G)$ denote the chromatic number of the graph *G*.

Theorem 1—Brook's Theorem

If the graph *G* is not an odd circuit or a complete graph, then $\chi(G) \leqslant d$, where *d* is the maximum degree of a vertex of *G*.

The maximum degree is for most graphs a poor upper bound on $\chi(G)$. The examples in Section 2.3 had $\chi(G)$ closely related to the size of the largest complete subgraph. Thus it seems natural that there should be a good bound on $\chi(G)$ in terms of the size of the largest complete subgraph. The following theorem shows that this approach fails.

Theorem 2

For any positive integer *k*, there exists a triangle-free graph *G* with $\chi(G) = k$.

Instead of coloring vertices, we can color edges so that edges with a common end vertex get different colors. A very good bound on the edge chromatic number of a graph in terms of degree is possible. All edges incident at a given vertex must have different colors, and so the maximum degree of a vertex in a graph is a *lower* bound on the edge chromatic number. Even better, one can prove:

Theorem 3—Vizing's Theorem

If the maximum degree of a vertex in a graph G is d, then the edge chromatic number of G is either d or $d + 1$.

The one vertex coloring problem that can be nicely characterized concerns 2-colorable graphs. Such graphs can be divided into two sets such that all edges go between a vertex in one set and a vertex in the other set.

Theorem 4

A graph can be 2-colored if and only if it does not contain a circuit of odd length.

Two-colorable graphs are also called bipartite graphs (see Example 3 in Section 1.1). This theorem gives a general explanation of why the wheel in Example 2 in Section 2.3 cannot be 3-colored. Three-coloring this wheel is equivalent to 2-coloring the outer circuit, but Theorem 4 says that since this circuit is of odd length it cannot be 2-colored.

Finally we would like to present and prove a theorem about coloring planar graphs. As noted earlier, Appel and Haken recently proved that all planar graphs can be 4-colored. But their proof is incredibly long and required a computer to do this case-by-case analysis. We will prove an easier "second best" theorem.

Theorem 5

Every planar graph can be 5-colored.

Proof

We only need to consider connected planar graphs, since we can 5-color unconnected planar graphs by 5-coloring each connected component. A key step in this proof uses a fact about planar graphs proved in Exercise 16 in Section 1.4: Any connected planar graph has a vertex of degree at most 5. We prove this theorem by induction on the number of vertices. Trivially a one-vertex graph can be 5-colored.

Next we assume that all connected planar graphs with $n - 1$ vertices ($n \geq 2$) can be 5-colored. We will prove that a connected planar graph G with n vertices can be 5-colored. As noted above, G has a vertex x of degree at most

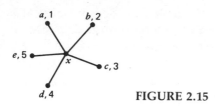

FIGURE 2.15

5. Delete x from G to obtain a graph with $n - 1$ vertices which by assumption can be 5-colored. Now we reconnect x to the rest of the graph and try to properly color x. If x has degree ≤ 4, then we can simply assign x a color different from the colors of its neighbors. The same approach works if the degree of x is 5 but two or more neighbors have the same color. Thus it remains to consider the case where x has five adjacent vertices each with a different color. See Figure 2.15, where we label the neighbors of x as a, b, c, d, e according to their clockwise order about x in some planar depiction of G. Let the color "names" be as shown in Figure 2.15.

First consider all paths from a whose vertices are colored 1 and 3. Suppose there is no path consisting of 1 and 3 vertices from a to c. Then we can change the color of a from 1 to 3, change the neighbors of a colored 3 to 1, and so on along all paths of 1 and 3 vertices emanating from a. This 1-3 interchange will not affect c since there is no path of 1's and 3's from a to c. After this 1-3 interchange from a, a and c are both color 3, and x can be properly colored with 1. On the other hand, if there is a 1-3 path from a to c, then we consider all paths of 2- and 4-colored vertices from b. The 1-3 path from a to c, together with edges (x,a), (x,c), forms a circuit that blocks the possibility of any 2-4 path from b to d. Thus we can perform a 2-4 interchange along all paths of 2's and 4's emanating from b without changing the color of d. After this 2-4 interchange, b and d are both color 4, and x can be properly colored with 2.

This completes the induction step of the proof that any n-vertex connected planar graph can be 5-colored. ∎

EXERCISES

SUMMARY OF EXERCISES   These exercises are proofs of results in coloring theory. Exercises 3 and 4 indicated two ways to prove Theorem 4 and Exercise 15 develops a proof of Theorem 2.

1. (a)   If the minimum degree of a vertex is d in an n-vertex graph G, then show that $\chi(G) \geq n/(n - d)$.
    (b)   Find a graph G for which $\chi(G) = n/(n - d)$.
2. A graph is *color critical* if the removal of any vertex decreases the graph's chromatic number. Show that every k-chromatic color critical graph G has the following properties:
    (a) G is connected.

    (b)   Every vertex of G has degree $\geq k - 1$.

    (c) G has no vertex whose removal disconnects G.

3.   (a)   Show that every k-chromatic graph contains a k-chromatic color critical subgraph (see Exercise 2).

    (b)   Show that to prove that each graph in a certain class S of graphs is k-colorable, it suffices to show that no graph in S contains a $(k + 1)$-chromatic color critical graph as a subgraph.

    (c)   Use the idea in part b to prove Theorem 4 (you may assume the results in Exercise 2 about color critical graphs).

4.   Show that every connected graph G without odd circuits can be 2-colored by picking a vertex x and coloring all vertices at the end of even-length paths from x color 1 and all vertices at the end of odd-length paths from x color 2. Show that this coloring is proper and well defined (i.e., a vertex cannot be assigned both colors), thus proving Theorem 4.

5.   Show that G can be edge k-colored if and only if $L(G)$, the line graph of G (see Exercise 16 in Section 2.1), can be vertex k-colored.

6.   If $\bar{G}$ is the complement of G, then show that:

    (a) $\chi(G) + \chi(\bar{G}) \leq n + 1$ (*Hint:* Use induction.)

    (b) $\chi(G)\chi(\bar{G}) \geq n$.

    (c) $\chi(G) + \chi(\bar{G}) \geq 2\sqrt{n}$

7.   Show that if every region in a planar graph has an even number of bounding edges, then the vertices can be 2-colored.

8.   Show that a planar graph G with 8 vertices and 13 edges cannot be 2-colored. (*Hint:* Use results in Section 1.4 to show that G must contain a triangle.)

9.   Show that no planar graph G has a chromatic polynomial of the form $P_k(G) = (k^2 - 6k + 8)Q(k)$ (see Exercise 7 in Section 2.3).

10.   Show that a graph with at most two odd-length circuits can be 3-colored.

11.   Show that a graph is k-colorable if and only if its edges can be directed so that its longest path has length $k - 1$.

12.   Use the fact that every planar graph has a vertex of degree ≤ 5 to give a simple induction proof that every planar graph can be 6-colored.

13.   Use the fact that every planar graph with fewer than 12 vertices has a vertex of degree ≤ 4 (Exercise 17 in Section 1.4) to prove that every planar graph with less than 12 vertices can be 4-colored.

14.   Show that if G is an interval graph (see Example 6 in Section 1.1), then $\chi(G)$ equals the size of the largest complete subgraph in G.

15.   Prove Theorem 2 using the following hints.

    (a)   The proof should be by induction on k, the chromatic number. Initially for $k = 3$, we use the graph G_3 consisting of a 5-circuit.

(b)   Assuming one can construct G_k, a triangle-free graph with $\chi(G_k) = k$, one constructs G_{k+1} by making k copies of G_k and then adding $(n_k)^k$ vertices, where n_k is the number of vertices in G_k. Each new vertex has as its set of neighbors a different k-tuple consisting of one vertex from each copy of G_k. Confirm that this new graph is the desired G_{k+1}.

16.   A *cut-set* of a graph G is a set C of vertices whose removal disconnects G while no proper subset of C disconnects G. Show that if all cutsets of G are complete subgraphs, then $\chi(G)$ equals the size of the largest complete subgraph in G. (*Hint:*   Use induction on number of vertices.)

17.   For any two nonadjacent vertices x,y in a graph G, define graphs G_{xy}^+ and G_{xy}^c as follows. G_{xy}^+ is obtained by adding the edge (x,y) to G, and G_{xy}^c is obtained from G by coalescing vertices x and y into a single vertex. Show that $\chi(G) = \min(\chi(G_{xy}^+), \chi(g_{xy}^c))$.

2.5   SUMMARY AND REFERENCES

This chapter presented two important graph-theoretic concepts: covering circuits and coloring. Section 2.1 discussed Euler circuits—circuits that traverse every edge exactly once. Section 2.2 discussed Hamilton circuits—circuits that visit every vertex exactly once. Both types of covering circuits arise naturally in operations research routing problems. Despite the similarity in the definitions of Euler and Hamilton circuits, determining the existence of Euler and Hamilton circuits in a graph are as different as graph-theoretic problems can be. Euler's theorem allows one quickly to decide whether an Euler circuits exists. On the other hand, except in special cases, the existence or nonexistence of a Hamilton circuit can only be determined by a laborious systematic search to try all possible ways of constructing a Hamilton circuit.

Section 2.3 introduced graph coloring with ad hoc coloring schemes and some applications of coloring. Section 2.4 gave a sampling of coloring theory, highlighted by a proof of the fact that any planar graph can be 5-colored. The stronger theorem proved in 1976 by Appel and Haken [1], that planar graphs are 4-colorable, was the motivation of much of the research in graph theory over the last one hundred years. The search for a proof of the Four Color Theorem led to reformulations of this theorem in terms of Hamilton circuits and other graph concepts whose properties were then examined.

As mentioned in Section 1.4, Euler's 1736 analysis of Euler circuits was the first paper on graph theory. Euler's paper (translated in Biggs, Lloyd, and Wilson [3]) makes very interesting reading. It is instructive to see how awkward Euler's writing was when he lacked the modern terminology of graph theory. The first use of the concept of a Hamilton circuit occurred in a 1771 paper by A. Vandermonde that presented a sequence of moves by which a knight could tour all positions of a chessboard (without repeating a position).

The name "Hamilton" refers to W. Hamilton whose algebraic research led him to consider special types of circuits and paths on the edges of a dodecahedron (see Exercise 2(c) in Section 2.2). Hamilton even had a game marketed that involved finding a Hamilton circuit on a dodecahedron (Hamilton's instructions for this game are reprinted in [3]). See Barnette [2] for a good history of the Four Color Problem, its restatements, and final solution by Appel and Haken.

1. K. Appel and W. Haken, "Every planar map is 4-colorable," *Bull. Am. Math. Soc.* **82** (1976), 711–712.

2. D. Barnette, *Map Coloring and The Four Color Problem*, Mathematical Association of America, Washington, D.C., 1984.

3. N. Biggs, E. Lloyd, and R. Wilson, *Graph Theory* 1736–1936, Clarendon Press, Oxford, 1976.

4. J. Bondy and U. Murty, *Graph Theory with Applications*, American Elsevier, New York, 1976.

Chapter Three
Trees and Searching

3.1 PROPERTIES OF TREES

The most widely used special type of graph is a tree. A **tree** is a graph with a designated vertex called a **root** such that there is a unique path from the root to any other vertex in the tree. Intuitively, a tree looks like a tree. See the examples of trees in Figure 3.1. The vertex labeled a is a root for each of these trees. In this chapter, we show how trees can be used to decompose and systematize the analysis of various search problems. We also present some graph connectivity algorithms based on trees. Finally, we analyze several common sorting techniques in terms of their underlying tree structure.

In this section, we present some basic properties of trees and introduce some convenient terminology for working with trees. We prove a few useful counting formulas about trees and apply these formulas to some decomposition problems. Note that trees are connected graphs, since there is a path connecting the root with all other vertices. Further, if a tree is an undirected graph (with no directed edges), then any vertex can be the root. For example, the middle tree in Figure 3.1, is drawn so that a appears to be a root, but the right tree in Figure 3.1, which is a redrawing of the middle tree, has no single vertex that is a natural root—that is, any vertex can be the root.

A tree has no circuits (in a directed graph, no circuits even when the directions of edges are ignored), because a circuit would provide two ways to get from the root to certain vertices. For example, in the left tree in Figure 3.1, the addition of an edge (g,h) would create a circuit (when edge directions are

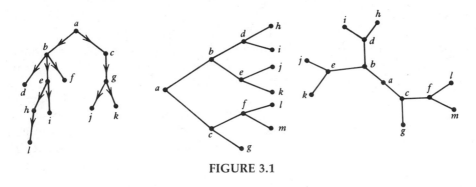

FIGURE 3.1

ignored) and a second path from a to h via g; the addition of an edge (d, a) would create a circuit that would constitute a second path from a to a, in addition to the trivial null (empty) path from a to a. If G is a connected undirected graph, then it is easy to show that G *is a tree if and only if it has no circuits*.

In most of this chapter we will be using trees with directed edges. Following common terminology, we call a directed tree a **rooted tree**. A rooted tree T has a unique root, for if vertices a and b were both roots of T, then there would be paths from a to b and from b to a forming a circuit. An undirected tree is unrooted in the sense that it has no one particular root. An undirected tree can be made into a rooted tree by choosing one vertex as the root and then directing all edges away from the root. For example, to root the undirected middle tree in Figure 3.1 at vertex a, we would simply direct all the edges from left to right.

The standard way to draw a rooted tree T is to place the root a at the top of the figure. Then the vertices adjacent from a are placed one level below a, and so on, as in the left tree in Figure 3.1. We say that the root a is at level 0, vertices b and c in that tree are at level 1, vertices d, e, f, and g in that tree are at level 2, and so forth. In general, the **level number** of a vertex x in T is the length of the (unique) path from a to x.

For any vertex x in T, except the root, the **parent** of x is the unique vertex y with an edge to x (the next-to-last vertex on the unique path from a to x). Conversely, x is a **child** of y. If two vertices have the same **parent**, they are **siblings**. The **parent–child** relationship extends to **ancestors** and **descendants** of a vertex. In the left tree in Figure 2.1, vertex e has b as its parent, h and i as its children, d and f as its siblings, a as its other ancestor, and l as its other descendant. Each vertex x in T is the root of the subtree of x and its descendants.

Theorem 1

A tree with n vertices has $n - 1$ edges.

Proof

Assume the tree is rooted; if undirected, make it rooted as described above. Each vertex except the root is at the lower end of a unique edge (from its parent). There are $n - 1$ nonroot vertices and hence $n - 1$ edges. ∎

Vertices of T with no children are called **leaves** of T; all other vertices (with children) are called **internal** vertices of T. If each internal vertex of a rooted tree has m children, we call T an **m-ary tree**. If $m = 2$, T is a **binary tree**.

Theorem 2

An m-ary tree T with i internal vertices has $n = mi + 1$ vertices in all.

Proof

Each internal vertex has m children. So there are mi children plus the one non-child vertex, the root. ∎

Corollary

Let T be an m-ary tree. Then we have:
(a) If T has i internal vertices, it has $l = (m - 1)i + 1$ leaves.
(b) If T has l leaves, it has $i = (l - 1)/(m - 1)$ internal vertices and $(ml - 1)/(m - 1)$ total vertices.
(c) If T has n total vertices, it has $(n - 1)/m$ internal vertices and $[(m - 1)n + 1]/m$ leaves.

The proof of the corollary's formulas follow directly from $n = mi + 1$ (Theorem 2) and the fact that $l + i = n$ (details left as an exercise).

Example 1

If 56 people sign up for a tennis tournament, how many matches will there be?

The tournament proceeds in a reverse binary treelike fashion. The entrants are leaves and the matches are the internal vertices. See Figure 3.2. By the corollary (part (b)), if there are $l = 56$ leaves and the tree is binary, then there are $i = (56 - 1)/(2 - 1) = 55$ matches. ∎

Example 2

Suppose a telephone chain is set up for an organization of 100 people. It is activated by a leader who calls a chosen set of three people. Each of these three people calls given sets of three other people, and so on. How many

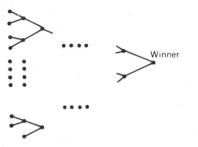

Winner

FIGURE 3.2

people will have to make any calls? Repeat the problem for an organization of 200 people.

Such a telephone chain is a rooted tree with 100 vertices. An edge corresponds to a call. Since the tree is ternary (3-ary), the corollary (part c) tells us that there will be $i = (100 - 1)/3 = 33$ internal vertices—that is, 33 people who make calls.

When we repeat the computation for an organization of 200 people, we get $i = (200 - 1)/3 = 66\frac{1}{3}$ internal vertices. By Theorem 2, a ternary tree must have a number of vertices n equal to $3k + 1$, for some k—that is, $n \equiv 1$ (mod 3). But $200 \equiv 2$ (mod 3), and so we do not have a true ternary tree. Either 199 or 202 would give a ternary tree. As a practical matter, with 200 people there will be 67 people who each make 3 calls and then one person who makes just one call (one internal vertex with one child). ∎

The **height** of a rooted tree is the length of the longest path from the root, or equivalently, the largest level number of any vertex. A rooted tree of height h is called **balanced** if all leaves are at levels h and $h - 1$. Balanced trees are "good" trees. The telephone chain tree in Example 2 should be balanced to get the message to everyone as quickly as possible. A tennis tournament's tree should be balanced to be fair; otherwise some players could reach the finals by playing several fewer matches than other players. Making an m-ary tree balanced will minimize its height. The middle tree in Figure 3.1, with a as root, is a balanced binary tree of height 3.

Theorem 3

An m-ary tree of height h has at most m^h leaves. An m-ary tree T with l leaves has height $h \geq \lceil \log_m l \rceil$; if T is a balanced m-ary tree, $h = \lceil \log_m l \rceil$.

Proof

Recall that $\lceil r \rceil$ is the smallest integer $\geq r$, and so $h = \lceil \log_m l \rceil$ means that $m^{h-1} < l \leq m^h$. Obviously an m-ary tree of height 1 has m leaves (children of the root). Now we use induction to show that an m-ary tree of height h has at most m^h leaves. An m-ary tree of height h can be broken into m subtrees rooted at the m

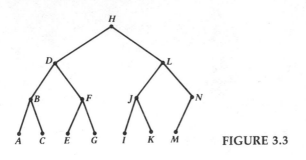

FIGURE 3.3

children of the root. These m subtrees have height at most $h - 1$. By induction they have at most m^{h-1} leaves each or at most $m \cdot m^{h-1} = m^h$ leaves all together. The second part of the theorem now follows immediately. ■

One of the most common uses of trees is in sequential testing procedures. The following two sequential testing examples, one a basic computer science problem and the other a logical puzzle, illustrate the variety of such applications of trees.

Example 3

Let us re-examine the compiler "dictionary look-up" problem discussed in Example 4 of Section 1.1. We want to identify an unknown word (number) X by testing it in a 3-way branch (less than, equal to, greater than) against "words" in a set (dictionary) to which X belongs. The test procedure can be represented by a binary, or almost binary, tree. If X were one of the first 14 letters of the alphabet, then Figure 3.3 is such a binary search tree. Each vertex is labeled with the letter tested at that stage in the procedure. The procedure starts by testing X against H. The left edge from a vertex is taken when X is less than the letter and the right edge when X is greater. Such a tree may have one internal vertex with just one child if the number of vertices is even (as is the case for vertex N in Figure 3.3).

To minimize the number of tests needed to recognize any X, that is, the height of the search tree, we should make the tree balanced. Suppose that X were to belong to a set of n "words." What is the maximum number of tests that would be needed to recognize X?

By the corollary, a binary search tree with n vertices has

$$\frac{(2 - 1)n + 1}{2} = \frac{n + 1}{2}$$

leaves. Then the maximum number of tests needed to recognize X is the height of a balanced $\frac{1}{2}(n + 1)$-leaf search tree,

$$h = \left\lceil \log_2 \frac{n + 1}{2} \right\rceil = \lceil \log_2(n + 1) \rceil - 1$$

■

Implicit in the tree model in the preceding example is the fact that there is 1-1 correspondence between binary search schemes and binary trees.

Example 4

A well-known logical puzzle has n coins, one of which is counterfeit, too light or too heavy, and a balance to compare the weight of any two sets of coins (the balance can tip to the right, to the left, or be even). The problem is for a given value of n to determine a procedure for finding the counterfeit coin in a minimum number of weighings. Sometimes one is told whether the counterfeit coin is too light or too heavy. If we are told that the fake coin is too light, how many weighings are needed for n coins?

Our testing procedure will form a tree in which the first test is the root, the other tests are the other internal vertices, and the solutions, that is, which coin is counterfeit, are the leaves. See the testing procedure in Figure 3.4 for eight coins. The coins are numbered 1 through 8, and the left edge is followed when the left set of coins in a test is lighter, the middle edge when both sets have the same weight, and the right edge when the right set of coins is lighter. Note that when weighing 4 and 5 (and already knowing that 1, 2, 3, 6, 7, 8 are not the light coin), the balance cannot be even. The test tree is ternary, and with n coins there will be n leaves—that is, n different possibilities of which coin is counterfeit. Theorem 3 tells us that the test tree must have height at least $[\log_3 n]$ to contain n leaves. It is not automatic that a testing procedure exists that can achieve the $[\log_3 n]$ height bound. For the light counterfeit coin problem, this bound can be achieved by successively dividing the current subset known to have the fake coin into three almost equal piles and comparing two of the piles of equal size, as in Figure 3.4.

If the counterfeit coin could be either too light or too heavy, then the problem is much harder. A particular coin will usually (but not always) appear at 2 leaves in the test tree, once when the coin is determined to be too light and another time when too heavy. ■

We conclude this section with an example of building a tree "bottom up," starting from a set of leaves.

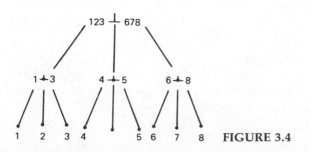

FIGURE 3.4

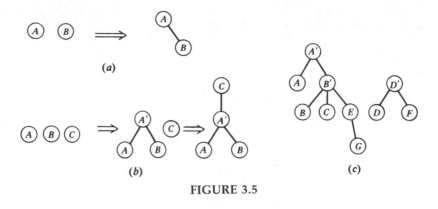

FIGURE 3.5

Example 5

In building a symbol table, a compiler initially lists a large number of different variable names used by the program (and extra names generated by the compiler). Later, the compiler repeatedly combines two names or two subsets of names that should be assigned the same location in memory (in various subsections of the program, different names are used for the same variable). Since a particular variable name may go through several rounds of being renamed and combined with other names, it is inefficient to change a name each time it is combined. Instead, one can build trees as shown in Figure 3.5*a* and 3.5*b*. If names *A* and *B* are equivalent and should both be a variable named *A*, we combine leaves *A* and *B* into a tree as shown in Figure 3.5*a* with *A* as root. If *A* and *B* are combined with the new name *A'* and then *A'* is combined with *C*, with both *A'* and *C* now named *C*, we construct the tree shown in Figure 3.5*b*. By traveling up to the top of its tree, one can determine the current name of any original name.

Let COMB(*W*;*X*,*Y*) be the operation of combining the sets of variables currently named *X* and *Y* into a variable named *W*. Suppose we started with variable names *A*, *B*, *C*, *D*, *E*, *F*, *G* and successively did the following operations: COMB(*E*;*E*,*G*), COMB(*B'*;*B*,*C*), COMB(*D'*;*D*,*F*), COMB(*B'*;*B'*,*E*), COMB(*A'*;*A*,*B'*). Then we would have the family of trees shown in Figure 3.5*c*. For example, the variable originally named *G* is now named *A'* (as are *A*, *B*, *C*, and *E*). See the exercises for information on the difference in computation times between simple re-naming and the tree method presented here. ■

EXERCISES

SUMMARY OF EXERCISES   Exercises 3–14 and 28–30 present some basic theory about trees. The remaining exercises involve various modeling problems with trees.

1. Draw all nonisomorphic trees with:
   (a) four vertices (b) five vertices (c) six vertices

2. Suppose a connected graph has 20 edges. What is the maximum possible number of vertices?

3. Show that all trees are 2-colorable.

4. Show that all trees are planar.

5. Show that an undirected connected graph is a tree if any one of the following conditions hold:
   (a) G has no circuits.
   (b) G has fewer edges than vertices.
   (c) Removal of any edge disconnects G.

6. Reprove Theorem 1 by using the fact that trees are planar (Exercise 4) and Euler's formula (Theorem 2 in Section 1.4).

7. Show that any tree with more than one vertex has at least two vertices of degree 1.

8. Prove the following parts of the corollary:
   (a) Part a (b) Part b (c) Part c

9. Reprove that $l \leqslant m^h$ in a m-ary tree of height h with a simple combinatorial "block-walking" type of argument.

10. What is the maximum number of vertices (internal and leaves) in an m-ary tree of height h?

11. Show that the fraction of internal vertices in an m-ary tree is about $1/m$.

12. A *forest* is an unconnected graph that is a disjoint union of trees. If G is an n-vertex forest of t trees, how many edges does it have?

13. Show that the sum of the level numbers of all l leaves in a binary tree is at least $[l(\log_2 l)]$, and hence the average leaf level is at least $\log_2 l$.

14. Show that the chromatic polynomial of an n-vertex tree is $k(k - 1)^{n-1}$ (see Exercise 7 in Section 2.3).

15. Any m-ary tree, $m \geqslant 3$, can be "converted" into a binary tree by the following substitution.

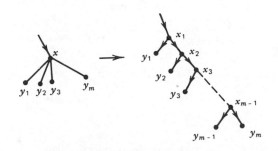

(a) Perform this conversion for the tree in Figure 3.4.

(b) If an m-ary tree has height h, what is the maximum possible height after conversion.

16. Build a minimal length tree in a coordinate grid with all edges parallel to the x- and y-axes and with leaves or internal vertices at the following coordinates: (0,0), (0,4), (1,1), (1,2), (2,3), (3,2), (4,1), (5,0). Internal "dummy" vertices at other coordinates are permitted. Minimal length means that the sum of the lengths of the edges is minimized.

17. Consider the problem of summing n numbers by adding together various pairs of numbers and/or partial sums, for example, $\{[(3 + 1) + (2 + 5)] + 9\}$.
    (a) Represent this addition process with a tree. What will internal vertices represent?
    (b) What is the smallest possible height of an "addition tree" for summing 100 numbers?

18. A tree can be used to represent a binary code, a left branch is a 0 and a right branch a 1. The path to a letter (vertex) is its binary code. To avoid confusion, one sometimes requires that the first k digits of one letter's code cannot be the code of another letter. Under this requirement, which vertices in a tree represent letters? How many letters can be encoded using n-digit binary sequences?

19. Suppose that each player in a tennis tournament brings a new can of tennis balls. One can is used in each match and the other can is taken by the match's winner along to the next round. Use this fact to show that a tennis tournament with n entrants has $n - 1$ matches.

20. Suppose that the losers in the first two rounds of a tennis tournament with 32 entrants qualify for a losers tournament. The people who lose in the first two rounds of this tournament qualify for another tournament, and so on until finally there is just one grand loser (the last tournament has two people). How many tournaments are required to determine this grand loser (all tournaments are balanced trees with a power of two people in each round after the first round; some people may have a first-round bye).

21. Suppose that a chain letter is started by someone in the first week of the year. Each recipient of the chain letter mails copies on to five other people in the next week. After six weeks, how much money in postage (20¢ a letter) has been spent on these chain letters?

22. What type of search procedure is represented by the search tree below?

23. (a) Repeat Example 3 assuming now that only a two-way branch (less than, greater than or equal to) is available. Draw a balanced search tree for the first 13 letters and determine the height of an n-letter search.

    (b) Suppose a two-way branching search tree for letters A, B, C, D, E is to take advantage of the following letter frequencies: A 10%, B 30%, C 10%, D 30%, E 20%. Build a two-way tree that minimizes the average number of tests required to identify a letter.

    (c) Repeat part a with the additional constraint that if both children of a vertex x are internal (branching) vertices, then the letter for branching at vertex x is exactly midway between the two letters at its children.

24. (a) Repeat Example 4 for 20 coins with one too light.

    (b) Prove by induction that 3^n or fewer coins with one too light can be tested in at most n weighings to find out which one is too light.

25. Suppose we have four coins and *possibly* one coin is either too light or too heavy (all four might be true).

    (a) Show how to determine which of the nine possible situations holds with just two weighings, if given one additional coin known to be true.

    (b) Show that two weighings are not sufficient without the extra true coin.

26. (a) How many total times would the names A, B, C, E, F, G be re-named in the combinations shown in Figure 3.5c? In comparison, only eight edges were created in Figure 3.5c.

    (b) Compare the number of renamings versus number of edges for a procedure with the names A_1, A_2, . . . , A_m in which the ith and ith-from-the-end letter are combined into the ith letter, and then the resulting $\frac{1}{2}m$ names are combined again and again in the same fashion. Assume m is a power of 2.

27. The tree method in Example 5 can be further optimized with respect to requests for the current name of an original name. FIND(X) is the operation of finding the current name of X. When a FIND(X) is performed and we trace up along a path to the root of the tree containing X, we remove the vertices along this path and attach these vertices as sons of the root. For example, if FIND(G) is performed with the trees in Figure 3.5c, the result will be the trees below. Give the resulting set of trees if the following sequence of COMB's and FIND's are performed on

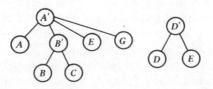

names A, B, C, D, E, F, G: COMB(A';A,B), COMB(C;A',C), FIND(B), COMB(E';E,F), COMB(G;E',G), COMB(D;C,D), COMB(D';D,G), FIND(A).

28. Describe a method, working from largest to smallest level, of rearranging the left-to-right orders of children of vertices in two rooted trees to facilitate a simple test to see if the two trees are isomorphic.

29. Let T be an undirected tree. If the choice of vertex x to be the root yields a rooted tree of minimal height, then x is call a *center* of T. Show that any undirected tree has at most two centers.

30. Show that there is the following one-to-one correspondence, due to H. Prufer, between all trees built on a given set of n labeled vertices x_1, x_2, . . . , x_n and all $(n-2)$-digit sequences using the numbers 1, 2, . . . , n. This correspondence shows that there are n^{n-2} different trees on n labeled vertices. The correspondence is formed as follows: if x_i is the vertex of degree one in the n-vertex tree T_n with the smallest index, then the first integer in the associated sequence is the index of the vertex adjacent to x_i; now delete x_i and its incident edge to obtain T_{n-1}; repeat the same procedure on T_{n-1} to obtain the second integer in the sequence; and so on until $n-2$ integers are obtained (and what is left is T_2, a single edge). For example, the sequence (8,8,7,5,5,5) is obtained from

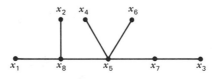

3.2 ENUMERATION WITH TREES

Trees provide a natural logical framework for finding solutions to problems that involve a (finite) sequence of choices. Finding one's way out of a maze, optimal strategies in a game, or a shortest tour in a routing problem are examples of such uses of trees. Most of the problems in the two preceding chapters—isomorphism, Hamilton circuits, minimal colorings, and the examples in Section 1.1, such as placing policemen on street corners—require tree-based searching for computerized solutions. Whether we want to find one solution, all solutions, or an optimal solution, the first and foremost challenge is to be sure that we check all possible ways to generate a solution, that is, the enumeration must be complete. By letting the sequential choices by internal vertices in a rooted tree and the solutions and "dead-ends" be the leaves, as in the coin balancing problem in Example 4 of the previous section, we can organize our enumeration of possible solutions. Trees also make it easier to discern and implement short-cuts, such as "pruning" subtrees that can be shown not to lead to a solution (or not to lead to an optimal solution).

In this section we present examples of tree enumeration that involve games rather than operations research applications. Most operations research tree enumeration problems involve very large trees and use special tree "pruning" algorithms. As an example of such applications, we solve a very small Traveling Salesperson problem in the next section. The next chapter, Network Algorithms, discusses three optimization algorithms that implicitly use trees to search through graphs.

There are two basic approaches to tree enumeration. The method called **depth-first search** or **backtracking**, builds a path from the root as far as possible in the tree, that is, to some leaf. If the leaf is not a solution or if we must continue to find all other solutions, we backtrack up this path one level to the parent of the leaf (the previous choice) and then branch along a different edge building a path to a new leaf. In backtracking, if all edges from the previous vertex (choice) have already been tried, then we backtrack up one level higher, and so forth. Eventually this method will generate paths to all leaves—that is, enumerate the whole tree of possible sequences of choices. If we only need one solution, a depth-first search will terminate as soon as it finds a leaf that is a solution.

There is one major difficulty that we must guard against—cycling. In games and real-world routing problems, it is usually inevitable that there will be two or more different paths (sequences of choices) that lead to the same "position," for example, the same corner in a maze. Such redundancy is acceptable. However, it is usually disastrous to visit the same position twice on one path. This could lead to an endless path that cycled around and around through this position forever. Thus we should always check that each successive position reached on a path does not appear earlier on the path. If it does, then treat the position as a dead-end leaf and backtrack.

Example 1

Consider the maze in Figure 3.6. We start at the location marked with an S and seek to reach the end marked with an E.

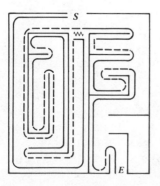

FIGURE 3.6

We use a depth-first search. For mazes, there is a convenient rule of thumb (whose verification is left as an exercise): stick to the right wall in the maze. When we come to a dead-end, we follow the right wall to the end wall, along the end wall, and then backtrack along the left wall (now the right wall as we leave the dead-end). When we cycle around to a previously visited corner, we put an artificial (wiggly) dead-end wall to stop us from actually reaching that corner (as at S in Figure 2.6). We use solid lines to indicate the (forward) path-building and dashed lines for backtracking in the maze in Figure 3.6. Because the maze is easily searched directly, we have not drawn the search tree for this problem (in which S would be the root, other corners internal vertices, and dead-ends and E the leaves). ∎

The other common method of tree enumeration, called **breadth-first search**, is to determine all edges leaving the root, that is, all possible children of the root; then determine all edges leaving these children; and so on. This procedure fans out uniformly from the root. Again, no single path should repeat a position. In a breadth-first search, it is sometimes possible to check that different paths do not use a common vertex. If the tree of possible paths is large, then the breadth-first method quickly becomes unwieldy. The depth-first method that only traces one path at a time is much easier to use by hand or to program. Further, in cases where we only need to find one of the possible solutions, it pays to go searching all the way down a path for a solution rather than to take a long time building a large number of partial paths, only one of which in the end will actually be used. On the other hand, when we want a solution involving a shortest path or when there may be very long dead-end paths (while solution paths tend to be relatively short), then the breadth-first method is better. All the network optimization algorithms in the next chapter use breadth-first searches.

Example 2

Suppose we are given three pitchers of water, of sizes 10 quarts, 7 quarts, and 4 quarts. Initially the 10-quart pitcher is full and the other two empty. We can pour water from one pitcher into another, *pouring until the receiving pitcher is full or the pouring pitcher is empty.* Is there a way to pour among pitchers to obtain exactly 2 quarts in the 7- or 4-quart pitcher? If so, find a minimum sequence of pourings to get two quarts.

The positions, or vertices, in this enumeration problem are ordered triples (a,b,c), the amounts in the three pitchers. Actually it suffices to record only (b,c), the 7- and 4-quart pitcher amounts, since $a = 10 - b - c$. A directed edge corresponds to pouring water from one pitcher to another. Let us draw the tree on a b,c-coordinate grid as shown in Figure 3.7. The grid is bounded by $b = 7$, $c = 4$, $b + c = 10$. Pouring between the 10- and 7-quart pitchers will be a horizontal edge, between 10- and 4-quart pitchers a vertical edge, and between 7- and 4-quart pitchers a diagonal edge with slope -1.

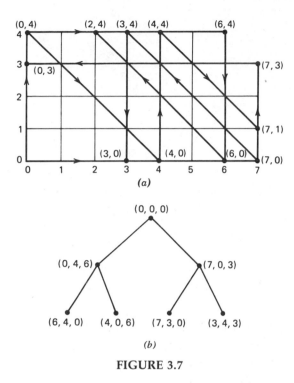

FIGURE 3.7

The root of this search tree is (0,0). We will use a breadth-first search. From the root, we can get to positions (7,0) and (0,4). From (7,0), we can get to new positions (7,3) and (3,4), and from (0,4), we can get to new positions (6,4) and (4,0). The tree built thus far is shown in Figure 3.7b. From (7,3), the only new position is (0,3), and from (3,4), the only new position is (3,0). From (6,4), the only new position is (6,0), and from (4,0), the only new position is (4,4). We have now checked all paths of length 3. The only new moves now are from (4,4) to (7,1) and from (6,0) to (2,4). But (2,4) has 2 quarts in one pitcher. So (0,0) to (0,4) to (6,1) to (6,0) to (2,4) is a minimum sequence of pourings to obtain 2 quarts. ∎

Example 3

Three jealous wives and their husbands come to a river. The party must cross the river (from near shore to far shore) in a boat that can hold at most two people. Find a sequence of boat trips that will get the six people across the river without ever letting any husband be alone (without his wife) in the presence of another wife.

Let the wives be represented by the letters A, B, and C, and their respective husbands by a, b, and c. The positions will be the possible partitions of people when the boat is at one of the two shores, and an edge will correspond to a

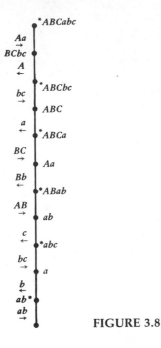

FIGURE 3.8

crossing of the river by the boat. We denote a position with the names of the people currently on the near shore plus a star (*) if the boat is on the near shore. An edge is labeled with the direction of the boat and the people in the boat. At each position, we must check that the people will not violate the "jealousy" condition of an unaccompanied husband in contact with another wife (or vice versa). Figure 3.8 shows one path of feasible positions that gets everyone across the river. There are other similar paths possible; for example, we could start with Bb or Cc instead of Aa, or start with ab or ac or bc, and so forth. ■

Example 4

Find all ways to place eight nontaking queens on an 8×8 chessboard. Recall that one queen can capture another queen if they are both in the same row or the same column or on a common diagonal.

We develop an algorithm using backtracking tree enumeration to print all solutions for placing eight nontaking queens. We will try all ways to place nontaking queens successively in column 1, column 2, and so forth, through column 8. Let a_k be the row of the queen in column k. For each i, $1 \leq i < k$, we require $a_k \neq a_i$ and $|a_k - a_i| \neq k - i$. When these conditions hold for a_k, we say that "a_k is compatible with $a_1, a_2, \ldots, a_{k-1}$." When a compatible a_k is found, we descend in the tree to the next level (next column). When no compatible a_k is found, we backtrack and try the next larger value (row) for a compatible a_{k-1}. The algorithm can then be written as follows:

$$k \leftarrow 1; \ a_1 \leftarrow 1;$$

DESCEND: $k \leftarrow k+1; \ a_k \leftarrow 1;$

ADDQUEEN: WHILE $a_k \leq 8$ AND a_k is not compatible with $a_1, \ \ldots, \ a_{k-1}$

                   DO $a_k \leftarrow a_k + 1;$

         IF $a_k = 9$ THEN

                         IF $k > 1$ THEN GOTO *BACKTRACK* ELSE GOTO *END*;

         IF $k < 8$ THEN GOTO *DESCEND*;

         PRINT solution $a_1, a_2, \ \ldots, \ a_8;$

BACKTRACK: $k \leftarrow k-1; \ a_k \leftarrow a_k+1;$ GOTO *ADDQUEEN*;

      END: STOP;

If $a_k = 9$, then there was no new a_k compatible with $a_1, \ \ldots, \ a_{k-1}$ found in the preceding WHILE loop and we must backtrack (or if $k = 1$, the search is finished).   ∎

EXERCISES

SUMMARY OF EXERCISES   The first 14 exercises are based on the four examples in this section. The last five exercises seek search algorithms for solving various games; use the depth-first method for searching positions in these games.

1.   (a)   Use a depth-first search to find your way through this maze from S to E.

       (b)   How many ways through this maze (without cycles) are there?

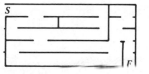

2.   Perform a breadth-first search of the graph in Figure 3.6 to find the shortest path from S to E.

3.   Repeat Example 2 using a depth-first search.

4.   (a)   Repeat Example 2 with pitchers of size 8, 5, and 3 with an objective of 4 quarts in one pitcher.

       (b)   Do Example 2 with pitchers of sizes 12, 8, and 5, and an objective of 7 quarts in one pitcher.

5.   Find another (actually shorter) way in Example 2 to get 2 quarts in one pitcher.

6.   Use a depth-first search in Example 2 to show that any amount between 0 and 10 quarts can be obtained in one of the pitchers.

7. Show that the stick-to-the-right-hand-wall rule will always get one out of a maze.

8. Suppose there is a dog, a goat, and a bag of tin cans to be transported across a river in a ferry that can only carry one of these three items at once (along with a ferry driver). If the dog and goat cannot be left alone on a shore (when the driver is not present) nor can the goat and tin cans be left alone, then find a scheme for getting all across the river.

9. (a) Repeat Example 3 for the problem of four jealous wives, if possible.
   (b) Do part a with a three-person boat.

10. Consider the following alternative approach to Example 3. Make a graph with a vertex for each legal configuration of people (on each side of the river) with the boat *on the near shore*. Add one more vertex for the configuration of all people on the far shore (and the boat on the far shore). Next draw an edge joining a pair of vertices (configurations) if there is a legal round-trip of the boat from one vertex to the other. Now simply find a path from the starting vertex (all on the near shore) to the end vertex (all on the far shore).

11. (a) Suppose three missionaries and three cannibals must cross a river in a two-person boat. Show how this can be done so that at no time the cannibals outnumber the missionaries on either side (unless there are no missionaries on a shore). (*Hint:* Try the idea in Exercise 10.)
    (b) Repeat part a with the additional condition that only one of the cannibals can row.

12. (a) Write a program for finding all ways to place eight non-taking rooks on a chessboard.
    (b) Write a program for finding all ways to place eight non-taking bishops on a chessboard.
    (c) What is the largest number of non-taking bishops that can be placed on a chessboard?

13. (a) Write a program for finding all ways to place four non-taking kings on a 4 × 4 chessboard.
    (b) Write a program for finding all ways to place four non-taking knights on a 4 × 4 chessboard.

14. Write a program for finding the minimal number of queens such that all squares on the 8 × 8 chessboard are under attack by one of the queens.

15. Consider the tic-tac-toe game after a first move by X and then a move by O as shown. Build a tree for the successive plays of this game and show how X can always win.

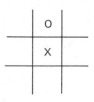

16. Write a program to find a solution to the Instant Insanity problem in Section 11.1.

17. Given a matching graph (Example 3 in Section 1.1) of people with jobs, write a program to find a minimal-sized subset of people who can collectively do all the jobs.

18. Write a program to find all 3×3 magic squares. An $n \times n$ *magic square* is an $n \times n$ array of the integers 1, 2, . . . , n^2 such that the sum of every row, of every column, and of the two main diagonals is the same. (*Hint:* For a 3×3 square, the sum is 15).

19. Write a program to find all ways to assemble the seven given pieces into a $3 \times 3 \times 3$ cube. (The game is called *Soma.*)

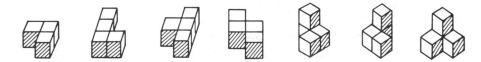

3.3 THE TRAVELING SALESPERSON PROBLEM

The traveling salesperson problem calls for a minimal-cost Hamilton circuit in a complete graph having an associated cost matrix C. Entry c_{ij} in C is the cost of using the edge from the ith vertex to the jth vertex. Minimal-cost means minimizing the sum of the costs of the edges.

In this section, we first present a tree enumeration approach that uses a "branch and bound" method to limit the number of different paths in the tree (each path would correspond to a Hamilton circuit) that must be inspected in search of a minimal solution. Note that a complete graph on n vertices has $(n - 1)!$ different Hamilton circuits, for example, a 12-city (vertex) problem has $11! = 39,916,800$ possible circuits.

In the latter part of this section, we discuss a faster technique that finds a near-minimal-cost Hamilton circuit. The huge amount of time required to find exact solutions to problems that involve enumeration of very large trees, such as the traveling salesperson problem, has led researchers in recent years to concentrate on heuristic, near-optimal algorithms for such problems. Finding a minimal coloring of any arbitrary graph or testing for isomorphism are other

To From	1	2	3	4
1	∞	3	9	7
2	3	∞	6	5
3	5	6	∞	6
4	9	7	4	∞

FIGURE 3.9

problems we have seen whose tree enumeration is similarly complex. Near-minimal coloring algorithms have received much attention, but unfortunately, a "near-isomorphism" is usually of no value.

Initially we consider a very small traveling salesperson problem with four vertices x_1, x_2, x_3, x_4. Let the cost matrix for this problem be the matrix in Figure 3.9 (note that we do not require $c_{ij} = c_{ji}$; the ∞'s on the main diagonal indicate that we cannot use these entries). A Hamilton circuit will use four entries in C, one in each row and in each column and such that no proper subset of entries (edges) forms a subcircuit. This latter constraint means that if we choose entry c_{ij}, then we cannot also use c_{ji}, for these two entries form a subcircuit of length 2. Similarly, if entries c_{ij} and c_{jk} are used, then c_{ki} cannot be used.

We first show how to obtain a lower bound for the cost of this traveling salesperson problem. Since every solution must contain an entry in the first row, the edges of a minimal tour will not change if we subtract a constant value from the first row of the cost matrix (of course, the cost of a minimal tour will change by this constant). Suppose then that we subtract as large a number as possible from the first row without making any entry in the row negative, that is, we subtract the value of the smallest entry in row 1, namely 3. Do this for the other three rows also.

We display the altered cost matrix in Figure 3.10. After subtracting a total of $3 + 3 + 5 + 4 = 15$ from the different rows, a minimal tour for Figure 3.10 will cost 15 less than for Figure 3.9. Still the edges of a minimal tour for one problem are the edges of a minimal tour for the other.

In a similar fashion, we can subtract a constant from any column without changing the set of edges of a minimal tour. The only column in Figure 3.10 without a 0 is column 4, whose smallest value is 1. So we subtract 1 from the last column in Figure 3.10 to get the matrix in Figure 3.11. The cost of a

To From	1	2	3	4
1	∞	0	6	4
2	0	∞	3	2
3	0	1	∞	1
4	5	3	0	∞

FIGURE 3.10

To	1	2	3	4	
From					
1	∞	0	6	3	Lower Bound = 16
2	0	∞	3	1	
3	0	1	∞	0	
4	5	3	0	∞	

FIGURE 3.11

minimal tour in Figure 3.11 has been reduced by a total of $15 + 1 = 16$ from the cost in Figure 3.9. We can use this reduction of cost to obtain a lower bound on the cost of a minimal tour: a minimal tour in Figure 3.11 must trivially cost at least 0, and hence a minimal tour in Figure 3.9 must cost at least 16.

Now we are ready for the branching part of the method. We look at an entry in Figure 3.11 that is equal to 0. Say c_{12}. Either we use c_{12} or we do not. We "branch" on this choice. In the case that we do not use c_{12}, we represent the no-c_{12} choice by setting $c_{12} = \infty$. The smallest value in row 1 of the altered Figure 3.11 is now $c_{14} = 3$, and so we can subtract this amount from row 1. Similarly we can subtract 1 from column 2. We obtain the cost matrix in Figure 3.12. Hence, any tour for Figure 3.9 that does not use c_{12} must cost at least $16 + (3 + 1) = 20$.

In the case that we use c_{12}, there is no immediate increase in the cost of the tour (since $c_{12} = 0$). We shall now consider partial tours using c_{12}, and continue to extend these partial tours until we get a lower bound of more than 20. If the lower bound were to exceed 20, then we would have to consider partial tours not using c_{12}. Our enumeration tree for this problem will be a binary tree whose internal vertices represent choices of the form: use c_{ij} or do not use c_{ij}. As long as the lower bound for possible tours using c_{ij} is less than the lower bound for tours not using c_{ij}, we do not need to look at the subtree of possible tours not using c_{ij}.

If we use c_{12} to build a tour for Figure 3.11, then the rest of the tour cannot use another entry in row 1 or column 2; also entry c_{21} must be set equal to ∞ (to avoid a subcircuit of length 2). The new smallest value of the reduced matrix in row 2 is 1, and so we subtract 1 from row 2 to obtain the cost matrix in Figure 3.13. The lower bound for a tour in Figure 3.13 (using c_{12}) is now $16 + 1 = 17$. Since 17 is less than 20 (the lower bound if we do not use c_{12}), we

To	1	2	3	4	
From					
1	∞	∞	3	0	Lower Bound = 20
2	0	∞	3	1	
3	0	0	∞	0	
4	5	2	0	∞	

FIGURE 3.12

To	1	3	4
From			
2	∞	2	0
3	0	∞	0
4	5	0	∞

Lower Bound = 17

FIGURE 3.13

continue our tour building with c_{12} by choosing another entry. The new entry need not connect with c_{12}, that is, need not be of the form c_{j1} or c_{2j}, but for simplicity we shall pick an entry in row 2. As before, we want to pick a 0 entry. The only 0 entry in row 2 of Figure 3.13 is c_{24}.

Again we have the choice of using c_{24} or not using c_{24}. Not using c_{24} will increase the lower bound by $2 + 0 = 2$ (the sum of the smallest entries in row 2 and column 4 of Figure 3.13 after we set $c_{24} = \infty$). Using c_{24} will not increase the lower bound, and so we first try the partial tour using c_{24} along with c_{12}. Again we delete row 2 and column 4 in Figure 3.13 and set $c_{14} = \infty$ (to block the subcircuit $x_1 - x_2 - x_4 - x_1$). Figure 3.14 shows the new remaining cost matrix (all rows and columns still have a 0 entry).

Now the way to finish the tour is clear: use entries c_{43} and c_{31} for a complete tour $x_1 - x_2 - x_4 - x_3 - x_1$. We actually have no choice since not using either of c_{43} or c_{31} forces us to use an ∞ (which represents a forbidden edge). Since $c_{43} = c_{31} = 0$, this tour has a cost of 17. Then this tour must be minimal since its cost equals our lower bound. Re-checking the tour's cost with the original cost matrix in Figure 3.9, we have (in Figure 3.9) $c_{12} + c_{24} + c_{43} + c_{31} = 3 + 5 + 4 + 5 = 17$. We summarize the preceding reasoning in Figure 3.15 with the binary tree of choices we made and their lower bounds (L.B.).

One general point should be made about how to take best advantage of the branch-and-bound technique. At each stage, we should pick as the next entry on which to branch (use or do not use the entry), the 0 entry whose removal maximizes the increase in the lower bound. In Figure 3.11, a check of all 0 entries reveals that not using entry c_{43} will raise the lower bound by $3 + 3 = 6$ (3 is the new smallest value in row 4 and in column 3). So c_{43} would theoretically have been a better entry than c_{12} to use for the first branching, since the greater lower bound for the subtree of tours not using c_{43} makes it less likely that we would ever have to check possible tours in that subtree.

Now consider the 6×6 cost matrix in Figure 3.16. We present the first few

To	1	3
From		
3	0	∞
4	∞	0

Lower Bound = 17

FIGURE 3.14

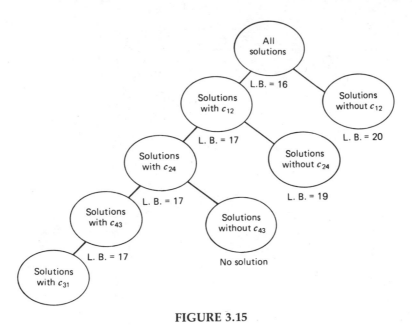

FIGURE 3.15

branchings in Figure 3.17 for this cost matrix. We circle the 0 entry on which we branch at each stage.

Now we present a quicker algorithm for obtaining good (near-minimal) tours, when the costs are symmetric (i.e., $c_{ij} = c_{ji}$) and the costs satisfy the triangle inequality—$c_{ik} \leq c_{ij} + c_{jk}$. These two assumptions are satisfied in most traveling salesperson problems. After describing the algorithm and giving an example of its use, we will prove that at worst the quick algorithm's tour is always less than twice the cost of a true minimal tour. A bound of twice the true minimum may sound bad, but in many cases where a "ballpark" figure is needed (and the exact minimal tour can be computed later if needed), such a bound is quite acceptable. Usually the quick algorithm finds a tour that is very close to the true minimum.

To	1	2	3	4	5	6
From						
1	∞	3	4	2	6	3
2	5	∞	3	4	4	5
3	4	1	∞	5	3	4
4	4	2	6	∞	4	5
5	3	3	3	5	∞	4
6	7	4	5	6	7	∞

FIGURE 3.16

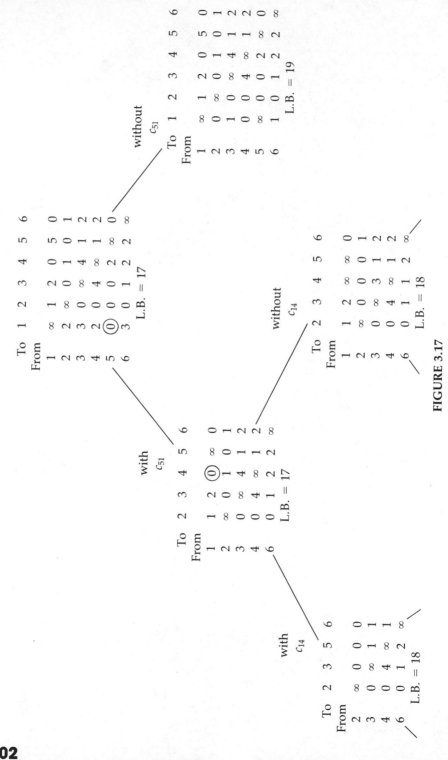

FIGURE 3.17

Quick Traveling
Salesperson Tour Construction

1. Pick any vertex as a starting circuit T_1 with one vertex (and 0 edges).

2. Given the k-vertex circuit T_k, find the vertex z_k not on T_k that is closest to a vertex, call it y_k, on T_k.

3. Let T_{k+1} be the $k + 1$-vertex circuit obtained by inserting z_k immediately in front of y_k in T_k.

4. Repeat Steps 2 and 3 until a Hamilton circuit (containing all vertices) is formed.

Example 1

Let us apply the preceding algorithm to the six-vertex traveling salesperson problem whose cost matrix is given in Figure 3.18. Name the vertices x_1, x_2, x_3, x_4, x_5, x_6. We will start with x_1 as T_1. Vertex x_4 is closest to x_1, and so T_2 is x_1-x_4-x_1. Vertex x_3 is closest to T_2 at x_4; thus $T_3 = x_1$-x_3-x_4-x_1. There are now two vertices, x_2 and x_6, three units from T_3. Suppose we pick x_2 and insert it before x_3 to obtain $T_4 = x_1$-x_2-x_3-x_4-x_1. Vertex x_6 is still three units from x_1, and so we insert x_6 before x_1 obtaining $T_5 = x_1$-x_2-x_3-x_4-x_6-x_1. Finally, x_5 is within four units of x_3 and x_6. Inserting x_5 before x_6, we obtain our near-minimal tour $T_6 = x_1$-x_2-x_3-x_4-x_5-x_6-x_1, whose cost we compute to be 19.

Looking at the numbers in Figure 3.18, it is clear that this value is quite close to the minimum (which happens to be 18). The length of the tour generally depends on the starting vertex; by applying the algorithm to each starting vertex and taking the shortest of the six tours obtained, we would get an improved estimate for the true minimal tour. ∎

Theorem

The cost of the tour generated by the quick construction is less than twice the cost of the minimal traveling salesperson tour.

Proof—optional

Suppose we are successively building the k-vertex circuits T_k according to the Quick Construction in the n-vertex complete graph G. Let S_k be a set of edges

To	1	2	3	4	5	6
From						
1	∞	3	3	2	7	3
2	3	∞	3	4	5	5
3	3	3	∞	1	4	4
4	2	4	1	∞	5	5
5	7	5	4	5	∞	4
6	3	5	4	5	4	∞

FIGURE 3.18

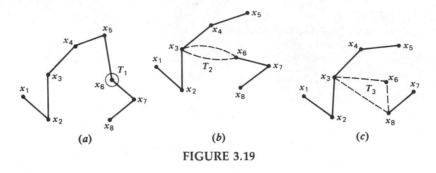

FIGURE 3.19

in the true minimal tour T^* that connect the vertices in T_k to the other vertices in G (S_k is described further below). When $k = 1$, we let S_1 consist of $T^* - e^*$, where e^* is the costliest edge in T^*. Since T_1 is a single vertex and $T^* - e^*$ is a Hamilton path, $S_1 = T^* - e^*$ will connect T_1 with the rest of G. Suppose that $S_1 = T^* - e^*$ is the Hamilton path shown in Figure 3.19a with T_1 the vertex x_6. Next we suppose that T_2 is x_6-x_3-x_6. Then we let $S_2 = S_1 - (x_5, x_6)$, as shown in Figure 3.19b. Further if T_3 is x_6-x_8-x_3-x_6, then let $S_3 = S_2 - (x_6, x_7)$, as shown in Figure 3.19c. In general, we will obtain S_{k+1} from S_k by removing the first edge on the path in S_k from T_k to the new vertex z_k (that is inserted in T_k to obtain T_{k+1}).

In Figure 3.19c, $z_3 = x_8$ is inserted into T_2 between x_3 and x_6 to obtain T_3. If c_{ij} is the cost of edge (x_i, x_j), then we know that c_{68} is the smallest cost among edges between T_2, that is, x_3 and x_6, and the rest of G. Thus $c_{68} \leq c_{67}$ ((x_6, x_7) was the edge removed from S_2 to get S_3). Using this fact and the triangle inequality, we shall now prove that inserting x_8 into T_2 to get T_3 has a net increase in cost $\leq 2c_{67}$. The increase of T_3 over T_2 is $c_{68} + c_{38} - c_{36}$, since edges (x_6, x_8) and (x_3, x_8) replace (x_3, x_6). But by the triangle inequality, $c_{38} \leq c_{36} + c_{68}$, or equivalently $c_{38} - c_{36} \leq c_{68}$. Thus $c_{68} + (c_{38} - c_{36}) \leq c_{68} + c_{68} \leq 2c_{67}$, as claimed.

This same argument can be applied to the insertion of z_k into T_k to prove that the increase of T_{k+1} over T_k is at most twice the cost of the edge dropped from S_k to get S_{k+1}. Starting from T_1 and repeating this bound on the insertion cost, we see that T_n, the final near-optimal Hamilton circuit, is bounded by twice the sum of the costs of the edges in $S_1 + T^* - e^*$. Thus we have proved the theorem. ∎

EXERCISES

1. Subtract the value of each nondiagonal entry in Figure 3.9 from 10. Now solve the traveling salesperson problem for this new cost matrix.

2. Solve the traveling salesperson problem for the cost matrix in Figure 3.16.

3. Write a computer program to solve a six-city traveling salesperson problem and use it for the cost matrix in Figure 3.18.

4. Use ad hoc arguments to show that the cost of a minimal tour for the cost matrix in Figure 3.18 is 18.

5. Every month a plastics plant must make batches of five different types of plastic toys. There is a conversion cost c_{ij} in switching from the producting of toy i to toy j shown in the matrix. Find a sequence of toy production (to be followed for many months) that minimizes the sum of the monthly conversion costs.

To From	T_1	T_2	T_3	T_4	T_5
T_1	—	3	2	4	3
T_2	4	—	4	5	6
T_3	5	3	—	4	4
T_4	3	5	1	—	6
T_5	5	4	2	3	—

6. The *assignment problem* is a matching problem with n people and n jobs and a cost matrix with entry c_{ij} representing the "cost" of assigning person i to job j. The goal is a one-to-one matching of people to jobs that minimizes the sum of the costs. Set all diagonal entries in Figure 3.9 equal to 5 and solve this 4×4 assignment problem by a branch-and-bound approach. How does an assignment problem differ from a traveling salesperson problem?

7. Use the quick construction to find approximate traveling salesperson tours for the cost matrices in (use just entries above the main diagonal):
   (a) Figure 3.9   (b) Figure 3.16   (c) Exercise 5

8. Consider the following rule for building approximate traveling salesperson tours. Starting with a single-vertex tour T_1, successively add the vertex whose insertion into T_k to form T_{k+1} minimizes the increase in cost, that is, if x_r is inserted between x_k and x_{k+1} then $c_{kr} + c_{r(k+1)} - c_{k(k+1)}$ should be minimal over all choices of x_r and x_k.
   (a) Can you prove an upper bound on this method similar to the one found for the Quick Construction method (making the same assumptions).
   (b) Apply this method to the cost matrix in (i) Figure 3.18 and (ii) Figure 3.16 (using just entries above the main diagonal).

9. Find a 5×5 cost matrix for which two different initial lower bounds can be obtained (with different sets of 0 entries) by subtracting from the rows and columns in different orders.

10. Make up a 5×5 cost matrix for which the quick construction finds:
   (a) An optimal tour.
   (b) A fairly costly tour (at least 50% over the true minimum).

11. Let $\chi(G)$ be the chromatic number of the graph G. For any two nonadjacent vertices x, y in G, define graphs G_{xy}^+ and G_{xy}^c as follows: G_{xy}^+ is obtained by adding the edge (x, y) to G, and G_{xy}^c is obtained from G by coalescing vertices x and y into a single vertex. As shown in Exercise 17

in Section 2.4, $\chi(G) = \min(\chi(G_{xy}^+), \chi(G_{xy}^c))$. With repeated use of this fact, design a branch-and-bound method to find $\chi(G)$. Apply your method to the graph in Figure 2.10 in Section 2.3. (*Hint:* Use the largest complete subgraph in your bound.)

3.4 SPANNING TREES AND GRAPH ALGORITHMS

In many applications, graph algorithms are needed to test whether a graph has a certain property, such as connectedness or planarity, or to count all occurrences of a given structure, such as circuits or complete subgraphs. In their search for these properties or structures, the algorithms usually employ a spanning tree to move through the graph. A **spanning tree** of a graph G is a subgraph of G that is a tree containing all vertices of G. Spanning trees can be constructed either by depth-first (backtrack) search or by breadth-first search.

To build a depth-first spanning tree, we pick some vertex as the root and begin building a path from the root. The path continues until it cannot go any further without repeating a vertex already on the path. The vertex where this path must stop is a leaf. We now backtrack to the parent of this leaf and try to build a path from the parent in another direction; and so on, as in the depth-first search method described in Section 3.2.

To build a breadth-first spanning tree, we pick any vertex x as the root and put all edges leaving x (along with the vertices at the ends of these edges) in the tree. Then we successively add to the tree the edges leaving the vertices adjacent from x, unless such an edge goes to a vertex already in the tree. We continue this process as in the breadth-first search method described in Section 3.2. It is important to note that if the graph is not connected, then no spanning tree exists. Our first basic graph algorithm is thus:

**Algorithm to Test Whether an
Undirected Graph is Connected**

Use a depth-first (or breadth-first) search to try to construct a spanning tree. If all vertices of the graph are reached in the search and a spanning tree is obtained, then the graph is connected. If the search does not reach all vertices, the graph is not connected.

A formal proof that depth-first (breadth-first) searching does search all vertices in a connected graph is left as an exercise.

Example 1

Is the undirected graph G whose adjacency matrix is given in Figure 3.20a connected (entry (i,j) is 1 if and only if x_i is adjacent to x_j)?

	x_1	x_2	x_3	x_4	x_5	x_6	x_7	x_8
x_1	0	1	1	1	0	1	0	0
x_2	1	0	1	0	1	0	1	0
x_3	1	1	0	0	0	0	0	0
x_4	1	0	0	0	0	0	1	0
x_5	0	1	0	0	0	0	1	0
x_6	1	0	0	0	0	0	0	0
x_7	0	1	0	1	1	0	0	1
x_8	0	0	0	0	0	0	1	0

(*a*)

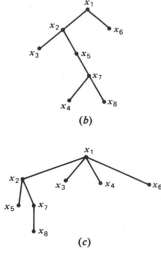

(*b*)

(*c*)

FIGURE 3.20

Let us perform a depth-first search of G starting with x_1 as the root. At each successive vertex, we pick the next edge on the tree to be the edge going to the lowest number vertex not already in the tree. So from x_1 we go to x_2 and x_3. Since x_3 is not adjacent to any other (new) vertex, we backtrack from x_3 to x_2 and re-continue the search from x_2 to x_5 to x_7 to x_4. At x_4 we backtrack to x_7 and re-continue to x_8. From x_8, we must backtrack all the way back to x_1 and then re-start to x_6. This finishes the search—all vertices have been visited. The spanning tree obtained is shown in Figure 3.20*b*. The result of a breadth-first search is shown in Figure 3.20*c*. ∎

The computation time required to make a depth-first search of a graph is proportional to the number of edges in the graph (each edge in the spanning tree is traversed twice, and edges that cannot be used are tried just twice). If an undirected graph is not connected, then we can find its components (connected pieces) by applying a depth-first (or breadth-first) search at any vertex to find one component; then apply this search starting at a vertex not on the previous tree to find another component; and continue finding additional components until no unused vertices are left.

If a graph is directed, then the term "connected" normally refers to the connectivity of the graph when directions of edges are ignored (all edges made undirected). A directed type of connectivity is strong connectivity. A directed graph is **strongly connected** if there is a directed path from any vertex x to any vertex y (and interchanging roles of x and y, from y to x). A natural way to test whether a directed graph G is strongly connected is to try to build spanning trees of G rooted at every vertex in G, for there will be a path from vertex x to all other vertices in G if and only if there is a spanning tree of G rooted at x. In Exercise 28 at the end of this section, we describe a faster algorithm to test for strong connectivity.

Suppose we have built a spanning tree to provide the framework for a search of a graph. Testing for the existence of a graph property or counting the number of special configurations in a graph normally involves a depth-first type of traversal of the spanning tree. However, there are several times during a traversal when internal vertices can be checked. A **preorder traversal** of a spanning tree is a depth-first search that examines an internal vertex when the vertex is first encountered in the search. A **postorder traversal** examines an internal vertex when last encountered (before the search backtracks away from the vertex and its subtree). If a tree is binary, we can define an **inorder traversal** that checks an internal vertex in between the traversal of its left and right subtrees. Figures 3.21a and 3.21b display numberings of the vertices of the tree in Figure 3.20b as checked in preorder and postorder traversals, respectively.

We now give examples of each type of traversal. In the binary search tree example of a compiler dictionary look-up (Example 3 in Section 3.1 and Example 4 in Section 1.1), the alphabetical order of the vertices corresponds to inorder traversal. In searching through a graph (along a spanning tree) for a vertex with a specified label, we should test each new vertex as it is encountered on a depth-first search to see if it has the desired label. There is no advantage to postponing such testing. Thus in searching for a special vertex, vertices should be examined according to a preorder traversal.

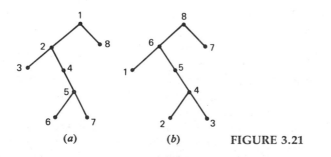

(a)   (b)   FIGURE 3.21

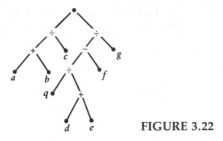

FIGURE 3.22

We demonstrate the need for postorder traversal with an arithmetic tree. The arithmetic expression

$$((a + b) \div c) \cdot (((q \div (d + e)) - f) \div g)$$

can be decomposed into a binary tree as shown in Figure 3.22. The internal vertices are arithmetic operations and the leaves variables. To evaluate this arithmetic expression, we need to execute the operations specified by internal vertices according to a postorder traversal of the arithmetic tree, that is, we cannot perform an operation until the subexpressions represented by the operation vertex's two subtrees are evaluated.

An **articulation point** in a connected undirected graph G is any vertex whose removal disconnects G. If G has no articulation points, then G is called **biconnected**. The graph in Figure 3.23 has c, e, and f as articulation points. The division of the edges of G into biconnected subgraphs, or **biconnected components**, is shown in Figure 3.23b. There are many graph properties and structures that need to be checked only within individual biconnected components. For example, circuits and complete subgraphs occur *within* biconnected components. It is not hard to show that a graph is planar if and only if each of its biconnected components are planar. So finding articulation points and breaking a graph into biconnected components are important steps in simplifying the work of many algorithms (of course, no such simplification occurs if the whole graph is biconnected).

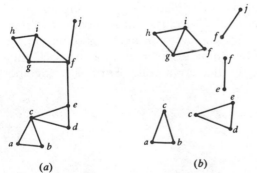

(*a*) (*b*) FIGURE 3.23

Before we can present our algorithm for finding articulation points and biconnected components, there is an important property of depth-first spanning trees that must be verified.

Lemma

Let G be a connected undirected graph and let x and y be two vertices of G. If (x,y) is an edge in G that is not part of the depth-first spanning tree T, then x is either an ancestor or a descendant of y in T.

Proof

Assume that (x,y) is an edge of G in T. See Figure 3.24. By symmetry, let x precede y in a preorder traversal of T, that is, in a depth-first search to build T, x is encountered before y. Then the depth-first search would use edge (x,y) before backtracking from x (leaving x), unless y was already incorporated in T. All vertices incorporated in T after first visiting x but before backtracking from x are descendants of x. Thus y is a descendant of x. ∎

If x and y are related, as in the lemma, but $(x,y) \notin T$, then (x,y) is a **back edge** of T. An edge between two vertices not related in T is called a **cross edge** of T. The lemma states that all out-of-tree edges in depth-first spanning trees are back edges. The work of many graph-theoretic algorithms is greatly simplified by the absence of cross edges.

Suppose that T is a depth-first spanning tree in an undirected graph G. A vertex w in G will be an articulation point if and only if the removal of w disconnects one (or more) of w's subtrees in T from the rest of G. Since G has no cross edges (by the lemma), the only way that the removal of w will not disconnect a subtree of w is if a vertex in the subtree has a back edge that reaches above w (to an ancestor of w) in T. In Figure 3.24, (x,y) is such a back edge.

Let $i(x)$ denote the index of vertex x according to a preorder traversal of G. Clearly if x is an ancestor of y, then $i(x) < i(y)$, that is, x is visited before y in a depth-first search. Let back(y) be the smallest value of $i(z)$ among all z connected to y or to one of y's descendants. If y is a child of w and back$(y) < i(w)$, then the subtree rooted at y has no back edge to an ancestor of w. On the other hand, if back$(y) = i(w)$, then the subtree rooted at y has no back edge reaching above w in T, and so w is an articulation point. Further, w together

FIGURE 3.24

with the subtree rooted at y will be a biconnected component of G (assuming that the subtree itself contains no articulation points).

It remains to find an efficient way to calculate back(y). The following recursive definition of back(y) permits us to determine back(y) for each vertex in one traversal through the tree T. If $y_1, y_2, \ldots, y_k$ are the children of vertex y in T, and if back*(y) is the smallest value of $i(z)$ among all vertices connected to y, then

$$\text{back}(y) = \min(\text{back*}(y), \text{back}(y_1), \text{back}(y_2), \ldots, \text{back}(y_k)) \qquad (*)$$

To use this recurrence, we must order vertices so that all children precede their parents. The required order is a reversed preorder traversal. That is, if G has n vertices $x_1, x_2, \ldots, x_n$ indexed by a preorder traversal, then we scan them in reverse order. First determine back(x_n), then back(x_{n-1}), and so forth. When we are at vertex x_j, we first determine back*(x_j) and then compute back(x_j). We can now test to see if back(x_j) = $i(w)$, where w is the parent of x_j. If so, the subtree rooted at x_j together with a copy of w can be split off from G and put in a list of biconnected components of G. Summarizing, we have:

Algorithm to Find Articulation Points and Biconnected Components of a Connected Undirected Graph G

(a) Build a depth-first spanning tree T and let the n vertices of G be indexed according to a preorder traversal of T, $x_1, x_2, \ldots, x_n$.

(b) Starting with x_n and progressing backwards to x_{n-1}, x_{n-2}, and so forth, perform the following steps for each x_j (continue backwards to x_3; omit x_2 and x_1):
   (i) determine back*(x_j) = min($i(z)$: x_j is adjacent to z);
   (ii) determine back(x_j) = min(back*(x_j), {back(y): y is a son of x_j});
   (iii) if back(x_j) = $i(w)$, where w is the parent of x_j, then w is an articulation point, and the current subtree of T rooted at x_j together with a copy of w is put in a list of biconnected components of G.

Example 2

Let us find the articulation points and biconnected components of the undirected graph used in Example 1. For convenience, we repeat its adjacency matrix in Figure 3.25a. For Step (a) of the algorithm, we use the depth-first spanning tree T found in Example 1. See Figure 3.25b; back edges in Figure 3.25b are drawn as dashed edges. The y_i's at the ends of the rows and columns are a re-indexing of the vertices according to a preorder traversal of T. Now we perform Step (b), starting with y_8. Vertex y_8 has no back edge, and so back*(y_8) = 1 (the index of y_8's parent). Since y_8 is a leaf, back(y_8) = back*(y_8)

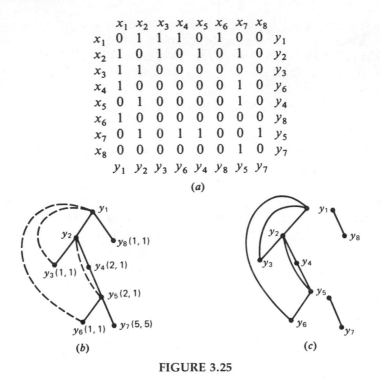

	x_1	x_2	x_3	x_4	x_5	x_6	x_7	x_8	
x_1	0	1	1	1	0	1	0	0	y_1
x_2	1	0	1	0	1	0	1	0	y_2
x_3	1	1	0	0	0	0	0	0	y_3
x_4	1	0	0	0	0	0	1	0	y_6
x_5	0	1	0	0	0	0	1	0	y_4
x_6	1	0	0	0	0	0	0	0	y_8
x_7	0	1	0	1	1	0	0	1	y_5
x_8	0	0	0	0	0	0	1	0	y_7

$$y_1 \ y_2 \ y_3 \ y_6 \ y_4 \ y_8 \ y_5 \ y_7$$

(a)

(b)   (c)

FIGURE 3.25

= 1. Since back (y_8) equals the index of y_8's parent y_1, then y_8 and y_1 form a biconnected component.

The algorithm proceeds to examine y_7, y_6, and so forth, computing the values shown in Figure 3.25b. Beside each vertex y_j in Figure 3.25b, we write (b^*,b), where $b^* = \text{back}^*(y_j)$ and $b = \text{back}(y_j)$. The articulation points of G are found to be y_1 and y_5. The biconnected components are displayed in Figure 3.25c.   ∎

EXERCISES

SUMMARY OF EXERCISES   The first seven exercises involve building spanning trees and different traversal orders. Exercises 8, 16–18, and 30 and 31 ask for computer programs. The remaining exercises examine properties of spanning trees and biconnected components.

1. Find depth-first spanning trees for each of these graphs;
   (a) K_8 (a complete graph on eight vertices).
   (b) The graph in Figure 2.9 in Section 2.3.
   (c) The graph in Figure 1.21b in Section 1.4.
2. Find breadth-first spanning trees of each of the graphs in Exercise 1.

3. Find all spanning trees (up to isomorphism) in the following graphs:

    (a) Figure 1.2   (b) K_4 (c)

4. Test the graph whose adjacency matrix is given below to see if it is connected.

	x_1	x_2	x_3	x_4	x_5	x_6	x_7
x_1	0	0	1	0	1	0	0
x_2	0	0	0	1	0	1	0
x_3	1	0	0	0	1	0	0
x_4	0	1	0	0	0	1	1
x_5	1	0	1	0	0	0	1
x_6	0	1	0	1	0	0	1
x_7	0	0	0	1	1	1	0

5. Consider an undirected graph with 25 vertices $x_2, x_3, \ldots, x_{26}$ with edges (x_i, x_j) if and only if i and j have a common divisor. How many components does this graph have? Find a spanning tree for each component.

6. List the vertices in order of a preorder and a postorder traversal of:
   (a) The left tree in Figure 3.1.
   (b) The middle tree in Figure 3.1.

7. Generalize the arithmetic tree in Figure 3.22 to include unary operations such as inverses or sin( ). Give the tree for the following expression:

$$\sin(((a + ((b \times c)^{-1} + ((a + d) \times v))) - (v + c)) \mid (a \quad b)^{-1})$$

8. (a) Write a program for building a depth-first spanning tree of a graph whose adjacency matrix is given.
   (b) Repeat part a for a breadth-first spanning tree.

9. Show that a connected undirected graph with just one spanning tree is a tree.

10. (a) Prove that a depth-first search reaches all vertices in an undirected connected graph.
    (b) Repeat (a) for breadth-first search.

11. (a) Show that the height of any depth-first spanning tree of a graph G starting from a given root a must be at least as large as the height of a breadth-first spanning tree of G with root a.

(b) By starting with different roots, it might be possible for a depth-first spanning tree of G to have a smaller height than a breadth-first spanning tree of G. Find a graph in which this can happen.

12. Let T be a depth-first spanning tree in a connected undirected graph G.
    (a) Show that when any non-tree edge is added to T a unique circuit results.
    (b) If C_1 and C_2 are circuits, show that $C_1 \oplus C_2 = C_1 \cup C_2 - C_1 \cap C_2$ is a circuit.
    (c) If $C_1, C_2, \ldots, C_n$ are circuits, show that $C_1 \oplus C_2 \cdots \oplus C_n$ is a circuit or collection of circuits.
    (d) Show that any circuit in G can be formed as a $\oplus$ sum of circuits of the type described in part a.

13. Prove that any spanning tree T' of a graph G can be converted into any other spanning tree T'' of G by a sequence of spanning trees $T_1, T_2,$ $\ldots, T_m$, where $T' = T_1$, $T'' = T_m$, and T_k is obtained from T_{k-1} by removing one edge of T_{k-1} that was in T' and adding one edge in T'' (m is the number of edges by which T' and T'' differ). (*Hint:* Use induction on m.)

14. Explain why in tests for the following graph properties it is sufficient to confine testing to biconnected components of a graph:
    (a) Largest complete subgraph.
    (b) Planarity.
    (c) Chromatic number.

15. Explain how using biconnected components can speed testing for graph isomorphism.

16. Write a program to determine whether a graph of n vertices has a circuit.

17. Write a program to enumerate all simple circuits (without repeated vertices) of a given graph.

18. Write a program to find, if possible, any occurrence of the following subgraph in a given undirected graph:

(a)   (b)

19. (a) Show that the lemma about all back edges in depth-first spanning trees is false for breadth-first spanning trees of most graphs. Specifically show that in a breadth-first spanning tree of any graph, all non-tree edges are cross edges.
    (b) Generalize part a to any graph that is not a tree.

20. (a) Show that the lemma is false when directed graphs are considered by finding a directed graph G and a depth-first rooted spanning tree of G that has a cross edge.

(b) Show that if $(x \overset{\rightarrow}{,} y)$ is a cross edge of a depth-first rooted spanning tree T in a directed graph G, then $i(y) < i(x)$, where $i(\ )$ is the preorder traversal indexing of vertices of T.

21. Apply the algorithm for finding articulation points and biconnected components to the graphs in:
(a) Exercise 4 (b) Exercise 5 (c) Figure S1.1 in Supplement I of Chapter One.

22. Use the biconnected components algorithm to show that the graph in Figure 1.3 in Section 1.1 is biconnected.

23. How many biconnected components does an n-vertex tree have?

24. Show how our biconnected components algorithm could have been based on a postorder traversal to compute back(x).

25. (a) If a directed graph is not strongly connected, show that its vertices can be partitioned into (disjoint) *strongly connected components*— vertices x and y are in the same strongly connected component if and only if there exist directed paths from x and y and from y to x.
(b) Find the strongly connected components of the directed graph in (i) Figure 1.9a in Section 1.1 and (ii) the right graph of Figure 1.15 in Section 1.2.
(c) How many strongly connected components does an m-vertex rooted tree have?

26. Why should the biconnected components algorithm not be used for x_2 or x_1 (see step (b))?

27. Show that a vertex x is an articulation point if and only if in a depth-first spanning tree rooted at x, there are at least two level-1 vertices.

28. Modify the biconnected components algorithm as follows to produce a strongly connected components algorithm. First, T is now a depth-first rooted tree that may not span the directed graph G; if T does not span G, repeat the algorithm on $G - T$. Second, use a postorder traversal of T (although $i(x)$ is still the preorder index). Third, in the definition of back*($\ $), "$x_j$ adjacent to z" means $(x \overset{\rightarrow}{,} z)$. Fourth, replace b (iii) by "if back(x_j) $\leqslant j$ ($=i(x_j)$), then the current subtree of T rooted at x_j is a strongly connected component of G—remove this component from G."
(a) Prove that this algorithm finds all strongly connected components of G. (*Hint:* Use the result of Exercise 20(b).)
(b) How can this algorithm be used to determine whether a directed graph is strongly connected?

29. A cutset is a set S of edges in a connected graph G whose removal disconnects G, but no proper subset of S disconnects G.
(a) Show that any cutset of G has at least one edge in common with any spanning tree of G.
(b) Suppose a set of edges in G contains exactly one edge of the spanning tree T and has an even number of edges in common with any circuit of G. Show that this set is a cutset.

30. Write a program for the Euler-circuit algorithm in Exercise 19 of Section 2.1; use a connectedness subroutine to test whether use of an edge disconnects the graph of remaining edges.

31. Use a breadth-first inverted spanning tree T (edges directed from children to parents) to build an Euler circuit in a directed graph possessing an Euler circuit as follows. Starting at the root of T, trace out a path such that at any vertex x we only choose the edge in T to the parent of x (in T) if there are no other unused edges leaving x at this stage. Verify the correctness of the algorithm. Program this algorithm.

32. Suppose that during a preorder traversal of a binary tree T, we write down a 1 for each internal vertex and a 0 for each leaf in the traversal, building a sequence of 1's and 0's. If T has n leaves, the sequence will have n 0's and $n - 1$ 1's. We call this sequence the *characteristic sequence of T*.

    (a) Find the binary tree with the characteristic sequence 110100110100100.

    (b) Prove that the last two digits in any characteristic sequence are 0's (assuming $n \geqslant 2$).

    (c) Prove that a binary sequence with n 0's and $n - 1$ 1's, for some n, is a characteristic sequence of some binary tree if and only if the first k digits of the sequence contain at least as many 1's as 0's, for $1 \leqslant k \leqslant 2n - 2$.

3.5 TREE ANALYSIS OF SORTING ALGORITHMS

One of the basic combinatorial procedures in computer science is sorting a set of items. Items are sorted according to their own numerical value or the value of some associated variable. Since names are stored in a computer as numbers, numerical arrangement is equivalent to alphabetical arrangement. To simplify our discussion, we will assume that all items have distinct numerical values.

Some of the sorting procedures we mention will use binary trees explicitly. All the procedures use binary testing trees implicitly, that is, the procedures use a sequence of tests that compare various pairs of items. A binary-testing tree model of sorting has one important consequence. The tree must have at least $n!$ leaves if there are n items, since there are $n!$ possible permutations, or rearrangements, of the set (if several test sequences in the tree lead to the same permutation, then there will be more than $n!$ leaves). By Theorem 3 in Section 3.1, the height of a binary testing tree must be at least $[\log_2 n!]$, which is approximately $n \log_2 n$.

Theorem

The maximum number of binary comparisons required to sort n items is at least $O(n \log_2 n)$. ∎

A function $g(n)$ is $O(f(n))$ when for large n, $g(n) \leq cf(n)$, c some constant. One can also show that the average number of binary comparisons required to sort n items is at least $O(n \log_2 n)$; see Exercise 13 in Section 3.1.

The best known sorting algorithm is called **bubble sort**, so named because small items move up the list the way bubbles rise in a liquid. It is compactly written as:

> FOR $m \leftarrow 2$ to n DO
> > FOR $i \leftarrow n$ step -1 to m DO
> > > IF $A_i < A_{i-1}$ THEN interchange items A_i and A_{i-1}.

This procedure will always require $(n-1) + (n-2) + \cdots + 1 = \frac{1}{2}n(n-1)$ binary comparisons. Thus this procedure requires $O(n^2)$ comparisons, as opposed to the theoretical bound of $O(n \log_2 n)$ comparisons. To see how much faster $O(n^2)$ grows than $O(n \log_2 n)$, observe that for $n = 50$, $n^2 = 2500$, and $n \log_2 n \cong 300$; and for $n = 500$, $n^2 = 250,000$, and $n \log_2 n \cong 4,500$.

The simplest to state sorting procedure that achieves the $O(n \log_2 n)$ bound is **merge sort**. It recursively subdivides the original list and successive sublists in half (or as close to half as possible) until each sublist consists of one item. Then it successively merges (in numerical order) the sublists. The subdivision process is naturally represented as a balanced binary tree and the merging as a reflected image of the tree.

Example 1

Sort the list 5, 4, 0, 9, 2, 6, 7, 1, 3, 8 using a merge sort.

The subdivision tree and subsequent ordered merges are shown in Figure 3.26. ∎

To analyze the number of binary comparisons in a merge sort, we make the simplifying assumption that $n = 2^r$, for some integer r. Then the subdivision tree will have sublists of size 2^{r-1} after the first splitting, that is, at the level 1 vertices. In general, there will be 2^{r-k} items in the sublists at level k. At level r, there are leaves each with one item.

In the merging tree, pairs of leaves are ordered at each vertex on level $r - 1$; this requires one binary comparison (to see which leaf item goes first). In general, at each vertex on level k we merge the ordered sublists (of 2^{r-k-1} items) of the 2 children into an ordered sublist of 2^{r-k} items; this merging will require $2^{r-k} - 1$ binary comparisons (verification of this number is left as an exercise). Finally the two ordered sublists of 2^{r-1} items at level 1 are merged at

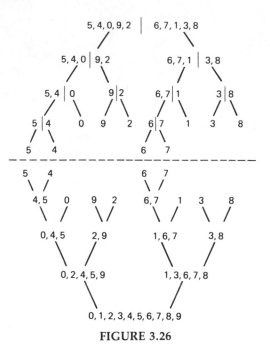

FIGURE 3.26

the root. The number of binary comparisons at all the vertices on level k is $2^k(2^{r-k} - 1)$, since there are 2^k different vertices on level k. Summing over all levels, we compute the total number of binary comparisons

$$\sum_{k=0}^{r-1} 2^k(2^{r-k} - 1) = \sum_{k=0}^{r-1} (2^r - 2^k)$$

$$= \sum_{k=0}^{r-1} 2^r - \sum_{k=0}^{r-1} 2^k$$

$$= r2^r - (2^r - 1) = (\log_2 n)n - (n - 1)$$

since $n = 2^r$ and so $r = \log_2 n$. Thus the number of binary comparisons in a merge sort is $O(n \log_2 n)$. However, extra computer time is required to implement the initial subdivision process. This extra work also requires only $O(n \log_2 n)$ steps. When $n = 15$, a bubble sort usually takes less time (and a lot less programming effort) than a merge sort.

A closely related sorting procedure is called **QUIK sort**. In QUIK sort, one takes the first item in the list L and uses it to divide the rest of the list into sublists L_1 and L_2 such that all items in L_1 are less than A_1 and all items in L_2 are greater than A_1. This subdivision would require $n - 1$ comparisons (of each item against A_1). Item A_1 is then put at the end of sublist L_1. Next the same subdivision procedure is used with the first item in each of the two sublists (L_1 is divided into L_{11} and L_{12}), and so on. Finally all sublists have only one item,

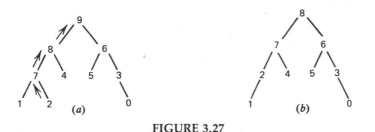

FIGURE 3.27

but the sublists are ordered and by concatenating the sublists in order we obtain a sorted list.

This procedure is simpler to implement than a merge sort, but it may not divide the lists evenly (for example, A_1 could be the largest item so that L_2 is empty). In the worst case, QUIK sort can be shown to require as many comparisons as a bubble sort and hence is an $O(n^2)$ algorithm. But when QUIK sort is used with many randomly arranged lists, the average number of comparisons is only $O(n \log_2 n)$ (see Exercise 6).

We conclude this section with a more complicated tree-based sorting procedure called **heap sort**. A **heap** is a binary, or almost-binary, tree with an item at each vertex such that each internal vertex's item is numerically greater than the items of its children. Each subtree of a heap is a heap. Figure 3.27a shows a heap involving the numbers 0 through 9. When a set of items has been arranged in a heap, the root of the heap must have the largest item in the set. If we remove the root of the heap (and place it at the end of the ordered list we are building), then we can re-establish a heap by making the larger child of the root the new root and recursively using this procedure to pick a new root for each subtree whose root was moved up a level. The arrows in Figure 3.27a show how vertices move up when the root 9 is removed. The new heap is shown in Figure 3.27b. Now we again remove the root of the heap (and make it the next-to-last item in the sorted list). Repeat this procedure until the heap is emptied and the sorted list complete. The one important missing step is creating the initial heap. This problem and further discussion of heap sort is left for the exercises.

EXERCISES

SUMMARY OF EXERCISES   The first exercises are based on the sorting methods presented in this section. The remaining exercises present other sorting schemes.

1.   Apply merge sort and QUIK sort to the following sequences:
     (a)   6, 9, 5, 0, 3, 1, 8, 4, 2, 7
     (b)   15, 27, 4, −7, 9, 13, 8, 28, 12, 20, −80
2.   Show that at most $n - 1$ comparisons are needed to merge two sorted sublists into a single sorted list of n items.

3. Use the result of Exercise 13 in Section 3.1 to show that the average number of binary comparisons required to sort n items is at least $O(n \log_2 n)$.

4. (a) Find a way to represent QUIK's sublist divisions with a binary tree.
   (b) Draw this tree for a QUIK sort of the list in Exercise 1(a).

5. Give an example of an n-item list for which QUIK sort requires $\frac{1}{2}n(n-1)$ comparisons.

6. Show by induction with a recurrence relation that for a random permutation, QUIK sort requires an average of $O(n \log_2 n)$ comparisons. (*Hint:* The ith item in the list is equally likely to be 1 or 2 or . . . or n).

7. (a) Describe how to build an initial heap from an unordered list of n items (the initial heap should be a balanced tree).
   (b) Use your method in part a to make a heap for the lists in Exercise 1.
   (c) Apply heap sort to the heaps in part b.

8. Show that heap sort requires $O(n \log_2 n)$ comparisons to sort n items (this includes the initial construction of a heap).

9. Given any heap-like n-vertex tree, show that the number of ways to reassign the numbers 1, 2, . . . , to the n vertices to form a heap (each parent's value is greater than its children's), is $n!/(r_1 r_2 r_3 \cdots r_n)$, where r_i is the number of vertices in the subtree rooted at the vertex with value i.

10. Modify the following sorting methods to allow for repeated (two or more equal) items:
    (a) bubble sort   (b) merge sort   (c) QUIK sort
    (d) heap sort

11. Write a computer implementation of (use a recursive language such as PASCAL):
    (a) merge sort   (b) QUIK sort   (c) heap sort

12. Describe a two-stage scheme for finding the largest and next-to-largest elements in a set of n items with $n + [\log_2 n] - 2$ comparisons.

13. Consider the following sorting scheme, which we call *tree sort*. We build a "dictionary look-up" binary tree recursively. The first item in the list to be sorted is the root. The second item is a left or right child of the root, depending on whether it numerically precedes or follows the root. We continue to add each successive item as a leaf to this growing tree. For example, if a list begins 8, 4, 11, 6, . . . , the tree after four items would be

When all items have been incorporated in the tree, the sorted order is obtained by an inorder traversal of the tree.

(a) Build the tree for the list in Exercise 1(a).

(b) Compare the tree in part a with the tree in Exercise 4(b) for this list.

(c) Generalize part b to show that tree sort is equivalent to QUIK sort.

(d) Show that if we rearrange each successive tree formed in tree sort to be a balanced tree, this rearrangement will require $O(n^2)$ operations for a list such as 0, 1, 2, . . . , n.

14. Consider the following sorting scheme for the list $A_1, A_2, . . . , A_n$ ($n = 2^r$). First do a sort (one comparison) of A_i and $A_{i+n/2}$; call this the ith ordered pair, for $i = 1, 2, . . . , n/2$. Next do a merge sort of the ith and the $(i + n/4)$th ordered pairs; call this the ith 4-tuple. Next do a merge sort of the ith and the $(i + n/8)$th ordered 4-tuples. Continue until there is just one sorted list of all n items.

(a) Apply this sorting method to the list 10, 12, 7, 0, 5, 8, 11, 15, 1, 6, 3, 9, 13, 4, 2, 14.

(b) How many comparisons does this sorting method require?

3.6 SUMMARY AND REFERENCES

This chapter examined a variety of search and data organization problems based on graph models. The common graph-theoretic tool for all these problems was trees. Section 3.1 presented basic properties and terminology of trees. Section 3.2 demonstrated the uses of depth-first and breadth-first search. Section 3.3 gave branch-and-bound and heuristic approaches to the traveling salesperson problem. Section 3.4 introduced spanning trees and graph algorithms. Section 3.5 looked at the decision trees underlying sorting algorithms.

The first paper implicitly using trees was Kirchhoff's 1847 fundamental paper about electrical networks. Cayley was the first person to use the term tree in an 1857 paper on counting ordered trees (using generating functions). Searching methods have been around for years (see Lucas [4]), but a systematic development of this subject came only in recent years with the advent of digital computers. There are several good computer science books about searching, sorting, and graph algorithms. See Aho, Hopcroft, and Ullman [1], Knuth [3], or Reingold, Nievergelt, and Deo [5]. For a good survey of the Traveling Salesperson Problem, see Bellmore and Nemhauser [2].

1. A. Aho, J. Hopcroft, and J. Ullman, *Data Structures and Algorithms*, Addison-Wesley, New York, 1983.

2. M. Bellmore and G. Nemhauser, "The Traveling Salesman Problem," *Operations Research* **16** (1968), 538–558.

3. D. Knuth, *The Art of Computer Programming, Vol. III: Sorting and Searching*, Addison-Wesley, New York, 1973.

4. E. Lucas, *Recreations Mathematiques*, Gauthier-Villars, Paris, 1891.

5. E. Reingold, J. Nievergelt, and N. Deo, *Combinatorial Algorithms*, Prentice-Hall, Englewood Cliffs, N.J., 1977.

Chapter Four
Network Algorithms

4.1 SHORTEST PATHS

In this chapter we present algorithms for the solution of several network optimization problems. By a **network**, we mean a graph with a positive integer $k(e)$ assigned to each edge e. This integer will typically represent the "length" of an edge, in units, or represent "capacity" of an edge, in units such as megawatts or gallons per minute. The optimization problems we shall discuss are all standard problems in operations research and have many practical applications. Thus good systematic procedures for their solution are essential. In the case of network flows, we shall see that the flow optimization algorithm can also be used to prove several important combinatorial theorems.

We begin with an algorithm for a relatively simple problem, finding a shortest path in a network from point a to point z. We say a shortest path because, in general, there may be more than one shortest path from a to z. For the rest of this section, let us assume that all networks are *undirected* and *connected*.

Let us immediately eliminate one possible shortest path algorithm: determine the lengths of all paths from a to z, and choose a shortest one. The computer is fast, but not that fast—such enumeration is already infeasible for most networks with 20 vertices. So when we find a shortest path, we must be able to prove it is shortest without explicitly comparing it with all other a–z paths. Although the problem is now starting to sound difficult, there still is a straightforward algorithmic solution.

The algorithm we present is due to Dijkstra. This algorithm gives shortest paths from a given vertex a to all other vertices. Let $k(e)$ denote the length of edge e. Let the variable m be a "distance counter." For increasing values of m, the algorithm labels vertices whose minimal distance from vertex a is m.

Shortest Path Algorithm

1.   Set $m = 1$ and label vertex a with $(-,0)$ (the "$-$" represents a blank).
2.   Check each edge $e = (p,q)$ from some labeled vertex p to some unlabeled vertex q. Suppose p's labels are $(r,d(p))$. If $d(p) + k(e) = m$, label q with (p,m).
3.   If all vertices are not yet labeled, increment m by 1 and go to Step 2. Otherwise go to Step 4. If we are only interested in a shortest path to z, then we go to Step 4 when z is labeled.
4.   For any vertex y, a shortest path from a to y has length $d(y)$, the second part of the label of y; and such a path may be found by backtracking from y (using the first part of the labels) as described below.

Observe that instead of concentrating on the distances to specific vertices, this algorithm solves the questions: How far can we get in 1 unit, how far in 2 units, in 3 units, . . . , in m units, . . .? Formal verification of this algorithm requires an induction proof (based on the number of labeled vertices).

The key idea is that to find a shortest path from a to any other vertex we must first find shortest paths from a to the "intervening" vertices. If $P_i = (s_1, s_2, \ldots, s_i)$ is a shortest path from s_1 to s_i, then $P_i = P_{i-1} + (s_{i-1}, s_i)$, where $P_{i-1} = (s_1, s_2, \ldots, s_{i-1})$ is a shortest path to s_{i-1}. Similarly $P_{i-1} = P_{i-2} + (s_{i-2}, s_{i-1})$, and so on.

To record a shortest path to s_i, all we need to store (as the *first* part of a label in the above algorithm) is the name of the next-to-last vertex on P_i, namely, s_{i-1}. Preceding s_{i-1} on the path is s_{i-2}, the next-to-last vertex on P_{i-1}. By continuing this backtracking process, we can recover all of P_i.

The algorithm given above has one significant inefficiency: If all sums $d(p) + k(e)$ in Step 2 have values of at least $m' > m$, then the distance counter m should be increased immediately to m'.

Example 1

A newly married couple, upon finding that they are incompatible, want to find a shortest path from point N (Niagara Falls) to point R (Reno) in the road network shown in Figure 4.1. We apply the shortest path algorithm. First N is labeled $(-,0)$. For $m = 1$, no new labeling can be done (we check edges (N,b), (N,d), and (N,f)). For $m = 2$, $d(N) + k(N,b) = 0 + 2 = 2$, and we label b with $(N,2)$. For $m = 3$, 4, no new labeling can be done. For $m = 5$, $d(b) + k(b,c) = 2 + 3 = 5$, and we label c with $(b,5)$. We continue to obtain the labeling shown

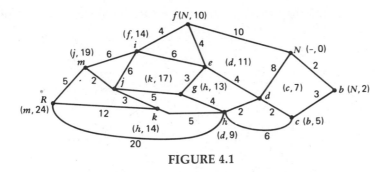

FIGURE 4.1

in Figure 4.1. Backtracking from R, we find the shortest path to be N-b-c-d-h-k-j-m-R with length 24. ■

If we want simultaneously to find shortest distances between all pairs of vertices (without directly finding all the associated shortest paths), we can use the simple algorithm due to Floyd. Let matrix D have entry $d_{ij} = \infty$ (or a very large number) if there is no edge from the ith vertex to the jth vertex; otherwise d_{ij} is the length of the edge from x_i to x_j. Then Floyd's algorithm is most easily stated with a computer program:

```
FOR k←1 TO n DO
  FOR i←1 TO n DO
    FOR j←1 TO n DO
      IF dik+dkj<dij THEN dij←dik+dkj;
```

When finished, d_{ij} will be the shortest distance from the ith vertex to the jth vertex.

EXERCISES

SUMMARY OF EXERCISES   The first six exercises involve shortest path calculations. The remaining exercises discuss associated theory (fairly easy theory), and the last exercise asks for a program.

1. Use the shortest path algorithm to find the shortest path between vertex c and vertex m in Figure 4.1.

2. Find the shortest path between the following pairs of vertices in the network in Figure 4.3 in the next section.
    (a) a and y (b) d and r (c) e and g

3. The network below shows the paths to success from L (log cabin) to W (White House). The first number on an edge is the time (number of years) it takes to traverse the edge; the second is the number of enemies

you make in taking that edge. Use the shortest path algorithm to answer the following:

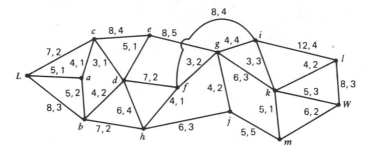

(a) Find the quickest path to success (From L to W).
(b) Find the quickest path to success that avoids points c, g, and k.
(c) Find the path to success (L to W) that minimizes the total number of enemies you make.
(d) Alter the order in which edges are checked in Step 2 of the shortest path algorithm to get another minimal-enemies path. How many such minimal-enemies paths are there?

4. With reference to Exercise 3, let the roughness index R of a path to success be $R = T + 2E$, where T is the time to get from L to W and E is the total number of enemies made.
(a) Find the smoothest (least rough) path to success.
(b) Find the smoothest path to success that includes edge (f,i); this edge can be traversed in either direction.

5. Ignore the numbers on the edges in Exercise 3 and use the shortest path algorithm to find the following shortest (least edges) paths:
(a) Shortest path from L to W.
(b) Shortest path from L to W including point d.
(c) Shortest path from L to W including both point e and point m.

6. Suppose that the edges in Exercise 3 are directed according to the alphabetical order of the endpoints, where L precedes a and W follows m (so edge (r,s) goes from r to s if s follows r in the alphabet). Find the quickest path from L to W.

7. Prove that the shortest path algorithm finds by induction the shortest path from a to every other vertex in the network.

8. Prove that Floyd's algorithm finds the shortest path between all pairs of vertices.

9. Make up an example to show that the algorithm in this section (by Dijkstra) fails if negative edge lengths are allowed.

10. Alter the shortest path algorithm so that in Step 2 we search among all edges from a labeled vertex p to an unlabeled vertex q for an edge that

minimizes $d(p) + k(p,q)$. The counter m is eliminated. Supply the rest of the details and show that this new algorithm is equivalent to the original one.

11. Show that in the shortest path algorithm the edges used in Step 2 to label new vertices form a spanning tree.

12. (a) Show that if the edges are properly ordered (and the edges are checked in this order in Step 2), the shortest path algorithm will produce any given shortest path from a to z.
    (b) Order the edges of the figure in Exercise 3 so that in Exercise 3(c), the "shortest" path found will be (L,a,c,d,f,g,k,m,W).

13. Consider the following physical model of a shortest path problem. Make pieces of string of length proportional to the length of each edge (e.g., an edge e with $k(e) = 7$ is a 7-in. string). Tie the string edges together at the appropriate nodes (vertices). Now take the start and end vertex of the network in your left and right hands, respectively, and pull them as far apart as possible. Show that any sequence of string-edges that form a taut path from a to z constitutes a desired shortest path.

14. Show that with the "short cut" in Exercise 10, the shortest path algorithm requires at most a number of steps proportional to the cube of the number of vertices in the network.

15. The *transitive closure* of a directed graph G is obtained by adding to G an edge (x_i, x_j) for each nonadjacent pair x_i, x_j with a directed path from x_i to x_j. Let $d_{ij} = 1$ if (x_i, x_j) is an edge in G and $= 0$ otherwise. Replace the IF statement in Floyd's algorithm with

$$\text{IF } d_{ij} \cdot d_{jk} > d_{ik} \text{ THEN } d_{ij} \leftarrow 1$$

Show that this revised Floyd's algorithm finds the transitive closure of G.

16. Write a computer program implementing the shortest path algorithm given in this section. Make the implementation as efficient as possible.

4.2 MINIMAL SPANNING TREES

Now we turn to the problem of finding a **minimal spanning tree**, a spanning tree in an (undirected) network such that the sum of the lengths of the tree's edges is as small as possible. This problem appears to be harder than the shortest path problem, but with the proper algorithm this problem is actually easier to do by hand than the shortest path problem. Surprisingly, the most straightforward "greedy" algorithms for finding a minimal spanning tree work. Let n be the number of vertices in the network.

Kruskal's Algorithm

Repeat the following step until the set T has $n - 1$ edges (initially T is empty): Add to T the shortest edge that does not form a circuit with edges already in T.

Prim's Algorithm

Repeat the following step until tree T has $n - 1$ edges: Add to T the shortest edge between a vertex in T and a vertex not in T (initially pick any edge of shortest length).

In both algorithms, when there is a tie for the shortest edge to be added, any of the tied edges may be chosen. That Kruskal's algorithm does indeed form a spanning tree follows from Exercise 5(a) of Section 3.1. Note that Prim's algorithm is very similar to the shortest path algorithm: If we consider vertices in T as labeled vertices, then both algorithms repeatedly search all edges from a labeled vertex to an unlabeled vertex (although the search is for different purposes). Indeed the edges used to label new vertices in Step 2 of the path algorithm form a spanning tree (see Exercise 11 in Section 4.1).

The difficult part in the minimal spanning tree problem is proving the minimality of the two algorithms. We give the proof for Prim's algorithm and leave Kruskal's as an exercise (Exercise 9).

Theorem

Prim's algorithm yields a minimal spanning tree.

Proof

Suppose $T_i = (e_1, e_2, \ldots, e_i)$ is the spanning tree constructed by i iterations of Prim's algorithm, and T' is a minimal spanning tree chosen to have as many edges in common with T_{n-1} as possible. We shall prove that $T_{n-1} = T'$.

If $T_{n-1} \neq T'$, then let $e_k = (a,b)$ be the first edge chosen by Prim's algorithm that is not in T'. Since e_k is not in T', then T' (a spanning tree) connects a to b via some path P. Some edge(s) of P are not in T_{k-1}, for otherwise $P \cup e_k$ would form a circuit in T_k. See Figure 4.2 (the thick edges are T_{k-1}). Let e^* be the first edge along P (starting from a) that is not in T_{k-1} (see Figure 4.2).

If e^* is shorter than e_k, then on the kth iteration, Prim's algorithm would have incorporated e^* not e_k. If e^* is the same length as (or greater than) e_k, we remove e^* from T' and replace it with e_k. It is not hard to show that the new T' is still a spanning tree that has the same minimal length (or less) and that has one more edge in common with T_{n-1}—this contradicts the choice of the original T'. ∎

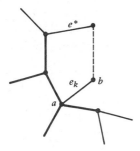

FIGURE 4.2

Example 1

We seek a minimal spanning tree for the network in Figure 4.3. Both algorithms start with a shortest edge. There are three edges of length 1: (a,f), (l,q), and (r,w). Suppose we pick (a,f). If we follow Prim's algorithm, the next edge we would add is (a,b) of length 2, then (f,g) of length 4, (g,l), then (l,q), then (l,m), and so forth. The next-to-last addition would be either (m,n) or (o,t) both of length 5 (suppose we choose (m,n)), and either one would be followed by (n,o). The final tree is indicated with darkened lines.

On the other hand, if we follow Kruskal's algorithm, we first include all three edges of length 1: (a,f), (l,q), (r,w). Next we would add all the edges of length 2: (a,b), (e,j), (g,l), (h,i), (l,m), (p,u), (s,x), (x,y). Next we would add almost all the edges of length 3: (c,h), (d,e), (k,l), (k,p), (q,v), (r,s), (v,w), but not (w,x) unless (r,s) were omitted (if both were present we get a circuit containing these two edges together with edges (r,w) and (s,x)). Next we would add all the edges of length 4 and finally either (m,n), or (o,t) to obtain the same minimal spanning tree(s) produced by Prim's algorithm. This similarity is no coincidence (see Exercise 11). ∎

a	1	f	7	k	3	p	2	u
2		4		3		5		7
b	5	g	2	l	1	q	3	v
8		6		2		7		3
c	3	h	4	m	10	r	1	w
4		2		5		3		3
d	7	i	8	n	8	s	2	x
3		5		4		4		2
e	2	j	9	o	5	t	6	y

FIGURE 4.3

EXERCISES

SUMMARY OF EXERCISES   The first six exercises involve minimal spanning tree computations; the seventh asks for programs. The remaining questions are theoretical.

1. Find a minimal spanning tree for the network in Figure 4.1 using:
    (a)   Prim's algorithm.
    (b)   Kruskal's algorithm.

2. Let us re-interpret the numbers in Exercise 3 of Section 4.1. Suppose the network shows possible routes for a freeway system linking a set of cities (the vertices) where the first number is the cost of building a freeway along the edge and the second is the number of trees (in thousands) that would have to be cut down.
    (a)   Use Kruskal's algorithm to find a minimal-cost set of freeways connecting all the cities together.
    (b)   Court action by conservationists rules out use of edges (c,e), (d,f), and (k,W). Now find the minimal-cost set of freeways.
    (c)   Find two nonadjacent vertices such that the tree in part a does not contain the cheapest path between them.
    (d)   Use Prim's algorithm to find a set of connecting freeways that minimizes the number of trees cut down.

3. With reference to Exercise 2, suppose that the set of freeways must include city c or city d (but not necessarily both) and all the other cities. Find the minimal-cost set of freeways.

4. With reference to Exercise 2, suppose that the governor's summer home is along edge (f,i). Find a minimal-cost set of freeways such that (f,i) is in that set.

5. Find a *maximal* spanning tree (whose sum of edge lengths is maximal) for the network in Figure 4.3.

6. Construct an undirected, connected network with 8 vertices and 15 edges that has a minimal spanning tree containing the shortest path between every pair of vertices.

7. Write a computer program to implement (as efficiently as possible):
    (a)   Kruskal's algorithm   (b)   Prim's algorithm

8. Let T be the minimal spanning tree found by Prim's algorithm in an undirected, connected network N.
    (a)   Prove that T contains all edges of shortest length in N unless such edges include a circuit.
    (b)   Prove that if $e^* = (a,b)$ is an edge of N not in T and if P is the unique path in T from a to b, then for each edge e in P, $k(e) \leq k(e^*)$.
    (c)   Prove that part (b) holds for any minimal spanning tree.

9. If each edge has a different cost, show that the minimal spanning tree is unique.

10. Verify the assertion in the last sentence in the proof of the theorem.

11. Suppose that the edges of the undirected, connected network N are ordered and that in both Prim's and Kruskal's algorithms, when there is a tie for the next edge to be added, the smaller indexed edge is chosen.
    (a) Prove that the edges can be ordered so that Prim's algorithm will yield any given minimal spanning tree.
    (b) Prove that the edges can be ordered so that Kruskal's algorithm will yield any given minimal spanning tree.
    (c) Prove that with ordered edges, both algorithms give the same tree.

12. Modify Prim's algorithm so that it finds a minimal spanning tree that contains a prescribed edge. Prove your modification works.

13. Modify Kruskal's algorithm so that it finds a maximal spanning tree.

14. Show that if the edges are first ordered by size, then Kruskal's algorithm requires at most a number of steps proportional to the square of the number of vertices in the network.

15. Prove that Kruskal's algorithm gives a minimal spanning tree.

16. Given an undirected, connected network N with n vertices, we form a graph G_N whose vertices correspond to minimal spanning trees of N with two vertices v_1, v_2 adjacent if the corresponding minimal spanning trees T_1, T_2 differ by one edge, that is, $T_1 = T_2 - e' + e''$ (for some e', e'').
    (a) Produce an eight-vertex network N such that G_N is a chordless 4-circuit.
    (b) Prove that if T_1 and T_2 are minimal spanning trees which differ by k edges, that is, $|T_1 \cap T_2| = (n - 1) - k$, then in G_N there is a path of length k between the corresponding vertices.

17. Show that the shortest network linking points a, b, and c in the plane is obtained by drawing a fourth point x whose lines to a, b, c form $120°$ angles (as shown) and using these lines to link a, b, c. (If the lines from a to b and a to c form an angle of $\geq 120°$, then $x = a$; similarly for b or c.)

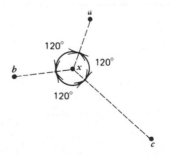

4.3 NETWORK FLOWS

In this section we interpret the integer $k(e)$ associated with edge e in a network as a capacity or upper bound. We seek to maximize a "flow" from vertex a to vertex z such that the flow in each edge does not exceed that edge's capacity. Many transport problems are of this general form, for example, maximizing the flow of oil from Houston to New York through a large pipeline network (here the capacity of an edge represents the capacity in barrels per minute of a section of pipeline), or maximizing the number of telephone calls possible between New York and Los Angeles through the vast Bell system. It is convenient to assume initially that *all networks are directed*.

Let $In(x)$ and $Out(x)$ be the sets of edges directed into and out from vertex x, respectively. Then we define an a–z **flow** φ in a directed network N to be an integer-valued function φ defined on each edge e—$\varphi(e)$ is the flow in e—together with a **source** vertex a and a **sink** vertex z satisfying the following three conditions:

(a) $0 \leqslant \varphi(e) \leqslant k(e)$.

(b) $\varphi(e) = 0$ if $e \in In(a)$ or $e \in Out(z)$.

(c) For $x \neq a$ or x, $\displaystyle\sum_{e \in In(x)} \varphi(e) = \sum_{e \in Out(x)} \varphi(e)$.

The second condition assures that the flow goes from a to z, not in the reverse direction. A sample flow is shown in Figure 4.4; the capacity and flow in each edge e are written beside the edge: $k(e)$, $\varphi(e)$. The assumption of integer capacities and flows is not restrictive, that is, the units we count could be thousandths of an ounce rather than barrels.

Let $(P, \bar{P})$ denote the set of all edges $(x \vec{,} y)$ with vertex $x \in P$ and $y \in \bar{P}$ (where $\bar{P}$ denotes the complement of P). We call such a set $(P, \bar{P})$ a **cut**. The cut $(\{a,b,c\}, \{d,e,z\})$ in Figure 4.4 consists of the edges $(b \vec{,} d)$, $(b \vec{,} e)$, $(c \vec{,} e)$; the edge $(d \vec{,} c)$ is not in the cut because it goes from $\bar{P}$ to P. We call $(P, \bar{P})$ an **a–z cut** if $a \in P$ and $z \in \bar{P}$. (Later in undirected graphs, $(P, \bar{P})$ will denote all edges between P and $\bar{P}$.)

Let φ be an a–z flow in the network N and let P be a subset of vertices in N not containing a or z. Summing together the conservation-of-flow equations in condition (c) for each $x \in P$, we obtain

$$\sum_{x \in P} \sum_{e \in In(x)} \varphi(e) = \sum_{x \in P} \sum_{e \in Out(x)} \varphi(e)$$

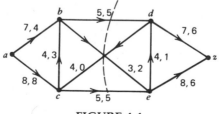

FIGURE 4.4

Certain $\varphi(e)$ occur on both sides of the preceding equality; namely, edges from one vertex in P to another vertex in P. After eliminating such $\varphi(e)$ from both sides, the left side becomes $\Sigma_{e\in(\bar{P},P)}\varphi(e)$ and the right side $\Sigma_{e\in(P,\bar{P})}\varphi(e)$. Thus we have:

(c') For each vertex subset P not containing a or z,

$$\sum_{e\in(\bar{P},P)} \varphi(e) = \sum_{e\in(P,\bar{P})} \varphi(e)$$

That is, the flow into P equals the flow out of P. We introduced condition (c') because it leads to simpler proofs. For example, the following intuitive result is readily verified with (c').

Theorem 1

For any a-z flow φ in a network N, the flow out of a equals the flow into z.

Proof

Assume temporarily that N contains no edge $(\vec{a,z})$. Let P be all vertices in N except a and z. So $\bar{P} = \{a,z\}$. Then

$$\text{flow out of } a = \sum_{e\in(\bar{P},P)} \varphi(e) \overset{\text{by (c')}}{=} \sum_{e\in(P,\bar{P})} \varphi(e) = \text{flow into } z$$

Remember that the only flow into P from $\{a,z\}$ must be from a, since condition (b) forbids flow from z. Similarly all flow out of P must go to z. The flow equality still holds if there is flow in an edge $(\vec{a,z})$. ■

We define $|\varphi|$, the **value** of the a–z flow φ, to be the sum of the flow out of a (or equivalently, into z). Let us consider the question of how large $|\varphi|$ can be. One obvious upper bound is the sum of the capacities of the edges leaving a, for

$$|\varphi| = \sum_{e\in\text{Out}(a)} \varphi(e) \leq \sum_{e\in\text{Out}(a)} k(e)$$

Similarly, the sum of the capacities of the edges entering z is an upper bound for $|\varphi|$. Intuitively, $|\varphi|$ is bounded by the sum of the capacities of any set of edges which cut all flow from a to z. For any cut $(P,\bar{P})$, we define $k(P,\bar{P})$, the **capacity** of $(P,\bar{P})$, to be

$$k(P,\bar{P}) = \sum_{e\in(P,\bar{P})} k(e)$$

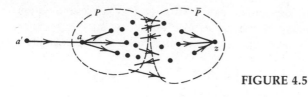

FIGURE 4.5

The capacity of the a–z cut $(P,\bar{P})$, where $P = \{a,b,c\}$, in Figure 4.4 is 13, and so no a–z flow in the figure can have a value greater than 13.

Theorem 2

For any a–z flow φ and any a–z cut $(P,\bar{P})$ in a network N, $|\varphi| \leq k(P,\bar{P})$.

Proof

Expand the network N by adding a new vertex a' with an edge $e' = (a',a)$ of immense capacity. Note that a' will be part of $\bar{P}$. Assign a flow value of $|\varphi|$ to e', and now φ becomes an a'–z flow in the expanded network (see Figure 4.5). In effect, the old source a now gets its flow from the "super source" a'. In the new network, we apply condition (c') to P (we could not use (c) in N because P contains N's source a). It says that the flow into P, which is at least $|\varphi|$ (other flow could come into P along edges from $\bar{P}$), equals the flow out of P. Thus

$$|\varphi| \leq \sum_{e\in(\bar{P},P)} \varphi(e) = \sum_{e\in(P,\bar{P})} \varphi(e) \leq \sum_{e\in(P,\bar{P})} k(e) = k(P,\bar{P}) \qquad (*)$$

∎

Corollary 2a

For any a–z flow φ and any a–z cut $(P,\bar{P})$ in a network N, $|\varphi| = k(P,\bar{P})$ if and only if:

(i)   For each edge $e \in (\bar{P},P)$, $\varphi(e) = 0$.
(ii) For each edge $e \in (P,\bar{P})$, $\varphi(e) = k(e)$.

Further, when $|\varphi| = k(P,\bar{P})$, then φ is a maximal flow and $(P,\bar{P})$ is an a–z cut of minimal capacity.

Proof

Consider the two inequalities in (*) of the preceding proof. The flow from $\bar{P}$ into P in the expanded network equals $|\varphi|$ if condition (i) holds; otherwise the flow into P is greater than $|\varphi|$. The flow out of P equals $k(P,\bar{P})$ if condition (ii) holds; otherwise it is less. Thus equality holds in (*) if and only if conditions

(i) and (ii) are both true. The last sentence in the corollary follows directly from Theorem 2. ∎

While we now have developed all the concepts needed to present our flow-maximizing algorithm, we first discuss an intuitive but faulty technique that can sometimes be used as a "short-cut" in place of the correct algorithm. Only after the fault in the short-cut is exposed can the real algorithm be fully appreciated.

All "normal" flows can be decomposed into a collection of **unit-flow paths** from a to z, for short, $a-z$ unit flows (a precise formulation of this assertion is left to Exercise 24). For example, in a telephone network, the flow from New York to Los Angeles can be decomposed into individual telephone calls. Similarly, flow of oil, in theory, can be decomposed into the paths of each individual petroleum molecule. Formally, an $a-z$ unit flow along path L is a flow (φ_L, a, z) with $\varphi_L(e) = 1$ if e is in L and $= 0$ if e is not in L. If we define the sum $\varphi = \varphi_1 + \varphi_2$ of two $a-z$ flows φ_1, φ_2 by $\varphi(e) = \varphi_1(e) + \varphi_2(e)$, then clearly φ is an $a-z$ flow, provided $\varphi(e) \leq k(e)$ for all edges.

This suggests a way to build a maximal flow. We build up the flow as much as possible by successively adding $a-z$ unit flows together, always being sure not to exceed any edge's capacity. Since an additional unit flow can only use **unsaturated** edges (where the present flow does not equal the capacity), we should build paths consisting of unsaturated edges. We define the **slack** $s(e)$ of edge e in flow φ by

$$s(e) = k(e) - \varphi(e)$$

If s is the minimum slack among edges in the $a-z$ unit flow φ_L, then we want an additional flow along L of $s\varphi_L = \varphi_L + \varphi_L + \varphi_L + \cdots$ (s times).

Example 1

Let us use the method just outlined to build a maximal $a-z$ flow for the network in Figure 4.6a. Note, as an upper bound, that the value of a flow cannot exceed 10, the capacity of edges going out of a.

We start with no flow, that is, $\varphi(e) = 0$, for all e. Now we find some path from a to z, for example, the $a-z$ path $L_1 = a\text{-}b\text{-}d\text{-}z$. The minimum slack on L_1 is 3 (at the start, the slack of each edge is just its capacity). So to our initial zero flow, we add the flow $3\varphi_{L_1}$. Now all edges except (b,d) are still unsaturated.

Suppose that we next find the $a-z$ path $L_2 = a\text{-}c\text{-}e\text{-}z$, also with minimum slack 3. Our new flow is $3\varphi_{L_1} + 3\varphi_{L_2}$, as shown in Figure 4.6b. Now all edges but (b,d) and (c,e) are unsaturated. The path $L_3 = a\text{-}b\text{-}e\text{-}z$ with minimum slack 2 can be used to get the augmenting flow $2\varphi_{L_3}$. Figure 4.6c shows the remaining unsaturated edges with their slacks.

Now the only possible path is $L_4 = a\text{-}c\text{-}d\text{-}z$ with minimum slack 1. After adding the flow φ_{L_4} (see Figure 4.6d), we can get no further than from a to c.

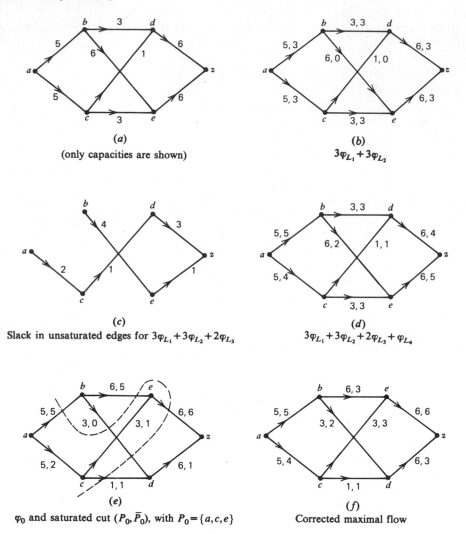

(a)

(only capacities are shown)

(b)

$3\varphi_{L_1} + 3\varphi_{L_2}$

(c)

Slack in unsaturated edges for $3\varphi_{L_1} + 3\varphi_{L_2} + 2\varphi_{L_3}$

(d)

$3\varphi_{L_1} + 3\varphi_{L_2} + 2\varphi_{L_3} + \varphi_{L_4}$

(e)

φ_0 and saturated cut $(P_0, \bar{P}_0)$, with $P_0 = \{a, c, e\}$

(f)

Corrected maximal flow

FIGURE 4.6

Observe that we have saturated the edges in the a–z cut $(P, \bar{P})$, where $P = \{a, c\}$. The value 9 of the final flow equals $k(P, \bar{P})$, and so by Corollary 2a the flow must be maximal. ∎

Example 2

Suppose the network in Figure 4.6a is redrawn as in 4.6e.

Let us again choose augmenting a–z flow paths across the top and bottom of the network, now having sizes 5 and 1, respectively. Since edges $(\vec{a,b})$ and $(\vec{c,d})$ are saturated by these flow paths, the only possible a–z path along unsaturated edges is $L_5 = a\text{-}c\text{-}e\text{-}z$ with minimum slack 1 (the minimum occurring in edge $(\vec{e,}z)$). After adding the a–z unit flow φ_{L_5}, we get the flow φ_0 shown in Figure 4.6e. The cut $(P_0, \bar{P}_0)$, where $P_0 = \{a,c,e\}$, is saturated, and so no more augmenting a–z unit flows exist. Yet $|\varphi_0| = 7$ and $k(P_0, \bar{P}_0) = 12$. Remember that a flow of size 9 was obtained for this same network in Example 1! What has happened? ∎

Example 2 shows that an arbitrary sequence of successive augmenting a–z unit flows need not inevitably yield a maximal flow. We are also faced with a flow φ_0 and a saturated a–z cut $(P_0, \bar{P}_0)$ such that $|\varphi_0| < k(P_0, \bar{P}_0)$. Corollary 2a implies that there must be some flow in an edge $e \in (\bar{P}_0, P_0)$. Looking at Figure 4.6e, we see that the flow path $L' = a\text{-}b\text{-}e\text{-}z$ crosses the a–z cut $(P_0, \bar{P}_0)$ backwards (from $\bar{P}_0$ to P_0) on the edge $(\vec{b,}e)$; or equivalently, L' crosses the cut forwards twice, on edge $(\vec{a,}b)$ and again on $(\vec{e,}z)$. Thus the 5 units of flow along L' use up 10 units of capacity in the cut, whence $k(P_0, \bar{P}_0)$ is 5 units greater than $|\varphi_0|$.

The reason that the sequence of augmenting flow paths in Example 2 did not lead to a maximal flow can be explained intuitively as follows. By sending 5 units of flow along L' (see Figure 4.6e), we have routed all the flow passing through b on to e and none of it to d. Then only 1 unit of flow passing through c can be routed on to e and then along edge $(\vec{e,}z)$. But, much of the flow through c must go to e, since the capacity of $(\vec{c,}d)$ is only 1. In sum, the initial 5-unit flow along $a\text{-}b\text{-}e\text{-}z$ was a "mistake" because some of the capacity in edge $(\vec{e,}z)$ should have been "reserved" for flow from c.

How can we avoid or correct such mistakes? If we understood where the mistakes were made, we could change some of the flow paths and try a new sequence of augmenting flow paths. However, we are likely to make other mistakes in subsequent constructions. Indeed, there may be certain networks in which it is impossible not to make such a mistake, no matter what sequence of flow paths is used. In terms of cuts, we may always end with a saturated a–z cut that one of our flow paths crosses twice.

Fortunately, there is a procedure to correct "mistakes" and thereby further increase the flow. The method will not look for edges that must be "reserved" for later flow paths, as suggested above (that is too hard a problem). Rather it looks for flow that is going the wrong way (backwards) across an a–z cut. Then it finds a way to decrease the backward flow without changing the forward flow across the cut, the result being more total flow across the cut (and through the network). The following example presents the idea behind this procedure.

Example 2 (continued)

The edge $(\vec{b,}e)$ contains 5 units of flow that is going backwards across the saturated cut $(P_0, \bar{P}_0)$. By how much can the flow in $(\vec{b,}e)$ be reduced?

Condition c of a flow—flow in equals flow out—requires that a reduction of the flow into e from b must be compensated by an increase to e from elsewhere in P_0 (if the compensating flow comes from $\bar{P}_0$ we would have a new backward flow). Such an increase must in the end come from a. Thus we need an a–e flow path in P_0. The only such path is $K_1 = a\text{-}c\text{-}e$. Similarly, a reduction of the flow out of b to e must be compensated by an increase in flow out of b to somewhere else in $\bar{P}_0$. So we need a b–z flow path in $\bar{P}_0$. The only such path is $K_2 = b\text{-}d\text{-}e$.

The minimum slack along K_1 is 2 and the minimum slack along K_2 is 3. Then we can decrease the flow in $(\vec{b,e})$ by 2 while increasing the flow in K_1 and K_2 by 2. Figure 4.6f shows the resulting maximal flow. Note that this new maximal flow is different from the maximal flow in Figure 4.6f, although both saturate the a–z cut $(\{a,c\},\{b,d,e,z\})$. ■

A **chain** in a directed graph is a sequence of edges that forms a path when the direction of the edges is ignored. **A unit-flow chain** from a to z along the a–z chain K is a "flow" φ_K with a value of 1 in each edge of K forwardly directed in K, a value of -1 in each edge of K backwardly directed in K, and a value of 0 elsewhere. Note that φ_K is not really a flow because it can assume a negative value on some edges. However, if φ is a flow that already has positive values in each backwardly directed edge of K and has slack in each forwardly directed edge of K, then $\varphi + \varphi_K$ is a flow.

The flow correction made in the continuation of Example 2 consisted of a (2-unit) a–z flow chain along chain $K = a\text{-}c\text{-}e\text{-}b\text{-}d\text{-}z$. When no backwardly directed edges occur in a flow chain, then it is simply a flow path. We shall see that any sequence of augmenting a–z flow chains leads to a maximal flow. Flow chains are the appropriate generalization of flow paths which both build additional flow and simultaneously correct possible "mistakes."

We now present a flow chain algorithm defined so that if the procedure fails to obtain an augmenting a–z flow chain, it will produce a saturated a–z cut whose capacity equals the value $|\varphi|$ of the current flow φ. Thus by Corollary 2a, φ would be maximal.

The algorithm recursively tries to build augmenting flow chains from a to all vertices in a manner reminiscent of the shortest path algorithm in Section 4.1. The algorithm assigns two labels to a vertex q: $(p^{\pm}, \Delta(q))$, where p is the previous vertex on a flow chain from a to q, the superscript of p is $+$ if the last edge of the chain is $(\vec{p,q})$ and is $-$ if the last edge is $(\vec{q,p})$, and $\Delta(q)$ is the minimum slack among the edges of the chain from a to q. On a backwardly directed edge e, the slack is the amount of flow that can be removed, namely $\varphi(e)$. A $+$ superscript on q means flow is being added to edge $(\vec{p,q})$; a $-$ superscript means flow is being subtracted from edge $(\vec{q,p})$.

Augmenting Flow Algorithm

1. Give vertex a the labels $(-, \infty)$. Let a be the first vertex to be scanned.
2. Call the vertex being scanned p with second label $\Delta(p)$.

(a)   Check each incoming edge $e = (q\overset{\rightarrow}{,}p)$. If $\varphi(e) > 0$ and q is unlabeled, then label q with $(p^-, \Delta(q))$, where $\Delta(q) = \min(\Delta(p), \varphi(e))$.

(b)   Check each outgoing edge $e = (p\overset{\rightarrow}{,}q)$. If $s(e) = k(e) - \varphi(e) > 0$ and q is unlabeled, then label q with $(p^+, \Delta(q))$, where $\Delta(q) = \min(\Delta(p), s(e))$.

3.   If z has been labeled, go to Step 4. Otherwise choose another labeled vertex to be scanned (which was not previously scanned) and go to Step 2. If there are no more labeled vertices to scan, let P be the set of labeled vertices and now $(P, \overline{P})$ is a saturated a–z cut. Moreover, $|\varphi| = k(P, \overline{P})$, and thus φ is maximal.

4.   Find an a–z chain K of slack edges by backtracking from z as in the shortest path algorithm. Then an a–z flow chain φ_K along K of $\Delta(z)$ units is the desired augmenting flow. Increase the flow in the edges of K by $\Delta(z)$ units (decrease flow if edge is backward directed in K).

Like the shortest path algorithm, the flow algorithm extends partial-flow chains from currently labeled vertices to adjacent unlabeled vertices. Before we prove that repeated application of our algorithm always leads to a maximal flow, let us give some examples.

Example 2 (continued)

Let us apply our augmenting flow algorithm to the flow in Figure 4.6e.

Vertex a is labeled $(-, \infty)$. The edge $(a\overset{\rightarrow}{,}b)$ is saturated, but $(a\overset{\rightarrow}{,}c)$ has slack $5 - 2 = 3$. So we label c $(a^+, 3)$. From c, we label e $(c^+, 2)$ (2 is the minimum of $\Delta(c)$, the extra flow we can get to c, and the slack in edge $(c\overset{\rightarrow}{,}e)$). From e, we label b $(e^-, 2)$. From b, we label d $(b^+, 2)$, and from d we label z $(d^+, 2)$. So we can get $\Delta(z) = 2$ more units of flow to z. Backtracking with the labels, the flow chain K (in backwards order) is z-d-b-e-c-a. The new flow $\varphi_0 + 2\varphi_K$ is the flow previously obtained in Figure 4.6f. ∎

Example 3

Consider the pipeline network shown in Figure 4.7a. We seek a maximal flow from a to z.

If a maximal flow were being found by a computer, it would have to start with a zero flow. When solving a flow problem by hand, we can speed the process by starting with a (nonzero) flow obtained by inspection (in small networks we can often obtain a maximal flow by inspection). Let the initial flow be $\varphi = 4\varphi_{K_1} + 4\varphi_{K_2} + 5\varphi_{K_3}$, where $K_1 = a$-b-e-z, $K_2 = a$-c-d-f-z, and $K_3 = a$-d-f-z. See Figure 4.7a. The reader can see that we actually sent too much flow from c to d. Some of it should have gone directly from c to f.

We now apply the labeling algorithm to the flow φ. Label vertex a $(-, \infty)$. Scanning edges at a, we find no incoming edges but there are three outgoing edges: edge $(a\overset{\rightarrow}{,}b)$ has slack $s(a\overset{\rightarrow}{,}b) = 2 > 0$ and b is unlabeled, so we label b

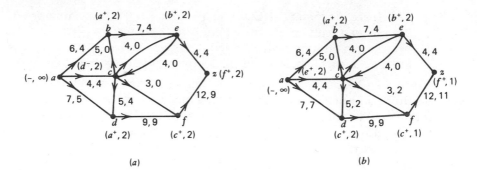

(a)

(b)

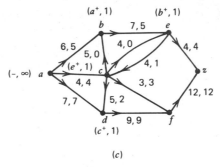

(c)

FIGURE 4.7

$(a^+,2)$, where 2 is the minimum of $\Delta(a)$ $(=\infty)$ and $s(a \vec{,} b)$; edge $(a \vec{,} c)$ has no slack; edge $(a \vec{,} d)$ has slack $s(a \vec{,} d) = 2$ and d is unlabeled, so we label it $(a^+,2)$.

Next we scan b (we will scan vertices in the order that they were labeled). There are two incoming edges at b; edge $(a \vec{,} b)$ has $\varphi(a \vec{,} b) = 4 > 0$ but a is already labeled; edge $(c \vec{,} b)$ has $\varphi(c \vec{,} b) = 0$. There is one outgoing edge at b: edge $(b \vec{,} e)$ has slack $s(b \vec{,} e) = 3$ and e is unlabeled, so we label it $(b^+,2)$ where 2 is min $(\Delta(b), s(b \vec{,} e))$.

Next we scan at d. There are two incoming edges at d: edge $(a \vec{,} d)$ comes from a labeled vertex; edge $(c \vec{,} d)$ has $\varphi(c \vec{,} d) = 4$ and c is unlabeled, so we label it $(d^-,2)$ where $2 = \min(\Delta(d), \varphi(c \vec{,} d))$. There is one outgoing edge at d: edge $(d \vec{,} f)$ is saturated.

No labeling can be done from e. At c, we label f with $(c^+,2)$. At f, we label z with $(f^+,2)$.

Since z is now labeled, the labeling procedure terminates. We can send 2 $(=\Delta(z))$ units in the augmenting $a-z$ flow chain $\varphi_{K_4} = a\text{-}d\text{-}c\text{-}f\text{-}z$, found by the backtracking procedure. Recall that since edge $(c \vec{,} d)$ is backwardly directed in K, the flow chain φ_{K_4} subtracts 2 units from edge $(c \vec{,} d)$. The new flow $\varphi' = \varphi + 2\varphi_{K_4}$ is shown in Figure 4.7b.

Next we repeat the labeling algorithm. At a, we label b with $(a^+,2)$. At b, we label e with $(b^+,2)$. At e, we label c with $(e^+,2)$. At c, we label d with $(c^+,2)$ and f

with $(c^+,1)$. At d we can make no new labels. At f, we label z with $(f^+,1)$. The augmenting $a-z$ flow chain is φ_{K_5} with $K_5 = a\text{-}b\text{-}e\text{-}c\text{-}f\text{-}z$, using forward edge (e,c). Our new flow is $\varphi'' = \varphi' + \varphi_{K_5}$, as shown in Figure 4.7c.

The reader has probably observed that the flow φ'' is maximal. The incoming edges at z are now saturated. Let us blindly apply the algorithm once more (as a computer would). At a, we label b; at b, we label e; at e, we label c; and at c, we label d. No more vertices can be labeled. We let P be the set of labeled vertices. Then $(P,\bar{P})$ is the saturated $a-z$ cut specified by the algorithm with $|\varphi''| = 16 = k(P,\bar{P})$. ∎

In applying our algorithm, we must always check for the possibility of minus labeling of vertices, even though this labeling is very infrequent. Recall that the minus labeling corresponds to correcting a mistaken flow assignment. A permissible short-cut would be to use only positive labeling (as in the faulty procedure discussed earlier) until no new flow paths can be found, and then apply the full algorithm to hunt for "mistakes." The use of this short-cut serves to increase the importance of having a rigorous proof that when repeatedly applied to any given flow, our algorithm will yield a maximal flow.

Theorem 3

For any given $a-z$ flow φ, a finite number of applications of the augmenting flow algorithm yields a maximal flow. Moreover, if P is the set of vertices labeled during the final (unsuccessful) application of the algorithm, then $(P,\bar{P})$ is a minimal $a-z$ cut set.

Proof

There are two main parts to the proof. First, if φ is the current flow and φ_K is the augmenting $a-z$ unit flow chain (along chain K) found by the algorithm with $m = \Delta(z)$, then we must show that the new flow $\varphi + m\varphi_K$ is indeed a legal flow in the network. Both φ and φ_K satisfy flow conditions (b) and (c) and are integer-valued. Hence $\varphi + m\varphi_K$ also satisfies (b) and (c) and is integer-valued. The labeling algorithm is designed so that $m = \Delta(z)$ is the minimum slack (of the appropriate kind) along chain K and hence $\varphi + m\varphi_K$ satisfies flow condition (a): $0 \leq \varphi(e) + m\varphi_K(e) \leq k(e)$.

Since m is a positive integer, each new flow is larger by an integral amount. The capacities and the number of edges are finite, and so the algorithm must eventually halt—fail to label z. Let P be the set of labeled vertices when the algorithm halts. Clearly $(P,\bar{P})$ is an $a-z$ cut, since a is labeled and z is not. Observe that there cannot be an unsaturated edge from a labeled vertex p to an unlabeled vertex q, or else at p we could have labeled q. Similarly, there cannot be a flow in an edge from an unlabeled vertex q to a labeled vertex p, or else again at p we could have labeled q. Thus both conditions of Corollary 2a are satisfied. Hence the value of the final flow is $k(P,\bar{P})$ and is maximal. Also $(P,\bar{P})$ is a minimal $a-z$ cut. ∎

Corollary 3a (Max Flow-Min Cut Theorem)

In any directed flow network, the value of a maximal a–z flow is equal to the capacity of a minimal a–z cut. ∎

Let us now indicate how flows in a directed network can be used to model a large variety of extensions in directed and undirected networks.

Example 4: Undirected Networks

Suppose the undirected network in Figure 4.8 represents a network of telephone trunk lines (the capacity of an edge is the number of calls the trunk line can handle). We wish to know the maximal number of calls that the network can simultaneously carry between locations a and z. That is, we seek the value of a maximal flow in this network.

To make the network directed, we can replace each undirected edge (x,y) by the two edges $(x\vec{\,}y)$ and $(y\vec{\,}x)$, each with the same capacity as (x,y). An equivalent approach is to allow a directed flow in an undirected edges. If $e = (x,y)$, $\varphi(e)$ would be a number with an "arrow" indicating whether the flow goes from x to y or from y to x. Step 2 of the flow algorithm is modified as follows: When checking edges at a labeled vertex p, edges with a flow directed inwards are treated like incoming edges and edges with no flow or flow away from p are treated like outgoing edges (in this case the edges can be checked in any order, not incoming first). ∎

Example 5: Flow Networks with Supplies and Demands

Consider the network of solid edges in Figure 4.9 with supplies (capacitated sources) and demands (capacitated sinks). Vertex b can supply up to 60 units of flow, and vertices c and d can each supply 40 units. Vertices h, i, and j have flow demands of 50, 40, and 40 units, respectively. Can we meet all the demands? The sources could be oil refineries and the sinks oil-truck distribution centers, or the sources factories and the sinks warehouses.

We model this network problem with a standard one-source, one-sink network as follows. Make b, c, and d regular nonsource vertices. Add a new

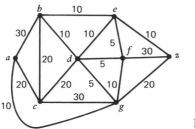

FIGURE 4.8

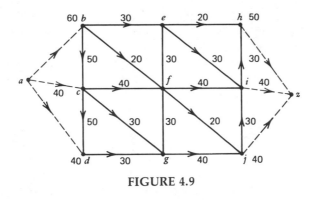

FIGURE 4.9

source vertex a and edges $(a\vec{,}b)$, $(a\vec{,}c)$, and $(a\vec{,}d)$ having capacities 60, 40, and 40, respectively. This construction simulates the role of the capacitated sources b, c, d. Next make h, i, and j regular nonsink vertices and add a new sink vertex z and edges $(h\vec{,}z)$, $(i\vec{,}z)$, and $(j\vec{,}z)$, having capacities 50, 40, and 40, respectively.

A flow satisfying the original sink demands is equivalent to a flow in the new network that saturates the edges coming into z, that is, a flow of value 130. Such a flow exists if an a–z cut has less capacity. ∎

Example 6: Edge-Disjoint Paths in a Graph

We are going to send messengers from a to z in the graph shown in Figure 4.10. Because certain edges (roads) may be blocked, we require each messenger to use different edges. How many messengers can be sent? That is, we want to know the number of edge-disjoint paths.

We convert this path problem into a network flow problem by assigning unit capacities to each edge. One could think of the flow as "flow messengers," and the unit capacities mean that at most one messenger can use any edge. The number of edge-disjoint paths (number of messengers) is thus equal to the value of a maximal flow in this undirected network. (See Example 4 for flows in undirected networks.)

Observe that we have implicitly shown that a maximal a–z flow problem for a unit-capacity network is equivalent to finding the maximal number of

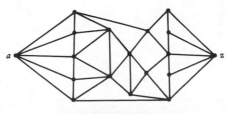

FIGURE 4.10

edge-disjoint a–z paths in the associated graph (where edge capacities are ignored). This equivalence can be extended using multigraphs to all networks by replacing each k-capacity edge (x,y) with k multiple unit-capacity edges and now proceeding with the above conversion. ∎

Example 7: Dynamic Network Flows

We want to know how many autos can be shipped from location a to location z in four days through the network in Figure 4.11a. We assume that each edge $(\vec{x,y})$ is the route of a train that leaves location x daily for a nonstop run to location y. The first number associated with an edge is the capacity of the trains (number of autos). The second number is the number of days the trip takes. Autos may be left temporarily at any location in the network.

We turn this dynamic problem into a static a–z flow problem by adding the dimension of time to our network: Each vertex x is replaced by five vertices x_0,

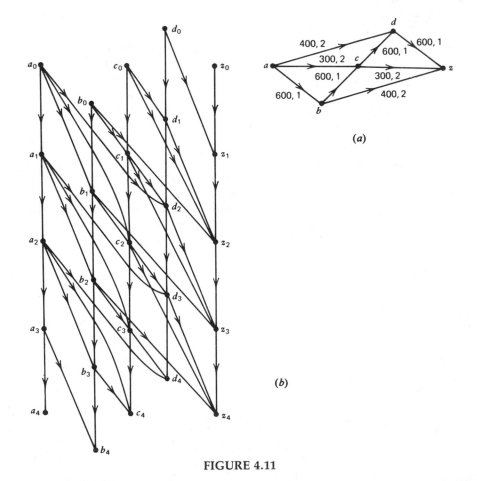

(a)

(b)

FIGURE 4.11

x_1, x_2, x_3, x_4, where the subscript refers to the ith day in the four-day shipping time (the starting day is day 0). For each original edge $(x,\vec{y})$ which takes k days to traverse, we make edges $(x_0,\vec{y_k})$, $(x_1,\vec{y}_{k+1})$, . . . , $(x_{4-k},\vec{y_4})$, each with the same capacity as $(x,\vec{y})$. For each vertex x, we make four edges of the form $(x_i,\vec{x}_{i+1})$ with very large (in effect, infinite) capacity; these edges correspond to the provision that permits autos to be left temporarily at any vertex. See Figure 4.11b.

A maximal flow in the new network gives the maximal dynamic flow in the original network. Note that for vertices other than a and z, the range of the subscripts of usable vertices will be at most 1 to 3. Nondaily trains could easily be incorporated in the model. ∎

It is hoped that the preceding four examples have impressed the reader with the versatility of our basic static a–z flow model. More examples are to be found in the exercises.

EXERCISES

SUMMARY OF EXERCISES   Exercises 1–17 mimic or extend the flow computations and models in Examples 3–7. Exercises 18–40 develop the theory of network flows. Exercises 41–45 present programming projects.

In the following problems, unless directed otherwise, the reader should route most of the flow by inspection and then when a near-optimal flow is obtained, use the augmenting flow algorithm to get further flow and afterwards a minimal cut.

1.   Apply the augmenting flow algorithm to the flow in Figure 4.4.

2.   Find a maximal a–z flow in the following networks:

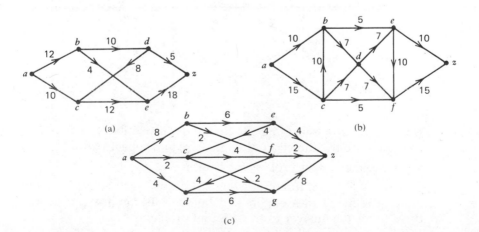

(a)

(b)

(c)

3. Find a maximal flow from a to z in the network of Figure 4.6 (using the associated directed network).

4. Treat the first number assigned to each edge in Exercise 3 of Section 4.1 as a capacity and let the edges be directed by the alphabetical order of their endpoints with L preceding a and W following n (e.g., edge (f,i) goes from f to i).
    (a) Find a maximal $L–W$ flow and a minimal $L–W$ cut in this network.
    (b) By inspection, find a different maximal $L–W$ flow.
    (c) Make a misrouted flow that yields a saturated $L–W$ cut whose capacity is greater than the flow. Now apply the algorithm.
    (d) In addition, let the second number of each edge be a lower bound on the flow in that edge. Try to find an $L–W$ flow satisfying both constraints.

5. Delete vertices p through y in Figure 4.3 in Section 4.2 and treat the remaining edge numbers as capacities (edges are still undirected). Use the undirected version of the flow algorithm suggested in Example 4.
    (a) Find a maximal flow from b to j and a minimal $b–j$ cut.
    (b) Build a $b–j$ flow that saturated edge (k,l) (in either direction) and now apply the algorithm.
    (c) Treat the edge numbers as lower bounds and modify the algorithm to find a minimal $b–j$ flow, using the whole network (a through y). Remember that there is no flow into b or out of j.

6. Is there a flow meeting the demands in Figure 4.9?

7. Suppose vertices b, c, d in Figure 4.9 have *unlimited* supplies. How much flow can be sent to the set $\{h, i, j\}$. Explain your model.

8. Vertices b, c, d have supplies 30, 20, 10, respectively, and vertices j, k have demands of 30 and 25 in this network.

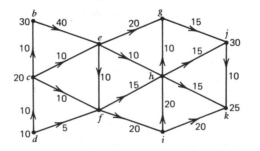

    (a) Find a flow satisfying the demands, if possible.
    (b) Reverse the direction of edge (h,g) and repeat part a.

9. Solve the messenger problem in Example 6.

10. Suppose that up to three messengers can use each edge in Example 6. Now how many messengers can be sent? Is the answer always just three times the answer to the original problem?

11. (a) Ignoring the numbers of the edges in Exercise 3 of Section 4.1, what is the size of the largest set of edge-disjoint paths from L to W?

    (b) What is the size of the largest set of paths from L to W such that no edge is used by more than five paths?

12. Suppose that no more than five units of flow can go through each intermediate vertex b, c, d, in Figure 4.7a. Now find a maximal flow in this revised network and associated minimal $a-z$ cut.

13. Suppose that no more than 20 units of flow can go through each intermediate vertex b, c, d, e, f, g in Figure 4.8. Now find the maximal flow in this revised network.

14. What is the size of a largest set of *vertex*-disjoint paths from a to z in Figure 4.10?

15. Solve the dynamic flow problem in Example 7.

16. Suppose that it takes one day to traverse each edge in Figure 4.4. How many units can be moved from a to z in five days in this network?

17. In Example 7, suppose that the trains do not run every day. Let trains depart from a and c on Monday, Wednesday, and Friday, and from b and d on Tuesday, Thursday, and Saturday. In one week, Monday through Sunday, how many cars can be sent from a to z in that network?

18. In the proof of Theorem 3, show that $\varphi = m\varphi_K$ satisfies the second part of condition (b).

19. (a) Prove that if a directed network contains edges $(x\vec{,}y)$ and $(y\vec{,}x)$ for some x, y, then the augmenting flow algorithm would never make assignments that would have flow occurring simultaneously in both edges.

    (b) Could edges $(x\vec{,}y)$ and $(y\vec{,}x)$ both get flow if the augmenting flow algorithm in step 2 checked outgoing edges before incoming edges?

20. (a) Restate the augmenting flow algorithm so that the labeling starts at z and "works back" to a.

    (b) Restate the augmenting flow algorithm for undirected networks (see Example 4). Sketch a proof of this algorithm.

21. Show that the set of edges used to label vertices in steps 2a and 2b of the augmenting flow algorithm form a tree rooted at a.

22. (a) Give a weakened replacement for condition (b) in the definition of an $a-z$ flow. The new condition should still insure that the net flow is from a to z.

    (b) Suppose condition (b) is eliminated. Can there exist maximal flows that violate (b)? Prove or give a counterexample.

23. Build a flow in the network in Figure 4.7a with the prescribed properties:

(a) Its value $|\varphi|$ is 0 but not all edges have 0 flow.

(b) Its value is 2 but it cannot be decomposed into a sum of a–z flow paths.

(c) Its value is maximal but it cannot be decomposed into a sum of a–z flow paths.

24. (a) Show that for a flow φ in a (directed or undirected) network, if φ is *circuit-free*, that is, there is no set of edges with flow that form a (directed) circuit, then φ can be decomposed into a sum of a–z flow paths. (*Hint:* Prove by induction on the value of φ.)

(b) Use the proof to get a decomposition algorithm for any such φ.

(c) Conclude that any flow φ can be decomposed into $|\varphi|$ a–z flow paths plus a set of circuits.

25. (a) Prove that starting from a circuit-free flow (see Exercise 24), perhaps a zero flow, the maximal flow generated by the augmenting flow algorithm is circuit-free and hence (by Exercise 24(b)) can be decomposed into a sum of a–z flow paths.

(b) Use part a and Exercise 24(b) to find the routings of a maximal set of phone calls in Example 4.

26. How many comparisons are required, as a function of the number of edges in a network, to perform one iteration of the augmenting flow algorithm?

27. Show that if a flow is decomposed into unit-flow paths and if each unit-flow path crosses a given saturated cut once, then the flow is maximal.

28. A *cut-set* in an undirected graph G is a set S of edges whose removal disconnects G but no proper subset of S disconnects G. Prove that in an undirected flow network, every cut-set that separates a and z is a cut and every minimal a–z cut is a cut-set.

29. Let G be a connected, undirected graph and a, b be any two vertices in G.

(a) Show that there are k edge-disjoint paths between a and b if and only if every a–b cut has at least k edges.

(b) Show that there are k vertex-disjoint paths between a and b if and only if every set of vertices disconnecting a from b has at least k vertices.

30. As mentioned in Example 6, we can model a flow network N (directed or undirected) by another multigraph flow network N' in which each edge has unit capacity; N' has the same vertices as N and for each edge e in N there are $k(e)$ edges in N' paralleling e. Since a flow in N' takes 0 or 1 values on the edges, we can drop the capacities in N' to get a multigraph G'. A flow in N' is just a subset of edges (with flow) in G'.

(a) Characterize the subsets of edges in G' that correspond to an a–z flow in N'.

(b) Restate the augmenting flow algorithm in terms of G'.

(c) Using the multigraph model, prove that any a–z flow φ in N' contains $|\varphi|$ a–z flow paths.

(d) Using the multigraph model and assuming the result in Exercise 29(a), prove Corollary 3a (max flow-min cut theorem).

31. Suppose the numbers on the edges of a directed network represent lower bounds for the flow. State a decreasing flow algorithm. Sketch a proof of this algorithm and deduce the counterpart of Corollary 3a.

32. (a) Explain how the algorithm in Exercise 31 can be applied to a bipartite graph to find a minimal set of edges incident with every vertex in the graph.

(b) Find such an edge set for the bipartite graph in Figure 4.12 of Section 4.4.

33. Suppose we have upper and lower bounds $k_1(e)$ and $k_2(e)$, respectively, on the flow in each edge e in a directed network and that we are given a feasible flow (satisfying these constraints).

(a) Modify the augmenting flow algorithm so that it can be used to construct a maximal flow from a given feasible flow.

(b) Prove that the maximal flow has value equal to the minimum of $k_1(P,\bar{P}) - k_2(\bar{P},P)$ among all a–z cuts $(P,\bar{P})$, where $k_1(P,\bar{P})$ and $k_2(P,\bar{P})$ are the sums of the upper and lower bounds of edges in $(P,\bar{P})$.

34. Consider a directed network with supplies and demands as in Example 5. Let $z(P)$ be the total demand of vertices in set P and $a(P)$ be the total supply of vertices in P.

(a) Prove that the demands can be met if and only if for all sets P, $z(P) - a(P) \leqslant k(\bar{P},P)$. (*Hint:* Generalize the reasoning in Example 5.)

(b) Prove that the supplies can all be used if and only if for all P, $a(P) - z(P) \leqslant k(P,\bar{P})$.

35. Suppose that the edges in Figure 4.11a in Example 7 were undirected. How would we construct a static flow model to simulate this dynamic flow problem so that a maximal static flow in the new network would correspond to a maximal dynamic flow? Are there any difficulties?

36. Suppose an undirected flow network is a planar graph and a (at the left side) and z (at the right side) are both on the unbounded region surrounding the network. Draw edges extending infinitely to the left from a and to the right from z; give them infinite capacity. This divides the unbounded region into two unbounded regions, an upper and a lower unbounded region. Now form the dual network (see Section 1.4) of this planar network with each dual edge assigned the capacity of the original edge it crosses.

(a) Show that a shortest path from the upper unbounded region's vertex to the lower unbounded region's vertex in the dual network corresponds to a minimal cut in the original network.

(b) Draw the dual network for the network in Figure 4.8 and find a shortest path corresponding to a minimal cut in the original network.

37. Suppose an undirected flow network N is a planar graph and a (at the left side) and z (at the right side) are both on the unbounded region surrounding the network. Consider the following flow path building heuristic. Starting from a, build an a–z flow path by choosing at each vertex x the first unsaturated edge (in clockwise order starting from the edge used to enter x).
    (a) Show that repeated use of this heuristic yields a maximal flow in N.
    (b) Can this heuristic be extended to directed planar networks?
    (c) Apply this heuristic to the network in Figure 4.8.

38. How many steps are required, as a function of the number of edges in a network, to find a maximal flow (starting from zero flow) in undirected planar networks using the heuristic in Exercise 37?

39. Prove that repeated use of the augmenting flow algorithm yields a maximal flow in a finite number of applications for networks with irrational capacities provided that vertices are ordered (indexed) and the next vertex scanned in Step 2 of the algorithm is the labeled vertex with lowest index.

40. Suppose we have no source or sink in a network with upper and lower bounds on the flow. We seek a *circulation*, a function on the edges satisfying the upper and lower bounds and condition (c) in the definition of a–z flows. Prove that a circulation exists if and only if for every cut $(P,\bar{P})$, $k_1(P,\bar{P}) \geq k_2(\bar{P},P)$, where $k_1(P,\bar{P})$ and $k_2(P,\bar{P})$ are the sum of the upper and lower bounds of the edges in $(P,\bar{P})$. (*Hint:* See Exercise 30.)

41. Write a program to find maximal flows:
    (a) In networks (the networds are input data).
    (b) In undirected networks.

42. Write a program to find the maximal number of paths in a graph between two given vertices such that:
    (a) The paths are edge disjoint.
    (b) The paths are vertex disjoint.

43. Write a program that, when given a network with an a–z flow of value k will extract from the flow k unit-flow paths from a to z.

44. Write a program to find for a given pair of vertices a and b in a given connected graph, a minimal set of vertices whose removal disconnects a from b (see Exercise 29).

45. Write a program to find a maximal flow in a planar undirected network:
    (a) Using the method in Exercise 36.
    (b) Using the method in Exercise 37.

4.4 ALGORITHMIC MATCHING

In this section, we apply the maximal flow algorithm to the theory of matchings. A **bipartite graph** $G = (X,Y,E)$ is an undirected graph with two specified vertex sets X and Y and with all edges of the form (x,y), $x \in X$, $y \in Y$. See Figure 4.12. Bipartite graphs are a natural model for matching problems. We let X and Y be the two sets to be matched and edges (x,y) represent pairs of elements that may be matched together.

A **matching** in a bipartite graph is a set of **independent** edges (with no common endpoints). The darkened edges in Figure 4.12 constitute a matching. An **X-matching** is a matching involving all vertices in X. A **maximal** matching is a matching of the largest possible size. As with network flows, we cannot always obtain a maximal matching in a bipartite graph by simply adding more edges to a nonmaximal matching. The matching indicated in Figure 4.12 cannot be so increased, even though it is not maximal.

A typical matching problem involves pairing off compatible boys and girls at a dance or the one-to-one assignment of workers to jobs for which they are trained. A closely related problem is to find a **set of distinct representatives** for a collection of subsets. We need to pick one element from each subset without using any element twice. In the bipartite graph model, we make one X-vertex for each subset, one Y-vertex for each element, and an edge (x,y) whenever element y is in subset x. Now an X-matching picks a distinct representative element for each subset. Conversely, any matching problem can be modeled as a set-of-distinct-representatives problem.

A $(0,1)$ matrix is an equivalent model for matching problems: We make a row for each element in X, a column for each element in Y, and place a 1 in the entry of row x and column y whenever x can be paired with y; other entries have a 0. See Figure 4.13. Such a matrix is merely the adjacency matrix for the associated bipartite graph. A matching in a $(0,1)$ matrix is a set of **independent** 1-entries (no two of which are in the same line, i.e., in the same row or column). The circled set of independent 1's in Figure 4.13 corresponds to the matching in Figure 4.12. Although we shall use bipartite graphs to prove our matching theorems, we shall later use $(0,1)$ matrices in a simplified matching algorithm.

We employ a modification of the trick in Example 5 of Section 4.3 to turn a matching problem into a network flow problem. Associate a supply of 1 at

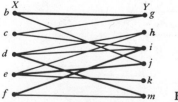

FIGURE 4.12

	g	h	i	j	k	m
b	①	0	0	1	0	0
c	1	0	1	0	0	0
d	0	1	1	0	0	①
e	0	1	0	①	1	0
f	0	0	①	0	0	1

FIGURE 4.13

each X-vertex and a demand of 1 at each Y-vertex. The capacities of the edges from X to Y can be any large positive integers, but it is convenient to pretend that these capacities are ∞. We assume edges are directed from X to Y. Now we apply the technique in Example 5 of Section 4.3 with source a connected by a unit-capacity edge to each X-vertex and sink z connected by a unit-capacity edge from each Y-vertex. We call such a network a **matching network**. See Figure 4.14. The X–Y edges used in an a–z flow constitute a matching. A maximal flow is a maximal matching. A flow saturating all edges from source a corresponds to an X-matching.

Example 1

Suppose the bipartite graph in Figure 4.12 represents possible pairings of boys b, c, d, e, f with girls g, h, i, j, m. A tentative matching indicated by darkened edges in Figure 4.12 was made. Although this matching cannot be increased, we still wonder whether a complete X-matching is possible.

As happened before in flow problems, we have made a "mistake" in this matching and now need to make some reassignments. We convert this matching into the corresponding flow in the associated matching network. See Figure 4.14. Now we apply the augmenting flow algorithm, as shown in Figure 4.14. The augmenting flow chain prescribed by the algorithm is a-c-g-b-j-e-k-z. Observe that subtracting a unit of flow from an X–Y edge means removing that edge from the matching.

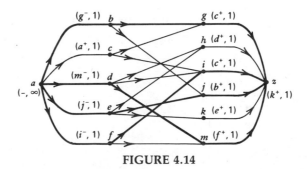

FIGURE 4.14

Our augmenting flow chain specifies that we add edge (c,g) to our matching, remove edge (b,g), add edge (b,j), remove edge (e,j), and add edge (e,k). The new flow corresponds to the matching b-j, c-g, d-m, e-k, f-i. Another application of the algorithm would only label a. The set of edges leaving a now form a (saturated) minimal a–z cut—a sign that we have an X-matching.

∎

Observe that the action of the augmenting flow algorithm in matching problems can be described as follows: Starting from an unmatched vertex x_1 in X, we go on a (nonmatching) edge to a matched vertex y_1; then we backtrack along the matching edge at y_1 to some matched vertex x_2; then we go on a nonmatching edge from x_2 to some matched vertex y_2; and so on until a nonmatching edge (x_k, y_k) goes from x_k to an unmatched Y-vertex y_k. In short, starting from an unmatched X-vertex, we create an odd-length alternating path L of nonmatching and matching edges to unmatched Y-vertex. Given such a path, we get a new, larger set of matching edges by interchanging the roles of matching and nonmatching edges on L.

We have seen that in bipartite graphs, matchings are analogous to flows in networks. What then is the bipartite graph counterpart to an a–z cut? Actually it is convenient to restrict our attention to finite-capacity a–z cuts. The corresponding concept is then an **edge cover**, a set S of vertices such that every edge has a vertex of S as an endpoint.

Lemma

Let $G = (X, Y, E)$ be a bipartite graph and let N be the matching network associated with G. For any subsets (possibly empty) $A \subseteq X$ and $B \subseteq Y$, $S = A \cup B$ is an edge cover if and only if $(P, \bar{P})$ is a finite capacity a–z cut in N, where $P = a \cup (X - A) \cup B$. In terms of P,

$$S = (\bar{P} \cap X) \cup (P \cap Y)$$

Further, $|S| = k(P, \bar{P})$.

Proof

A finite capacity a–z cut cannot contain an edge between X and Y (whose capacity is ∞). Then a possible finite-capacity cut $(P, \bar{P})$ must consist of edges of the form (a, x), $x \in A$ and (y, z), $y \in B$ for some sets $A \subseteq X$ and $B \subseteq Y$. These edges block all flow (and thus are an a–z cut) if and only if any X–Y edge starts at some $x \in A$ or ends at some $y \in B$, that is, if and only if $S = A \cup B$ is an edge cover. In terms of A and B, $P = a \cup (X - A) \cup B$, and in terms of P, $A = \bar{P} \cap X$ and $B = P \cap Y$. See Figure 4.15. Also, $|S| = k(P, \bar{P})$, since the edges in $(P, \bar{P})$ all have unit capacity. ∎

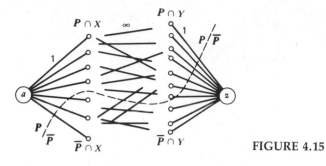

FIGURE 4.15

We now prove two famous theorems about matchings in bipartite graphs. In Theorem 2, $R(A)$, the **range** of A, denotes the set of vertices adjacent to a vertex in A.

Theorem 1

In a bipartite graph $G = (X, Y, E)$, the size of a maximal matching equals the size of a minimal edge cover.

Proof

Matchings correspond to flows in the associated matching network and, by the lemma, edge covers correspond to (finite capacity) a–z cuts. So this result is simply a bipartite-graph restatement of the max flow-min cut theorem (Corollary 3a in Section 4.3). ■

Theorem 2 (Hall's Marriage Theorem)

A bipartite graph $G = (X, Y, E)$ has an X-matching if and only if for each subset $A \subseteq X$, $|R(A)| \geq |A|$.

Proof

The condition is necessary, for if $|R(A)| < |A|$, a matching of all vertices in A is clearly impossible, and so an X-matching is impossible.

By Theorem 1, an X-matching exists if and only if each edge cover S has size $|S| \geq |X|$. If S contains only X-vertices—that is, $S = X$—then the result is immediate. We must show that S cannot be made smaller by dropping some X-vertices and in their stead using a smaller number of Y-vertices. If A is the set of X-vertices not in S, then $S \cap Y$ must contain the vertices in $R(A)$ to cover the edges between A and $R(A)$. Then we have

$$|S| = |X - A| + |S \cap Y| \geq |X| - |A| + |R(A)| \geq |X|$$

since $|R(A)| \geq |A|$. So $|S| \geq |X|$ and there is a matching of size X. ■

Theorems 1 and 2 are the starting point for a large family of matching theorems (for example, see Exercises 15, 22, 23).

Example 2

Country A would like to spy on all meetings in its territory between diplomats from Country X and from Country Y. We know which pairs of diplomats are likely to meet. We can make a bipartite graph expressing this likely-to-meet relationship. Country A cannot afford to assign spies to every X diplomat or to every Y diplomat. Instead it wants to find a minimal set S of diplomats such that every possible meeting would involve a diplomat of S.

Country A hires a graduate of this course who immediately sees that this minimal spying problem is really a minimal edge cover problem in disguise! The problem can thus be solved by using the augmenting flow algorithm to find a minimal a–z cut in the associate matching network and, from the a–z cut, obtain a minimal edge cover using the lemma. ■

Example 3

Suppose there are n people and n jobs, each person is qualified for k jobs, and for each job there are k qualified people. Is it possible to assign each person to a (different) job they can do?

We model this problem with a bipartite graph $G = (X, Y, E)$ with X-vertices for people and Y-vertices for jobs. Note that each vertex will have degree k. The question is now, is there an X-matching. From Theorem 2, we can adduce a yes answer as follows: Let A be any subset of X. Since each vertex has degree k, there will be $k|A|$ edges leaving A. Since at most k of these edges can go to any one vertex in Y, it follows that $R(A)$ has at least $|A|$ vertices. Then by Theorem 2, there is an X-matching. ■

We return now to the problem of simplifying our flow chain algorithm for matching networks. The second label assigned by the algorithm is superfluous, since it is always 1. The signed superscript can also be dropped, since the sign can be inferred from the type of the label, that is, if a Y-vertex is the label from an X-vertex, the superscript is minus; otherwise all superscripts are plus (check Figure 4.14).

A $(0,1)$ matrix is in many ways an easier setting than a graph for solving matching problems, and so we not restate the simplified flow algorithm for matching networks in terms of $(0,1)$ matrices. Recall that rows correspond to X-vertices and circled 1's correspond to edges used in a matching. We will place row labels at the right end of the rows and column labels at the bottom of the columns.

Augmenting Matching Algorithm

(An algorithm for a $(0,1)$ matrix M with a given set I of independent 1's.)

1. Circle the 1's in I and label with an * every row without a circled 1.

2. Scan every newly labeled row i; label with an i each unlabeled column with an uncircled 1 in row i.

3. Scan every newly labeled column j; label with a j the row with a circled 1 in column j. If there is no circled 1, go to Step 5.

4. If new rows were labeled in Step 3, go to Step 2. Otherwise the given set of circled 1's is maximal and the unlabeled rows together with the labeled columns constitute a minimal 1's cover (equivalent to an edge cover).

5. A "breakthrough" has occurred. In the column j_0 just scanned (with no circled 1), place a circle around the 1 in row i_1 designated by the label for column j_0; in row i_1 remove the circle around the 1 in column j_1 designated by the label for row i_1; now in column j_1 place a circle around the 1 in row i_2 designated by j_1's label; in row i_2 remove the circle around the 1 in column j_2; and so forth. Continue this process until a row with an * label is reached.

Step 1 corresponds to the initial network labeling of X-vertices from the source a. Step 2 corresponds to the plus labeling of Y-vertices done from X-vertices. Step 3 corresponds to the minus labeling of X-vertices done at Y-vertices; a breakthrough in Step 3 (no circled 1) corresponds to labeling z from a Y-vertex. Step 4 giving a minimal 1's cover corresponds to finding a saturated (minimal) a–z cut. Step 5 corresponds to backtracking to find an augmenting flow chain and modifying the flows along the chain. Because of its newness, this algorithm may seem more difficult than the familiar flow algorithm. However, with a little practice the matrix method is easier to use and to program.

Example 4

In Figure 4.16 we apply the augmenting matching algorithm to the same problem handled by the flow algorithm in Figure 4.14. The reader should check that the row and column labels are the same as the X-vertex and Y-

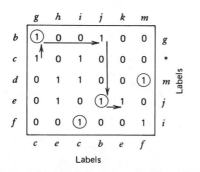

FIGURE 4.16

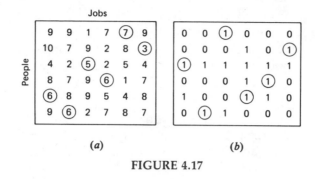

FIGURE 4.17

vertex labels in Figure 4.14 (the * corresponds to a). The arrows correspond to the augmenting flow chain in Example 1. Following the arrows, we circle entry (c,g), uncircle (b,g), circle (b,j), uncircle (e,j), and circle (e,k). If we were to repeat this algorithm, no new rows would be labeled in Step 1 and we would go straight to Step 4 where rows b, c, d, e, f would be designated as a minimal 1's cover. ∎

Example 5

The matrix in Figure 4.17a gives the level of dissatisfaction that person i has at job j. We wish to assign each person to a different job so as to keep the level of dissatisfaction as low as possible, that is, to minimize the worst case of dissatisfaction in the assignment. The problem of minimizing the largest "cost" in an assignment is called the **Bottleneck Assignment Problem**.

The following procedure, due to O. Gross, is quite efficient. By inspection, pick an assignment, that is, a set of six entries each in a different row and column, which appears to keep the highest dissatisfaction number fairly low. We have circled entries of such an assignment in Figure 4.17a. The highest number in the assignment is 7. Then a better assignment must consist of entries with numbers ≤6. The set of these entries is indicated by the 1's in Figure 4.17b. We guess at a maximal independent set of 1's and then use the augmenting matching algorithm to increase, if possible, the number of circled 1's.

If the algorithm succeeds in getting us six circled 1's, as is possible in this case (see Figure 4.17b), we have a better assignment. In this case the new assignment has maximal dissatisfaction of 6. Now we repeat the procedure to see if an assignment using only entries with values less than 6 exists. If the algorithm fails to get six circled 1's (as it would the next time with entries <6), then the current assignment is the best possible.

A nice aspect of this procedure is that while the problem is too complex to be solved by inspection, the procedure still rewards clever initial guesses. ∎

EXERCISES

SUMMARY OF EXERCISES   Exercises 1–9 involve matching computations. Exercises 10–14 apply matching models to obtain results similar to Example 3. Exercises 15–25 develop the theory of matching.

1. Bill is liked by Ann, Diana, and Lolita; Fred is liked by Bobbie, Carol, and Lolita; George is liked by Ann, Bobbie, and Lolita; John is liked by Carol and Lolita; and Larry is liked by Diana and Lolita. We want to pair each boy with a girl who likes him.
   (a) Set up the associated matching network and maximize its flow to solve this problem.
   (b) Make a flow corresponding to a partial pairing that has Bill with Diana and Fred with Carol (add other pairs). Now repeat part (a).
   (c) Do part b in terms of a (0,1) matrix and the augmenting matching algorithm.

2. Suppose there are five committees: committee A's members are a, c, e; committee B's members are b, c; committee C's members are a, b, d; committee D's members are d, e, f; and committee E's members are e, f. We wish to have each committee send a different representative to a convention.
   (a) Set up the associated matching network and maximize its flow to solve this problem.
   (b) Make a flow corresponding to the partial assignment A sends e, B sends b, C sends a, and D sends f. Now repeat part a.
   (c) Do part b in terms of a (0,1) matrix and the augmenting matching algorithm.

3. Suppose that in Exercise 2, e must be the representative of A. Model this situation with a (0,1) matrix. Now use the augmenting matching algorithm repeatedly to add more 1's until a full selection has been made (no 1's may be added by inspection).

4. Let us repeat Exercise 1 but this time Bill gets a total of 5 dates, Fred 4 dates, George 3, John 5, and Larry 3, while Ann gets 4, Bobbie 3, Diana 5, Carol 4, and Lolita 4. Compatible pairs may have any number of dates together. Model with a network to find a possible set of pairings.

5. Let us repeat Exercise 1 but this time we want to pair each boy twice (with two different girls) and each girl twice. Find the pairings using an appropriate network flow model.

6. Suppose that there are 6 universities and each will produce 5 mathematics Ph.D.'s this year, and there are 5 colleges that will be hiring 7, 7, 6, 6, 5 math Ph.D.'s, respectively. No college will hire more than one Ph.D. from any given university. Will all the Ph.D.'s get a job? Explain.

7. For the matrix in Figure 4.17a, find a minimal set of lines (rows or columns) such that every entry with a prime number (or 1) is in one of the lines. Use the augmenting matching algorithm.

8. The rows below represent people and the columns represent jobs. The entries tell the distance person i is from job j. Assign each person to a different job so as to minimize the farthest distance that anyone must travel. One job gets omitted.

$$
\begin{array}{cccccc}
1 & 2 & 3 & 4 & 3 & 5 \\
3 & 4 & 2 & 1 & 5 & 3 \\
3 & 1 & 3 & 2 & 4 & 3 \\
3 & 3 & 3 & 4 & 2 & 4 \\
3 & 4 & 3 & 5 & 4 & 3
\end{array}
$$

9. Suppose that the folks in Exercise 8 were bad and that their boss will punish them with a job assignment chosen to maximize the minimum distance anyone must go to his job. Find such an assignment.

10. Let M be a $(0,1)$ matrix with n rows and m columns, and suppose that every row sum is larger than any column sum. Show that M has a set of n independent 1's.

11. There are n boys and n girls in a computer dating service. The computer has made nm pairings so that each boy dates m different girls and each girl dates m different boys $(m < n)$.
    (a) Show that it is always possible to schedule the nm dates over m nights, that is, the pairings may be partitioned into m sets of complete pairings.
    (b) Find such a partition for the $(0,1)$ matrix of pairings below.
    (c) Show that in part (a), no matter how the first k complete pairings are selected $(0 < k < m)$, the partition can always be completed.

$$
\begin{array}{cccccc}
1 & 1 & 0 & 1 & 0 & 1 \\
1 & 0 & 1 & 0 & 1 & 1 \\
0 & 1 & 0 & 1 & 1 & 1 \\
1 & 0 & 1 & 1 & 1 & 0 \\
1 & 1 & 1 & 0 & 1 & 0 \\
0 & 1 & 1 & 1 & 0 & 1
\end{array}
$$

12. We want to construct an $n \times m$ matrix whose entries will be nonnegative integers such that the sum of the entries in row i is r_i, and the sum of the entries in column j in c_j. Clearly the sum of the r_i's must equal the sum of the c_j's.
    (a) What other constraints (if any) should be imposed on the r_i's and c_j's to assure such a matrix exists?

(b)   Construct such a 5 × 6 matrix with row sums 20, 40, 10, 13, 25 and column sums all equal to 18.

13.   The edges in a bipartite graph $G = (X,Y,E)$ are to be colored with edges having a common endpoint getting different colors. If r is the maximal degree of a vertex in G, prove that the edges of G can be r-colored.

14.   Give necessary and sufficient conditions for the existence of a circuit or a set of vertex-disjoint circuits which pass through each vertex once in a directed graph $G = (V,E)$. (*Hint:*   Make a bipartite graph $G' = (X, Y, E)$ with X and Y copies of V and for each edge (v_1, v_2) in G, G' has an edge (x_1, y_2); restate the problem.)

15.   Prove that a bipartite graph $G = (X,Y,E)$ has a matching of size t if and only if for all A in X, $|R(A)| \geq |A| + t - |X| = t - |X - A|$. (*Hint:*   Add $|X| - t$ new vertices to Y and join each new Y-vertex to each X-vertex.)

16.   Show that every bipartite-graph matching problem can be modeled as a set-of-distinct-representatives problem.

17.   Prove that the augmenting matching algorithm is equivalent to the augmenting flow algorithm for matching networks.

18.   Restate Theorems 1 and 2 in terms of (0,1) matrices.

19.   Prove Theorem 2 from Exercise 34 in Section 4.3, which applies directly to the bipartite network, not the augmented a–z matching network.

20.   Prove Theorem 2 for complete Y-matchings (without simply interchanging the roles of X and Y). By symmetry, the same condition is required but the set A in the reproof is chosen differently from the A in the text's proof.

21.   In this exercise we prove the equivalence of Theorems 1 and 2. In the text Theorem 2 was proved from Theorem 1. Now prove Theorem 1 using solely Theorem 2. Actually it is easier to prove the (0,1) matrix version. Proceed thus. Let R be the lines in the minimal 1's cover that are rows and C be the lines that are columns. Let R' be the set of other rows and C' the set of other columns. Then the submatrix of $R' \times C'$ is all 0's. Now apply the (0,1)-matrix version of Theorem 2 to the submatrix of $R \times C'$ and the submatrix of $R' \times C$.

22.   Let $\delta(G) = \max_{A \subseteq X}(|A| - |R(A)|)$. $\delta(G)$ is called the *deficiency* of the bipartite graph $G = (X,Y,E)$ and gives the worst violation of the condition in Theorem 2. Note that $\delta(G) \geq 0$ because $A = \varnothing$ is considered a subset of X.
(a)   Use Exercise 15 to prove that a maximal matching of G has size $|X| - \delta(G)$.
(b)   Given a maximal matching of size $t = |X| - \delta(G)$ (assume $\delta(G) > 0$), describe how the associated minimal edge covering of Theorem 1 can be used to find an A such that $|A| - |R(A)| = \delta(G)$.

23.   (a)   Show that the size of the largest independent set of vertices (mutu-

ally non-adjacent) in $G = (X,Y,E)$ is equal to $|Y| + \delta(G)$ (see Exercise 22). Describe how to find such an independent set.

(b)   Use part a to find such an independent set in Figure 4.12.

24.   (due to J. Hopcroft) Suppose each vertex of a bipartite graph G has degree 2^r, for some r. Partition the edges of G into circuits, and delete every other edge in each circuit. Repeat this process on the new graph (where each vertex now has degree 2^{r-1}), and continue repeating until a graph is obtained with each vertex of degree one. The edges in this final graph constitute a matching of the vertices of G.

(a)   Show that such a partition of edges into circuits exists in each successive graph and can be found in a number of steps proportional to the number of edges in the current graph.

(b)   Show that the total number of steps to find the vertex matching is proportional to the number of edges in G.

25.   In a bipartite graph with n X vertices and n Y vertices, show that a maximal matching can be found in a number of steps in proportional to n^3.

26.   Suppose we are given a partial matching to an arbitrary graph (such a matching is a set of edges with no common endpoints).

(a)   Prove that a generalization of the interchange method along a path alternating between nonmatching and matching edges (described in the text following Example 1) can be used to increase the size of the partial matching until it is a maximal matching.

(b)   Randomly pick a partial matching for the graph in Figure 4.9 and use this method to get a maximal matching.

27.   Write a computer program to find a maximal matching and a minimal edge cover in a bipartite graph (the graph is input data).

4.5   SUMMARY AND REFERENCES

In this chapter we presented algorithms for three basic network optimization problems: shortest path, minimal spanning tree, and maximal flow. Principal emphasis was placed on a thorough discussion of maximal flows. We showed how these flows could be applied to a wide variety of other network problems. In Section 4.4 we used flow models to develop a combinatorial theory of matching. All the material about flows in this chapter is discussed in greater detail in the pioneering work *Flows in Networks* by Ford and Fulkerson [2]. We have omitted discussion of the speed of these algorithms. The interested reader is referred to Rheingold, Nievergelt, and Deo [4] for efficient implementations of the shortest path and minimal spanning tree algorithms. A recent hybrid maximal flow algorithm of Karzanov requires only $O(n^3)$ operations for an n-vertex network (see Even [1]).

It is often natural in network flow problems to have costs associated with edges so that when many possible maximal flows exist, one can ask for a least-cost maximal flow. Such problems are called **trans-shipment** and **transportation** problems. Similarly, in a matching problem with many solutions (X-matchings), one can ask for a least-cost matching. Such a problem is called an **assignment** problem. Efficient algorithms exist for all these minimization problems (see any operations research text, such as [3]). Furthermore, any flow optimization problem, with or without the above-mentioned minimization, is a problem of optimizing a linear function of the edge flows subject to linear equalities and inequalities, such as the flow constraints (a), (b), and (c) of Section 3.3. Such a constrained linear optimization problem is called a **linear program**. Linear programming is a principal tool of operations research, and good algorithms exist for solving linear programs. However, it is much more efficient to solve network problems with the network-specific algorithms presented in this chapter.

1. S. Even, "The Max Flow Algorithm of Dinic and Karzanov: An Exposition," *Computer Science* 5 (1978).

2. L. Ford and D. Fulkerson, *Flows in Networks*, Princeton University Press, Princeton, N.J., 1964.

3. F. Hillier and G. Lieberman, *Introduction to Operations Research*, Holden-Day, San Francisco, 1980.

4. E. Reingold, J. Nievergelt, and N. Deo, *Combinatorial Algorithms*, Prentice-Hall, Englewood Cliffs, N.J., 1977.

PART TWO
ENUMERATION

Chapter Five

General Counting Methods for Arrangements and Selections

5.1 TWO BASIC COUNTING PRINCIPLES

In this chapter we discuss counting problems for which no specific theory exists. We present a few basic formulas, involving permutations and combinations, most of which the reader has seen before. Then we examine several "word problems" and show how they can be broken down into sums and products of simple numerical factors. Having read through the examples as passive readers, students next must assume the active role of devising solutions on their own to the exercises. The first exercises at the end of each section are similar to the examples discussed in the section. The later exercises, however, have little in common with the examples except that they require the same general types of logical reasoning, clever insights, and elementary mathematical modeling. Facility with these three basic skills in problem solving, as much as an inventory of special techniques, is the key to success in most combinatorial applications.

In order to motivate our problem solving, we mention applications to probability and statistics, computer science, operations research, and other disciplines. However the details of the counting problems arising from such applications often appear tedious to those not actively working in the area of application. So instead, we will base many of the worked-out examples (and exercises) on recreational problems, such as poker probabilities. The solution of these problems requires the same mathematical skills used in more practical applications.

This section starts with two elementary counting principles whose simplicity masks both their power and the ease with which they can be misused.

The Addition Principle

If there are r_1 different objects in the first set, r_2 objects in the second set, . . . , and r_m objects in the mth set, and *if the different sets are disjoint,* then the number of ways to select an object from one of the m sets is $r_1 + r_2 + \cdots + r_m$.

The Multiplication Principle

Suppose a procedure can be *broken into m successive (ordered) stages,* with r_1 outcomes in the first stage, r_2 outcomes in the second stage, . . . , and r_m outcomes in the mth stage. If the composite outcomes are all distinct, then the total procedure has $r_1 \cdot r_2 \cdots \cdots r_m$ different composite outcomes.

Rigorous proofs of these two principles for an arbitrary m require mathematical induction. Remember that the addition principle requires disjoint sets of objects and the multiplication principle requires that the procedure break into ordered stages and that the composite outcomes be distinct.

Example 1

Professor Mindthumper has 40 students in an algebra class and 40 students in a geometry class. How many different students are in these two classes?

By the addition principle, the answer is 80 students, provided that no students are in both classes. Suppose 10 students are in both classes. To obtain disjoint sets of students, we categorize students as just in algebra, just in geometry, and in both classes. Since 10 students are in both classes, then $40 - 10 = 30$ algebra students are just in algebra. Similarly 30 students are just in geometry. Now we can safely use the addition principle to sum the numbers of students in these three disjoint sets. The total number of students is $30 + 30 + 10 = 70$. ∎

Example 2

Two dice are rolled, one green and one red. How many different outcomes of this procedure are there? How many different outcomes are there with different values on the two dice (no doubles)?

There are six outcomes of a single die. So, by the multiplication principle, there are $6 \cdot 6 = 36$ outcomes of the procedure. In the case of no doubles, it is important to think of one die being rolled first, say, the red die. Once the red die is rolled, then there are five other permissible values for the green die. So there are $6 \cdot 5 = 30$ outcomes with no doubles. ∎

Generalizing the second part of Example 2, we note that the logic of counting requires that "make the two outcomes different" really means "make the second outcome different from the first." The no-doubles problem in Example 2 could also be answered by subtracting the number of doubles from all outcomes $36 - 6 = 30$.

Example 3

There are five different Spanish books, six different French books, and eight different Transylvanian books. How many ways are there to pick an (unordered) pair of two books not both in the same language?

If one Spanish and one French book are chosen, the multiplication principle says that the selection can be done in $5 \cdot 6 = 30$ ways; if one Spanish and one Transylvanian book, $5 \cdot 8 = 40$ ways; and if one French and one Transylvanian book, $6 \cdot 8 = 48$ ways. These three types of selections are disjoint, and so by the addition principle there are $30 + 40 + 48 = 118$ ways in all.   ■

The preceding example typifies a basic way of thinking in combinatorial problem solving: always try first to break a problem into a moderate number of manageable subproblems. There may be cleverer ways, but if we can reduce the original problem to subproblems with which we are familiar, then we are less likely to make a mistake.

Example 4

How many ways are there to form a three-letter sequence using the letters a, b, c, d, e, f: (a) with repetition of letters allowed? (b) without repetition of any letter? (c) without repetition that contain the letter e? (d) with repetition that contain e?

(a) With repetition, we have six choices for each letter in the sequence. So by the multiplication principle there are $6 \cdot 6 \cdot 6 = 216$ three-letter sequences with repetition.

(b) Without repetition, there are six choices for the first letter. For the second letter, there are five choices, the five remaining letters (no matter what the first choice was). Similarly for the third letter, there are four choices. Thus there are $6 \cdot 5 \cdot 4 = 120$ three-letter sequences without repetition.

(c) There are three choices for which position in the sequence is e. For any of the three choices for e, there are $5 \cdot 4 = 20$ choices for the first and second remaining positions in the three-letter sequence. Thus there are $3 \cdot 20 = 60$ three-letter sequences with e. A second equally simple approach to this problem is given in the exercises.

(d) Let us try the approach used in (c). As before, there are three choices for e's position. For any of these places for e, there are $6 \cdot 6 = 36$ choices for the other two positions, since e and the other letters can appear more than once. But the answer of $3 \cdot 36 = 108$ is not correct.

The Multiplication Principle has been violated because the outcomes are not distinct. Consider the sequence *ece*. It was generated two times in our procedure: once when *e* was put in the first position followed by *ce* as one of the 36 choices for the latter two positions, and a second time when *e* was put in the last position with *ec* in the other positions.

We must use an approach that insures distinct outcomes. Let us break the problem into disjoint cases based on where the first *e* in the sequence occurs. First suppose the first *e* is in the first position; then there are six choices each for the second and third positions. Next suppose the first *e* is in the second position; then there are five choices for the first position (cannot be *e*) and six choices for the last position. Finally, if the first (and only) *e* is in the last position, there are five choices each for the first two positions. The correct answer is then $6 \cdot 6 + 5 \cdot 6 + 5 \cdot 5 = 91$. ∎

The hardest part about solving most counting problems is finding a structure in the problem that allows it to be broken into subcases or stages, as was done in the preceding examples. In other words, the difficulty is in "getting started."

Typically, each counting problem requires its own special insights. Knowing the solution to problem A is of little help in solving problem B. Learning how to devise original solutions to counting problems is what a combinatorics course is all about. This skill cannot be acquired in reading lots of textbook examples. It is only gained by working lots of problems oneself.

The following example illustrates this type of problem-solving.

Example 5

How many nonempty different collections can be formed from five (identical) apples and eight (identical) oranges?

Readers with some experience in combinatorial problem-solving may want to break the problem into subcases based on the number of objects in the collections. Anyone of these subcases can be counted quite easily, but there are 13 possible subcases; that is, collections can have 1 or 2 or . . . up to 13 fruit.

In counting different possibilities, we must concentrate on what makes one collection different from another collection. The answer is, the number of apples and/or the number of oranges will be different in different collections. Then we can characterize any collection by a pair of integers (a, o), where a is the number of apples and o is the number of oranges.

Now the number of collections is easy to count. There are 6 possible values for a (including 0): 0, 1, 2, 3, 4, 5, and 9 possible values for o. Together there are $6 \cdot 9 = 54$ different collections. (Note that we multiply 6 and 9, not add them, because a collection combines any number of apples and any number of oranges; we add if we want to count the ways to get some amount of apples *or* some amount of oranges, but not both.)

Since the problem asked for nonempty collections and one of the possibilities allowed was (0,0), the desired answer is $54 - 1 = 53$. ∎

One piece of advice to consider when you are stuck and cannot get started with a problem. Try writing down in a systematic fashion some (a dozen or so) of the possible outcomes you want to enumerate. In listing outcomes, you should start to see a pattern emerge. Think of your list as being part of one special subcase. How many outcomes would your list need to include to complete that subcase?

EXERCISES

SUMMARY OF EXERCISES   The first eight exercises are straightforward. Then they become more challenging and require some analysis that will be different for each problem. For the harder problems, readers must devise their own method of solution rather than mimic a method used in one of the text's examples. All exercises should be read carefully two times to avoid misinterpretation.

The word "between" is always used in the inclusive sense; that is, "integers between 0 and 50" means 0, 1, 2, . . . , 49, 50. The exercises in Appendix A.5, Game of Mastermind, are a supplementary source of problems involving combinatorial logic.

Some problems include possible answers for comment. Try to infer from each expression the reasoning that would have generated such an answer. If the answer is wrong, point out the mistake in the reasoning.

1. (a) How many ways are there successively to pick a sequence of two different letters of the alphabet from the word BOAT? From MATHEMATICS?
    (b) How many ways are there to pick first a vowel and then a consonant from each of these words?

2. (a) How many integers are there between 0 and 50 (inclusive)?
    (b) How many of these integers are divisible by 2?
    (c) How many (unordered) pairs of these integers whose difference is 5?

3. A store carries eight styles of pants. For each style, there are 10 different possible waist sizes, six pants lengths, and four color choices. How many different types of pants could the store have?

4. How many secret codes are possible in the standard six-color, four-position Mastermind game (see Appendix A.5)? Comment on the answers: (a) 6^4, (b) 4^6, (c) $6 \cdot 5 \cdot 4 \cdot 3$, and (d) $6 + 6 + 6 + 6$.

5. How many different sequences of heads and tails are possible if a coin is flipped 100 times? Using the fact that $2^{10} = 1024 \simeq 1000$, give your answer in terms of an (approximate) power of 10.

6. How many four-letter "words" (sequence of any four letters with repetition) are there? How many with no repeated letters?

7. How many ways are there to pick a man and a woman who are not husband and wife from a group of n married couples?

8. Given eight different English books, seven different French books, and five different German books:
   (a) How many ways are there to select one book?
   (b) How many ways are there to select three books, one of each language?
   (c) How many ways are there to make a row of three books in which exactly one language is missing (the order of the three books makes a difference)?

9. There are four different roads from town A to town B, three different roads from town B to town C, and two different roads from town A to town C.
   (a) How many different routes from A to C altogether?
   (b) How many different routes from A to C and back (any road can be used once in each direction)?
   (c) How many different routes from A to C and back in (b) that visit B at least once?
   (d) How many different routes from A to C and back in (b) that do not use any road twice?

10. How many ways are there to pick 2 successive cards from a standard 52-card deck such that:
    (a) The first card is an Ace and the second card is not a Queen?
    (b) The first card is a spade and the second card is not a Queen?
        (*Hint:* Watch out for the Queen of spades.)

11. How many nonempty collections of letters can be formed from three A's and five B's?

12. Show that half of the 6^n outcomes of rolling n dice have an even sum. (*Hint:* For any outcome of the first $n - 1$ dice, how many possibilities on the last die will yield an even sum?)

13. How many ways are there to roll two dice to yield a sum divisible by 3?

14. How many four-letter "words" (sequences of letters with repetition) are there in which the first and last letter are vowels? In which vowels appear only (if at all) as the first and last letter? Comment on the answers to the second part: (a) $5^2 21^2$, (b) $5^2 26^2$, (c) $21^2 26^2$, and (d) $26^4 - 21^2 26^2$.

15. (a) How many different five-digit numbers are there (leading zeros, e.g., 00174, not allowed)?
    (b) How many even five-digit numbers are there?
    (c) How many five-digit numbers are there with exactly one 3?

(d) How many five-digit numbers are there that are the same when the order of their digits is inverted (e.g., 15251)?

16. How many different numbers can be formed by various arrangements of the five digits 1, 1, 1, 1, 2, 3?

17. What is the probability that the top two cards in a shuffled deck do not form a pair?

18. (a) How many different outcomes are possible when a pair of dice, one red and one white, are rolled two successive times?
    (b) What is the probability that each die shows the same value on the second roll as on the first roll?
    (c) What is the probability that the sum of the two dice is the same on both rolls?
    (d) What is the probability that the sum of the two dice is greater on the second roll?

19. A rumor is spread randomly among a group of 10 people by successively having one person call someone who calls someone, and so on. A person can pass the rumor on to anyone except the individual who just called.
    (a) How many different paths can a rumor travel through the group in three calls? In n calls?
    (b) What is the probability that if A starts the rumor, A receives the third call?
    (c) What is the probability that if A does not start the rumor, A receives the third call?

20. (a) How many different license plates involving three letters and three digits are there if the three letters appear together either at the beginning or end of the license?
    (b) How many license plates involving one, two, or three letters and one, two, or three digits are there if the letters must appear in a consecutive grouping.

21. Resolve the problem in Example 4, this section, of counting the number of three-letter sequences without repetition using a, b, c, d, e, f that have an e by first counting the number with no e.

22. What is the probability that the sum of two randomly chosen integers between 20 and 40 inclusive is even (the possibility of the two integers being equal is allowed)?
    Comment on the answers: (a) 1/2, (b) 11/21, and (c) $11^2/21^2$.

23. How many three-letter sequences without repeated letters can be made using a, b, c, d, e, f in which either e or f (or both) is used?
    Comment on the answers: (a) $3 \cdot 2 \cdot 4 \cdot 3$, (b) $3 \cdot 2 \cdot 5 \cdot 4$, (c) $3 \cdot 2 \cdot 4 \cdot 4 - 3 \cdot 2 \cdot 4$, and (d) $6 \cdot 5 \cdot 4 - 4 \cdot 3 \cdot 2$.

24. What is the probability that an integer between 1 and 10,000 has exactly one 8 and one 9?

Comment on the answers: (a) $4 \cdot 3/10^4$, (b) $4 \cdot 3 \cdot 8 \cdot 7/10^4$, and (c) $2 \cdot 4 \cdot 3 \cdot 8^2/10^4$.

25. How many different five-letter sequences can be made using the letters A, B, C, D with repetition such that the sequence does not include the word BAD—that is, the sequence A̲B̲A̲D̲D is excluded.

26. (a) How many election outcomes are possible with 20 people each voting for one of seven candidates (the outcome includes not just the totals but also who voted for each candidate)?
    (b) How many election outcomes are possible if only one person votes for candidate A and only one person votes for candidate D?

27. There are 15 different apples and 10 different pears. How many ways are there for Jack to pick an apple or a pear and then for Jill to pick an apple and a pear?

28. How many times is the digit 5 written when listing all numbers from 1 to 100,000?
    Comment on the answers: (a) 4, (b) $5 \cdot 10^4$, and (c) $1 + 10 + 100 + 1000 + 10000$.

29. What is the probability that if one letter is chosen at random from the word RECURRENCE and one letter is chosen from RELATION, the two letters are the same?

30. How many four-digit numbers are there formed from the digits 1, 2, 3, 4, 5 (with possible repetition) that are divisible by 4?

31. How many nonempty collections of letters can be formed from n A's, n B's, n C's and n D's?

32. There are 50 cards numbered from 1 to 50. Two different cards are chosen at random. What is the probability that one number is twice the other number?

33. If two different integers between 1 and 100 inclusive are chosen at random, what is the probability that the difference of the two numbers is 15?

34. If three distinct dice are rolled, what is the probability that the highest value is twice the smallest value?

35. How many different numbers can be formed by the product of two or more of the numbers 3, 4, 4, 5, 5, 6, 7, 7, 7?

36. How many n-digit binary sequences are there without any pair of consecutive digits being the same?

37. A chain letter is sent to five people in the first week of the year. The next week each person who received a letter sends letters to five new people, and so on. How many people have received letters in the first five weeks?

38. How many ways are there to place two identical rooks in a common row or column of an 8×8 chessboard? an $n \times m$ chessboard?

39. How many ways are there to place two identical kings on an 8×8 chessboard so that the kings are not in adjacent squares? On an $n \times m$ chessboard?

40. How many ways are there to place two identical queens on an 8×8 chessboard so that the queens are not in a common row, column, or diagonal?

41. How many different positive integers can be obtained as a sum of two or more of the numbers 1, 3, 5, 10, 20, 50, 82?

42. How many ways are there for a man to invite some (nonempty) subset of his 10 friends to dinner?

43. How many different rectangles can be drawn on an 8×8 chessboard (the rectangles could have sides of length 1 through 8; two rectangles are different if they contain different subsets of individual squares)?

44. How many ways are there to place a red checker and a black checker on two black squares of a checkerboard so that the red checker can jump over the black checker?

45. Use induction to verify formally:
    (a) The addition principle.
    (b) The multiplication principle.

46. On the real line, place n white pegs at positions 1, 2, . . . , n and n blue pegs at positions $-1, -2, . . . , -n$ (0 is open). Whites move only to the left, blues to the right. When beside an open position, a peg may move one unit to occupy that position (provided it is in the required direction). If a peg of one color is in front of a peg of the other color that is followed by an open position (in the required direction), a peg may jump two units to the open position (the jumped peg is not removed). By a sequence of these two types of moves (not necessarily alternating between white and blue pegs), one seeks to get the positions of the white and blue pegs interchanged. (See the article on this game in *Mathematics Teacher*, January 1982.)
    (a) Play this game for $n = 3$ and $n = 4$.
    (b) Use a combinatorial argument to show that, in general, $n^2 + 2n$ moves (unit steps and jumps) are required to complete the game.

5.2 SIMPLE ARRANGEMENTS AND SELECTIONS

A **permutation** of n distinct objects is an arrangement, or ordering, of the n objects. An **r-permutation** of n distinct objects is an arrangement using r of the n objects. An **r-combination** of n distinct objects is an unordered selection, or *subset*, of r out of the n objects. We use $P(n,r)$ and $C(n,r)$ to denote the number of r-permutations and r-combinations, respectively, of a set of n objects. From the multiplication principle we obtain

$$P(n,2) = n(n-1), \qquad P(n,3) = n(n-1)(n-2), \qquad \text{and}$$
$$P(n,n) = n(n-1)(n-2) \cdots 3 \cdot 2 \cdot 1$$

In enumerating all permutations of n objects, we have n choices for the first position in the arrangement, $n-1$ choices (the $n-1$ remaining objects) for the second position, . . . , and finally one choice for the last position. Using the notation $n! = n(n-1)(n-2) \cdots 3 \cdot 2 \cdot 1$ ($n!$ is said "n factorial"), we have the formulas

$$P(n,n) = n!$$

and

$$P(n,r) = n(n-1)(n-2) \cdots (n-(r-1)) = \frac{n!}{(n-r)!}$$

Our formula for $P(n,r)$ can be used to derive a formula for $C(n,r)$. All r-permutations of n objects can be generated by first picking any r-combination of the n objects and then arranging these r objects in any order. Thus $P(n,r) = C(n,r) \cdot P(r,r)$, and solving for $C(n,r)$ we have

$$C(n,r) = \frac{P(n,r)}{P(r,r)} = \frac{n!/(n-r)!}{r!} = \frac{n!}{r!(n-r)!}$$

The numbers $C(n,r)$ are frequently called **binomial coefficients** because of their role in the expansion of $(x+y)^n$. We will study identities involving binomial coefficients in Section 5.5. It is common practice to write the expression $C(n,r)$ as $\binom{n}{r}$ and to say "n choose r" [we usually write $C(n,r)$ in this book so that lines of text will not have to be widely spaced].

Note that $C(n,r) = C(n,n-r)$ — the number of ways to pick a subset of r out of n distinct objects equals the number of ways to throw away a subset of $n-r$.

Example 1

How many secret codes can be formed in Mastermind when no repetition of colors is allowed (Mastermind is described in Appendix A5)?

There are six colors and four positions. So there are $P(6,4) = 6!/2! = 360$ possibilities. ∎

Example 2

How many ways are there to rank n candidates for the job of chief wizard? If the ranking is made at random (each ranking is equally likely), what is the probability that the fifth candidate, Mr. Gandalf, is in second place?

A ranking is simply an ordering, or arrangement, of the n candidates. So there are $n!$ rankings. Intuitively, a random ranking should randomly position Mr. Gandalf, and so we would expect that he has probability $1/n$ of being second, or being in any given position (or, equivalently, by symmetry, each of the n candidates should have the same probability of being in second place). To confirm this suspicion, let us formally apply the formula for the probability of this event (see Appendix A3).

$$\text{Prob(Gandalf second)} = \frac{\text{no. of rankings with Gandalf second}}{\text{total no. of rankings}}$$

We know that the total number of rankings is $n!$ The number of rankings with Gandalf second can easily be obtained by first putting Gandalf in second place. Then there are $P(n - 1, n - 1) = (n - 1)!$ ways to assign the remaining $n - 1$ candidates to the remaining $n - 1$ positions in the ranking. So Mr. Gandalf is second in $(n - 1)!$ rankings and prob(Gandalf second) = $(n - 1)!/n! = 1/n$. ∎

Example 3

How many different ways are there to position 12 elves around a circular table with 12 equally spaced seats?

Assuming there is no preassigned orientation of the table (e.g., no certain seat faces north), a circular order around a table has no first position. It is instead a relative cyclic order that could "start" with any position. There are two approaches to this problem. One approach is to let one chosen elf sit anywhere. Now there are 11 ways to pick the first elf on the chosen elf's left, 10 ways to pick the second elf on the left, and so forth, yielding 11! circular arrangements.

The other approach is first to consider the circle to be oriented so that there are 12 ordered positions and 12! ways to seat the elves. Now we "unorient" the table by allowing it to rotate around, and we consider any two seatings to be equivalent if a rotation transforms one into another. Clearly, the set of all 12! seatings divides into equivalence classes with 12 cyclically equivalent seatings in each class. So there are $12!/12 = 11!$ equivalence classes, that is, 11! unoriented seatings. ∎

Example 4

How many ways are there to arrange the seven letters in the word SYSTEMS?

The difficulty of multiple S's can be avoided by first picking the positions that the other four letters E, M, T, Y will occupy in the seven letter arrangement (then the remaining three positions must be S's). Thus there are $P(7,4) = 7!/3! = 840$ ways. (A general formula for counting arrangements with repeated objects is given in the next section.) ∎

Example 5

How many 5-card hands (subsets) can be formed from a standard 52-card deck? If a 5-card hand is chosen at random, what is the probability of obtaining a flush (all 5 cards in the hand are in the same suit)? What is the probability of obtaining 3, but not 4, Aces?

A 5-card hand is a subset of 5 cards, and so there are $C(52,5) = 52!/(47!5!) = 2,598,960$ different 5-card hands. To find the probability of a flush, we need to find the number of such 5-card hands. There are 4 suits and a subset of 5 cards from the 13 in a suit can be chosen in $C(13,5) = 13!/(5!8!) = 1287$ ways. There are $4 \cdot 1287 = 5148$ flushes, and

$$\text{Prob(5-card hand is a flush)} = \frac{5,148}{2,598,960} = 0.00198 \ (=0.2\%)$$

To count the number of hands with exactly 3 Aces, we must pick 3 of the 4 Aces, done in $C(4,3) = 4$ ways, and then fill out the hand with 2 other non-Ace cards, done in $C(48,2) = 1128$ ways. So there are $4 \cdot 1128 = 4512$ hands with exactly 3 Aces, and

$$\text{Prob(5-card hand has exactly 3 Aces)} = \frac{4,512}{2,598,960} = .00174$$

Slightly less likely than a flush.   ∎

Note that to count hands with 3 Aces in Example 5, we implicitly used the multiplication principle to multiply the ways to pick 3 Aces by the ways to fill out the hand with 2 non-Ace cards. However, a hand is an unordered collection and by ordering it into two parts, we might generate two outcomes that are really the same set, violating the distinctness condition of the Multiplication Principle. Part of the next example shows how the distinctness condition could be violated. In this problem, hands could safely be decomposed into an Aces part and a non-Aces part. To be able to multiply parts of sets together, the parts must be distinct from each other.

Example 6

A committee of k people is to be chosen from a set of 7 women and 4 men. How many ways are there to form the committee if:

(a)   The committee has 5 people, 3 women, and 2 men.

(b)   The committee can be any positive size but must have equal numbers of women and men.

(c)   The committee has 4 people and at least 2 are women.

(d)   The committee has 4 people and one of them must be Mr. Baggins.

(e)   The committee has 4 people, two of each sex, and Mr. and Mrs. Baggins cannot both be on the committee.

(a)   By the set composition principle, a committee of 3 women and 2 men can be chosen by composing any subset of 3 women with any subset of 2 men, $C(7,3) \cdot C(4,2) = 35 \cdot 6 = 210$ ways.

(b)   To count the possible subsets of women and men of equal size on the committee, we must know definite sizes of the subsets. That is, we must break the problem into the four disjoint subcases of: 1 woman and 1 man, 2 each, 3 each, and 4 each (there are only 4 men). So the total number is the sum of the possibilities for these four subcases, $C(7,1) \cdot C(4,1) + C(7,2) \cdot C(4,2) + C(7,3) \cdot C(4,3) + C(7,4) \cdot C(4,4) = 7 \cdot 4 + 21 \cdot 6 + 35 \cdot 4 + 35 \cdot 1 = 329$.

(c)   One approach is to pick 2 women first, $C(7,2) = 21$ ways, and then pick any 2 of the remaining set of 9 people (5 women and 4 men). However, counting all committees in this fashion counts some outcomes more than once, since any woman in one of these committees could either be chosen as one of the first 2 women or one of the 2 remaining people. For example, if W_i denotes the ith woman and M_i the ith man, then (W_1, W_3) composed with the 2 remaining people (W_2, M_3) yields the same set as (W_1, W_2) composed with (W_3, M_3).

A correct solution to this problem must use a subcase approach, as in part b. That is, break the problem into 3 subcases: 2 women and 2 men, 3 women and 1 man, and 4 women. The answer is thus $C(7,2) \cdot C(4,2) + C(7,3) \cdot C(4,1) + C(7,4) = 21 \cdot 6 + 35 \cdot 4 + 35 = 301$.

(d)   If Mr. Baggins must be on the committee, this simply means that the problem reduces to picking 3 other people from the remaining 10 people (7 women and 3 men). So the answer is $C(10,3) = 120$.

(e)   There are three subcases involving whether or not each Baggins, but not both, is on the committee. If Mrs. Baggins is on it and Mr. Baggins is not, then one more woman must be chosen from the remaining 6 women and 2 more men must be chosen from the remaining 3 men (Mr. Baggins is excluded). This can be done $C(6,1) \cdot C(3,2) = 6 \cdot 3 = 18$ ways. If Mr. Baggins is on it and Mrs. Baggins is off, a similar argument yields $C(6,2) \cdot C(3,1) = 15 \cdot 3 = 45$ ways. Finally if neither is on the committee, we have $C(6,2) \cdot C(3,2) = 15 \cdot 3 = 45$ ways. The total answer is $18 + 45 + 45 = 108$.

A simpler approach to this problem is, we can have any of the $C(7,2) \cdot C(4,2)$ 2 women-2 men committees except the forbidden $C(6,1) \cdot C(3,1)$ committees with both Bagginses. Subtracting the excluded committees yields $21 \cdot 6 - 6 \cdot 3 = 108$. ∎

The problem of counting the same outcome twice, that arose in part c of Example 6, arises in many guises. The following principle should help the reader avoid this problem.

The Set Composition Principle

Suppose a set of distinct objects is being enumerated using the Multiplication Principle: multiplying the number of ways to form some *first part* of the set by the number of ways to form a *second part* (for a given first part). Given any set S thus constructed, one must be able to tell uniquely which elements of S are in the first part of S counted first.

In the three-Aces problem of Example 5, the collection A of different triples of three Aces and the collection NA of different pairs of non-Ace cards are disjoint, and so by the set composition principle the number of three-Ace hands is the size of A times the size of NA. In the "committee with at least two women" problem in Example 6, the method of choosing two of the seven women first and then picking any remaining two women or men violates the Set Composition Principle because given a committee with three women and one man, it is impossible to say which two women are chosen first. Thus all possible pairs of two of the three women in the committee could be and would be enumerated, counting this committee three times.

Example 7

How many different eight-digit binary sequences are there with six 1's and two 0's?

We use the one-to-one correspondence between the set of such binary sequences and the set of subsets of six out of eight objects. The objects are the positions in the sequence where the 1's go. Thus there are $C(8,6) = 28$ such binary sequences. (There is also a correspondence with subsets of two of the eight positions for 0's.) ∎

Example 8

How many different ways are there to form a row of 5 people out of a set of 10 people such that 5 people are lined up from left to right in order of increasing Social Security number?

Observe that there is only one way to arrange any given set of 5 people in order of increasing Social Security number. So the problem reduces to picking a subset of 5 out of the 10 people. There are $C(10,5) = 252$ ways. ∎

Example 9—Optional if Section 2.3 is not previously covered

The chromatic polynomial $P_k(G)$ of a graph G is a polynomial in k that gives the number of k-colorings of G. What is the chromatic polynomial of:

(a) A complete graph K_5 on five vertices (all vertices adjacent to each other).

(b) The graph C_4 of a circuit of length 4.

(a) $P_k(K_5) = k(k - 1)(k - 2)(k - 3)(k - 4) = P(k,4)$, since there are k possible choices for the first vertex to be colored; then that color cannot be used again, and so the second vertex to be colored has $k - 1$ choices, and so on.

(b) Let the vertices on the circuit be names x_1, x_2, x_3, x_4 with edges (x_1,x_2), (x_2,x_3), (x_3,x_4), (x_4,x_1). We break the computation of $P_k(C_4)$ into two cases, depending on whether or not x_1 and x_3 are given the same color.

If x_1 and x_3 have the same color, there are k choices for the color of these two vertices. Then x_2 and x_4 each must only avoid the color of x_1 and x_3, $k - 1$ color choices each. So the total number of k-colorings of C_4 in this case is $k(k - 1)^2$.

If x_1 and x_3 have different colors, there are $k(k - 1)$ choices for the two different colors for x_1 and then x_3. Now x_2 and x_4 each have $k - 2$ color choices. So in this case the total number of k-colorings of C_4 is $k(k - 1)(k - 2)^2$. Combining the two cases, we obtain $P_4(C_4) = k(k - 1)^2 + k(k - 1)(k - 2)^2$. ∎

Example 10

What is the probability that a 4-digit campus telephone number has one or more repeated digits?

There are $10^4 = 10,000$ different 4-digit phone numbers. We break the problem of counting 4-digit phone numbers with repeated digits into the 4 different classes of repetitions: (1) 2 digits the same, the other 2 digits each different (e.g., 5105); (2) 2 digits the same, the other 2 digits also the same (e.g., 2828); (3) 3 digits the same, the other different; and (4) all 4 digits the same. Let us do the easiest first. Trivially class 4 has 10 numbers (there are 10 digits, any of which can be repeated four times).

To count class 3 numbers, we pick which digit appears once, 10 choices, then where it occurs in the number, four positions, and finally which other digit appears in the other three positions, nine choices, for a total in class 3 of $10 \cdot 4 \cdot 9 = 360$ numbers.

To count class 2 numbers, we pick which 2 digits are each to appear twice, $C(10,2) = 45$ choices, and then how to arrange these 4 digits—pick which two positions for the smaller of the digits in $C(4,2) = 6$ ways—for a total in class 2 of $45 \cdot 6 = 270$ numbers.

To count class 1 numbers, we pick which pair of digits appear once, $C(10,2) = 45$ choices, then pick a position for one of these 2 digits (say, the smaller digit) and a position for the other digit, $4 \cdot 3 = 12$ choices, and finally pick which other digit appears in the remaining two positions, eight choices, for a total of $45 \cdot 12 \cdot 8 = 4320$ numbers. In sum, there are $4320 + 270 + 360 + 10 = 4960$ 4-digit phone numbers with a repeated digit. The probability of a repeated digit is thus $4960/10000 = 0.496 \cong 0.5$. Note that we were constantly using the set composition principle in the preceding computations of class sizes.

One point of caution, in the cases where 2 different digits both occur once or both occur twice, we pick those digits as an unordered pair (in $C(10,2)$

ways) and then arrange those digits in all ways in a phone number, rather than pick a first digit, position it, then pick a second digit, and position it. In the latter (wrong) approach, we cannot tell which digit in a resulting phone number was chosen first, violating the set composition principle.

Needless to say, there is also a simple way to count phone numbers with repeated digits: first count numbers with no repeated digits. These are just the $P(10,4) = 5040$ four-permutations of the 10 digits. So the remaining repeated-digit numbers amount to $10000 - 5040 = 4960$. ■

We close this section by noting that for large values of n, $n!$ can be approximated by the number $s_n = \sqrt{2\pi n}\left(\frac{n}{e}\right)^n$, where e is Euler's constant ($e = 2.718\ldots$). This approximation is due to Stirling and its derivation is given in most advanced calculus texts (see Buck [1]). The error $|n! - s_n|$ increases as n increases, but the relative error $|n!/s_n - 1|$ is always less than $1/11n$. The following table gives some sample values of $n!$, s_n, and $n!/s_n$.

n	$n!$	s_n	$n!/s_n$
1	1	.922	1.085
2	2	1.919	1.042
5	120	118.02	1.017
10	3,628,800	3,598,600	1.008
20	2.433×10^{18}	2.423×10^{18}	1.004
100	9.333×10^{157}	9.328×10^{157}	1.0008

EXERCISES

SUMMARY OF EXERCISES   As in the previous section, most of these exercises require individual analysis, different for each problem. Remember to read problems carefully to avoid misinterpretation. Pay special attention to whether a problem involves arrangements or subsets.

Problems assume that people are distinct objects (no identical people). Related programming problems are given in Section 5.6.

1. How many ways are there to arrange the cards in a 52-card deck?

2. How many different 10-letter "words" (sequences) are there with no repeated letters formed from the 26-letter alphabet?

3. How many ways are there to distribute nine different books among 15 children if no child gets more than one book?
   Comment on the answers: (a) $C(15,9)$, (b) 15^9, (c) $C(15,9)9!$

4. How many ways are there to seat five different boys and five different girls around a circular table with 10 seats? How many ways if boys and girls alternate seats?

5. How many arrangements are there of the 8 letters in the word VISIT-ING?

6. How many ways are there to pick a subset of 4 different letters from the 26-letter alphabet?

7. How many ways are there to pick a 5-person basketball team from 10 possible players? How many teams if the weakest player and the strongest player must be on the team?

8. There are 10 white balls and 5 red balls in an urn. How many different ways are there to select a subset of 5 balls, assuming the 15 balls are different? What is the probability that the selection has 3 whites and 2 reds?

9. If a coin is flipped 10 times, what is the probability of 8 or more heads?

10. What is the probability that a five-card poker hand has the following?
    (a) four Aces
    (b) four of a kind
    (c) two pairs (not four of a kind or a full house)
    (d) a full house (a three-of-a-kind and a pair)
    (e) a straight (a set of five consecutive values)
    (f) no pairs (possibly a straight or flush).

11. How many ways can a committee be formed from four men and six women with:
    (a) At least two men and at least twice as many women as men?
    (b) Four members, at least two of which are women, and Mr. and Mrs. Baggins will not serve together?

12. There are eight applicants for the job of dog catcher and three different judges who each rank the applicants. Applicants are chosen if and only if they appear in the top three in all three rankings.
    (a) How many ways can the three judges produce their three rankings?
    (b) What is the probability of Mr. Dickens, one of the applicants, being chosen in a random set of three rankings?
    Comment on the answers (numerator only): (a) $3(n - 1)!$, (b) $(3!(n - 3)!)^3$, (c) $3((n - 1)!)^3$.

13. There are six different French books, eight different Russian books, and five different Spanish books. How many ways are there to arrange the books on a shelf with all books of the same language grouped together?

14. If 13 players are each dealt 4 cards from a 52-card deck, what is the probability that each player gets one card of each suit?

15. How many 10-letter (sequences) are there using five different vowels and five different consonants (chosen from the 21 possible consonants)? What is the probability that one of these words has no consecutive pair of consonants?

16. What is the probability that an arrangement of a, b, c, d, e, f has
    (a) a and b side-by-side?
    (b) a occurring somewhere before b?

17. How many ways are there to pair off 10 women at a dance with 10 out of 20 available men?

18. There are 3 women and 5 men who will split up into two 4-person teams. How many ways are there to do this so that there is (at least) one woman on each team?

19. A man has n friends and invites a different subset of four of them to his house every night for a year (365 nights). How large must n be?

20. Suppose a subset of 60 different days of the year are selected at random in a lottery. What is the probability that there are five days from each month in the subset? (For simplicity, assume a year has 12 months with 30 days each.)

21. Suppose a subset of eight different days of the year are selected at random. What is the probability that each day is from a different month? (For simplicity, assume a year has 12 months with 30 days each.)

22. What is the probability of randomly choosing a permutation of the 10 digits $0, 1, 2, \ldots, 9$ in which:
    (a) An odd digit is in the first position and 1, 2, 3, 4, or 5 in the last position?
    (b) 5 is not in the first position and 9 is not in the last position?

23. What is the probability that an arrangement of INSTRUCTOR has:
    (a) Three consecutive vowels?
    (b) Two consecutive vowels?

24. What is the probability that a random 4-digit campus telephone number has no repeated digits?

25. What is the probability among the arrangements of GRACEFUL of having:
    (a) No pair of consecutive vowels?
    (b) F and G appear side-by-side?

26. (a) What is the probability that k is the smallest integer in a subset of four different numbers chosen from 1 through 20, for $k = 1, 2, \ldots, 17$?
    (b) What is the probability that k is the second smallest?

27. What is the probability that the difference between the largest and the smallest numbers is k in a subset of four different numbers chosen from 1 through 20, for $k = 3, 4, \ldots, 19$?

28. How many arrangements of JUPITER are there with the vowels occurring in alphabetic order?
    Comment on the answer: (a) $3!4!$, (b) $7!/3!$, (c) $C(7,3)4!$.

29. In standard six-color, four-position Mastermind (see Appendix A5) how
     many secret words are possible if:
     (a) The first guess is RWYG and its score is (1) ○ (2) ○● (3) ○○●?
     (b) The first guess is RRWY and its score is (1) ○ (2) ○● (3) ○○?

30. How many ways are there to place eight identical black pieces and eight
     identical white pieces on an 8 × 8 chessboard? How many ways that are
     symmetric, that is, so that the layout looks the same if the chessboard is
     rotated 180°?

31. How many different spanning trees are there in the following graph:

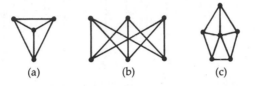

(a)   (b)   (c)

32. How many different 4-colorings do each of the graphs in Exercise 31
     have?

33. Determine the chromatic polynomial for the following graphs:
     (a) K_n, a complete graph on n vertices.
     (b) any n-vertex tree

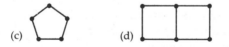

(c)   (d)

34. How many Euler circuits do each of the following graphs have:

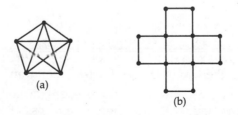

(a)

(b)

35. How many Hamilton paths do each of the graphs in Exercise 34 have?

36. How many ways are there for a man to invite different subsets of three
     of his five friends on three successive days? How many ways if he has n
     friends?

37. How many arrangements of INSTRUCTOR are there in which there are
     two consonants between successive pairs of vowels?

38. What is the probability that a random arrangement of REPLETE has no
     consecutive E's?

39. What is the probability that if a die is rolled a sequence of five times only two different values appear?

40. What is the probability that two letters selected from MISSISSIPPI are the same? That three letters are all different?

41. If a coin is tossed eight times, what is the probability of getting:
    (a) Exactly four heads in a row?
    (b) At least four heads in a row?

42. For the given map of roads between city A and city B,

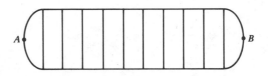

    (a) How many routes are there from A to B that do not repeat any road (a road is a line segment between two intersections)?
    (b) How many ways are there for two people to go from A to B without both ever traversing the same road in the same direction?

43. The U.S. Senate randomly creates a special committee with k Senators. What is the probability that the committee has a Senator from Kansas?

44. How many n-digit ternary (0,1,2) sequences are there with k 1's?

45. How many triangles can be formed by joining different sets of three corners of an octagon (with distinct corners)? How many triangles if no pair of adjacent corners are permitted?

46. What is the probability that 2 (or more) people in a random group of 25 people have a common birthday? (This is the famous *Birthday Paradox Problem*.)

47. A family has two boys and three girls to send to private schools. There are five boys' schools, eight girls' schools, and three coed schools. If each child goes to a different school, how many different *subsets* of five schools can the family choose for their children?

48. (a) How many points of intersection are formed by the chords of an n-gon (assuming no 3 of these lines cross at one point)?
    (b) Into how many line segments are the lines in part a cut by the intersection points?
    (c) Use Euler's Formula $r = e - v + 2$ and parts a and b to determine the number of regions formed by the chords of an n-gon.

49. If one quarter of all 3-subsets of the integers 1, 2, . . . , m contain the integer 5, determine m.

50. Ten fish are caught in a lake, marked, and then returned to the lake. Two days later 20 fish are again caught, two of which have been marked.

(a) Find the probability of two of the 20 fish being marked if the lake has k fish (assuming the fish are caught at random).

(b) What value of k maximizes the probability?

51. A batch of 50 different automatic typewriters contains exactly 10 defective machines. What is the probability of finding:

(a) At least 1 defective machine in a random group of 5 machines?

(b) At least 2 defective machines in a random group of 10 machines?

(c) The first defective machine to be the kth (in a random sequence of machines) machine taken apart for inspection?

(d) The last defective machine to be the kth machine taken apart?

52. How many ways are there to form an (unordered) collection of four pairs of two people chosen from a group of 30 people?

53. What is the probability that a random five-card hand has:

(a) Exactly one pair (no three-of-a-kind or two pairs)?

Comment on the answers: (a) $\binom{13}{1}\binom{4}{2}\binom{48}{3}\Big/\binom{52}{5}$,

(b) $13\binom{4}{2}48 \cdot 44 \cdot 40\Big/\binom{52}{5}$, (c) $\dfrac{52 \cdot 3 \cdot 48 \cdot 44 \cdot 40}{52 \cdot 51 \cdot 50 \cdot 49 \cdot 48}$.

(b) One pair or more (three-of-a-kind, two pairs, four-of-a-kind, full house)?

(c) At least one card of each suit?

(d) The cards dealt in order of decreasing value?

(e) At least one of each of the four values: Ace, King, Queen, and Jack?

(f) The same number of hearts and spades?

(g) At least one spade and at least one heart and the values of the spades are all greater than the values of the hearts?

54. Given a collection of $2n$ objects, n identical and the other n all distinct, how many different subcollections of n objects are there?

55. How many ways are there to pick a group of n people from 100 people (each of a different height) and then pick a selected group of m other people such that all people in the first group are taller than the people in the second group?

56. How many subsets of three different integers between 1 and 90 inclusive are there whose sum is:

(a) An even number?

(b) Divisible by 3?

(c) Divisible by 4?

57. Given 14 positive integers, 12 of which are even, and 16 negative integers, 11 of which are even, how many ways are there to pick 12 numbers from this collection of 30 integers such that 6 of the 12 numbers are positive and 6 of the 12 numbers are even?

58. A man has seven friends. How many ways are there to invite a different subset of three of these friends for a dinner on seven successive nights such that each pair of friends are together at just one dinner?

59. There are eight scientists who work at a secret research station. There are many doors into the station and each door has four locks. Each scientist is given a set of keys chosen so that a group of four scientists have to be present in order for the group collectively to have four keys that fit the four different locks on one of the doors. Give a lower bound on how many doors there must be. How many keys must each person have?

60. How many triangles are formed by (assuming no three lines cross at a point):

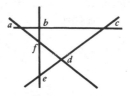

(a) Pieces of n nonparallel lines; for example, the four lines above form four triangles: acd, abf, efd, and ebc?

(b) Pieces of three chords of a convex 10-gon such that the triangles are wholly within the 10-gon (a corner of the 10-gon cannot be a corner of any of these triangles)?

(c) Pieces of n lines, m of which are parallel and the others mutually nonparallel?

(d) Pieces of three chords or outside edges of a convex n-gon?

5.3 ARRANGEMENTS AND SELECTIONS WITH REPETITIONS

In this section we discuss arrangements and selections with repetition—arrangements of a collection of objects with repeated objects, such as the collection b, a, n, a, n, a, and selections from a set when an object can be chosen more than once, such as ordering six hot dogs chosen from three varieties. We motivate the formulas for these counting problems with two examples.

Example 1

How many arrangements are there of the six letters b, a, n, a, n, a?

We form the arrangements by choosing first the three positions in the arrangement where the a's will go, $C(6,3) = 20$ ways, then the two positions (out of the remaining three) where the n's will go, $C(3,2) = 3$ ways, and finally the last remaining position gets the b, 1 way. Thus there are $20 \cdot 3 \cdot 1 = 60$ arrangements. ∎

Theorem 1

If there are n objects, r_1 of type 1, r_2 of type 2, . . . , and r_m of type m, where $r_1 + r_2 + \cdots + r_m = n$, then the number of arrangements of these n objects, denoted $P(n; r_1, r_2, \ldots, r_m)$, is

$$P(n; r_1, r_2, \ldots, r_m) = \binom{n}{r_1}\binom{n - r_1}{r_2}\binom{n - r_1 - r_2}{r_3}$$

$$\cdots \binom{n - r_1 - r_2 \cdots - r_{m-1}}{r_m} = \frac{n!}{r_1! r_2! \cdots r_m!} \quad (1)$$

Proof 1

First pick r_1 positions for the first types, then r_2 of the remaining positions for the second types, and so on. A mathematically precise proof of the "etc." part requires induction (see Exercise 34). The formula (1) is just a simplification that results from cancelling factorials, for example,

$$P(6; 3, 2, 1) = \binom{6}{3}\binom{3}{2}\binom{1}{1} = \frac{6!}{3!3!} \cdot \frac{3!}{2!1!} \cdot \frac{1!}{1!} = \frac{6!}{3!2!1!}$$

∎

Proof 2

This proof is similar to our derivation of $C(n,r)$ through the equation $P(n,r) = C(n,r) \cdot P(r,r)$. Suppose that for each type, the r_i objects of type i are given subscripts numbered 1, 2, . . . , r_i to make each object distinct. Then there are $n!$ arrangements of the n distinct objects. Let us enumerate these $n!$ arrangements of distinct objects by enumerating all $P(n; r_1, r_2, \ldots, r_m)$ patterns (without subscripts) of the objects, and then for each pattern placing the subscripts in all possible ways. For example, the pattern *baanna* can have subscripts on a's placed in the 3! ways:

$$ba_1a_2nna_3 \qquad ba_2a_1nna_3 \qquad ba_3a_1nna_2$$

$$ba_3a_2nna_1 \qquad ba_1a_3nna_2 \qquad ba_2a_3nna_1$$

For each of these 3! ways to subscript the a's, there are 2! ways to subscript the n's. Thus, in general a pattern will have $r_1!$ ways to subscript the r_1 objects of type 1, $r_2!$ ways for type 2, and $r_i!$ ways for type 1. Then

$$n! = P(n;r_1,r_2, \ldots ,r_m)r_1!r_2! \cdots r_m!$$

or

$$P(n;r_1,r_2, \ldots ,r_m) = \frac{n!}{r_1!r_2! \cdots r_m!}$$

■

Example 2

How many different orders for six hot dogs are possible if there are three varieties of hot dog?

To solve selection-with-repetition problems we require a one-to-one correspondence with a related selection-without-repetition problem. Suppose the three varieties are regular dog, chili dog, and super dog. Then let the order be jotted down on an order form

Regular	Chili	Super
x	xxxx	x

Each x represents a hot dog. The order shown on the form above is one regular, four chili, and one super. Since all workers know the sequence of dogs on the form is regular, chili, super, the order can simply be written as $x/xxxx/x$ without column headings.

Any order of r hot dogs will consist of some sequence of r x's and two /'s. Conversely, any sequence of r x's and two /'s represents an order: the x's before the first / count the number of regular dogs; the x's between the two /'s count chilis; and the final x's count supers. So there is a one-to-one correspondence between orders and such sequences. But the number of sequences of six x's and two /'s is just the number of different subsets of two positions in the sequence for the /'s. So the answer is $C(8,2) = 28$. ■

Theorem 2

The number of selections with repetition of r objects chosen from n types of objects is $C(r + n - 1,r)$.

Proof

We make an "order form" for a selection just as in Example 2, with an x for each object selected. As before, the x's before the first / count the number of the first type of object, the x's between the first and second /'s count the number of the second type, . . . , and the x's after the $(n - 1)$st / count the number of the nth type ($n - 1$ slashes are needed to separate n types). The number of sequences with r x's and $n - 1$ /'s is $C(r + (n - 1),r)$. ■

Example 3

Nine students, three from Mrs. Brown's class, three from Mr. Hill's class, and three from Mrs. Simon's class, have bought a block of nine seats for their school's homecoming game. If the nine seats are randomly divided into three for each class, what is the probability that each class will get a block of three consecutive seats?

There are $P(9;3,3,3) = 9!/3!3!3! = 1680$ ways to arrange the three types of students in the row of nine seats. If the three students of each class are to sit together, the nine seats must be divided into three blocks of three seats, one block for each class. There are $3! = 6$ ways to arrange three such blocks. Then the probability that each class sits together is 6/1680. ■

Example 4

How many ways are there to form a sequence of 10 letters from 4 a's, 4b's, 4 c's, and 4 d's if each letter must appear at least twice?

We have to break this problem into a set of subproblems that involve sequences with given numbers of a's, b's, c's, and d's, since to apply Theorem 1 we need to know exactly how many a's, b's, c's, and d's will be in the arrangement. There are two categories of letter frequencies that sum to 10 with each letter appearing two or more times. The first category is four appearances of one letter and two appearances of each other letter. The second is three appearances of two letters and two appearances of the other two letters.

In the first category, there are four subcases for choosing which letter occurs four times and $P(10;4,2,2,2) = 18,900$ ways to arrange four of one letter and two of the three others. In the second category, there are $C(4,2) = 6$ subcases for choosing which two letters occur three times and $P(10;3,3,2,2) = 25,200$ ways to arrange three of two letters and two of the two others. So the answer is $4 \cdot 18,900 + 6 \cdot 25,200 = 226,800$ ways. ■

Example 5

A domino is a thin rectangular piece of wood whose top face is divided into two squares. Each of the squares is either blank or contains between one and six dots. How many different domino faces are possible?

A domino face can be considered to represent an unordered pair of two numbers between 0 and 6 (with doubles allowed). By Theorem 2, there are $C(2 + 7 - 1,2) = 28$ different dominoes. ■

Example 6

How many ways are there to fill a box with a dozen doughnuts chosen from five different varieties with the requirement that at least one doughnut of each variety is picked?

We solve this problem by following the typical procedure of people making such a doughnut selection. They would first pick one doughnut of each variety and then pick the remaining seven doughnuts any way they pleased. There is no choice (only one way) in picking one doughnut of each type. The choice occurs in picking the remaining seven doughnuts from the five types. So the answer by Theorem 2 is $C(7 + 5 - 1,7) = 330$. (Note that the set composition principle does not apply here, because the objects are not all distinct.) ∎

Example 7

How many ways are there to pick a collection of exactly 10 balls from a pile of red balls, blue balls, and purple balls if there must be at least 5 red balls? If at most 5 red balls?

This problem is similar to Example 6. First pick 5 red balls and then pick the remaining 5 balls arbitrarily (possibly including more red balls). The remaining 5 balls can be chosen in $C(5 + 3 - 1,5) = 21$ ways.

To handle the constraint of at most 5 red balls, we count the complementary set. Of all $C(10 + 3 - 1,10) = 66$ ways to pick 10 balls from the three colors without restriction, there are $C(4 + 3 - 1,4) = 15$ ways to choose a collection with at least 6 red balls (first pick 6 reds and then any choice of 4 more balls). So there are $66 - 15 = 51$ ways to choose 10 balls without more than 5 red balls. ∎

EXERCISES

SUMMARY OF EXERCISES   Most of these problems are not too difficult variations on the section's Examples. Be careful to distinguish between arrangements and subsets problems. Related programming problems are given in Section 5.6.

1. How many ways are there to roll a die six times and obtain a sequence of outcomes with one "1", three "5"'s, and two "6"'s?

2. How many ways are there to arrange the letters in MISSISSIPPI?

3. (a) How many numbers between 100,000 and 1,000,000 contain only the digits 3, 5, and 7?
    (b) What fraction of the numbers in part a have two 3's, two 5's, and two 7's?

4. How many ways are there to invite one of three different friends over

for dinner on six successive nights such that no friend is invited more than three times?

Comment on the answers: (a) $\dfrac{6!}{3!2!1!} + \dfrac{6!}{2!2!2!}$, (b) $3^6 - 3\binom{6}{4}2^2$,

(c) $\left[\binom{6}{1} + \binom{6}{2} + \binom{6}{3}\right]^3$.

5. Ten different people walk into a delicatessen to buy a sandwich. Four always order tuna fish, two always order chicken, two always order roast beef, and two order any of the three types of sandwich.
   (a) How many different sequences of sandwiches are possible?
   (b) How many different (unordered) collections of sandwiches?

6. How many ways are there to pick a collection of 10 coins from piles of pennies, nickels, dimes, and quarters?

7. If three identical dice are rolled, how many different outcomes can be recorded?

8. How many ways are there to select a committee of 15 politicians chosen from a room full of indistinguishable Democrats, indistinguishable Republicans, and indistinguishable Independents if every party must have at least two members on the committee? If, in addition, no group may have a majority of the committee members?

9. How many ways are there to pick a selection of coins from $1 worth of identical pennies, $1 worth of identical nickels, and $1 worth of identical dimes if:
   (a) You select a total of 10 coins?
   (b) You select a total of 15 coins?

   Comment on the answers:

   (a) $\dbinom{15 + 3 - 1}{15}$, (b) $\displaystyle\sum_{k=0}^{10} \binom{15}{k}\binom{(15 + k) + 2 - 1}{(15 - k)}$,

   (c) $\dbinom{15 + 3 - 1}{15} - \displaystyle\sum_{k=11}^{15}\binom{(15 + k) + 2 - 1}{(15 - k)}$.

   (c) You select an equal (positive) number of each denomination of coin?

10. How many ways are there to pick 10 balls from large piles of (identical) red, white, and blue balls plus 1 pink ball, 1 lavender ball, and 1 tan ball?

11. How many numbers greater than 3,000,000 can be formed by arrangements of 1, 2, 2, 4, 6, 6, 6?

12. How many arrangements are possible with five letters chosen from MISSISSIPPI?

13. How many secret codes are there in Mastermind with:
    (a) Exactly two white pegs?

Comment on the answers (a) 5^2, (b) $\binom{4}{2} 5 \cdot 4$, (c) $5 \cdot 4 \cdot 5 \cdot 3/2$.

(b)   At most one color that occurs exactly one time (other colors appear at least twice or not at all).

14.   How many different rth-order partial derivatives does $f(x_1, x_2, \ldots, x_n)$ have?

15.   How many 8-digit sequences are there involving exactly six different digits?

16.   How many 9-digit numbers are there with twice as many different odd digits involved as different even digits (e.g., 945222123 with 9, 3, 5, 1 odd and 2, 4 even)?

17.   How many ways are there to select an unordered group of eight numbers between 1 and 25 inclusive with repetition? In what fraction of these ways is the sum of these numbers even?

18.   How many arrangements of 5 0's and 10 1's are there with no pair of consecutive 0's?

19.   How many ways are there to split a group of $2n$ α's $2n$ β's, and $2n$ γ's in half (into two groups of $3n$ letters)? *Note:* The halves are unordered, there is no first half.

20.   How many ways are there to place nine different rings on the four fingers of your right hand (excluding the thumb) if:
(a)   The order of rings on a finger does not matter?
(b)   The order of rings on a finger is considered?

21.   Show that $\Sigma P(10; k_1, k_2, k_3) = 3^{10}$, where k_1, k_2, k_3 are nonnegative integers ranging over all possible triples such that $k_1 + k_2 + k_3 = 10$.

22.   How many arrangements of six 0's, five 1's, and four 2's are there in which:
(a)   The first 0 precedes the first 1?
(b)   The first 0 precedes the first 1, which precedes the first 2? (Leave answer is the form of a summation.)

23.   How many arrangements are there of $4n$ letters, four of each of n types of letters, in which each letter is beside a similar letter?

24.   How many arrangements are there of eight α's, six β's, and seven γ's in which each α is beside (on at least one side) another α?

25.   When a coin is flipped n times, what is the probability that:
(a)   The first head comes after exactly m tails?
(b)   The ith head comes after exactly m tails?

26.   How many arrangements are there of TINKERER with two but not three consecutive vowels?

27.   How many arrangements of MISSISSIPPI are there with no pair of consecutive S's?

28. What is the probability that a random arrangement of a deck of 52 cards has exactly k instances where two consecutive cards are both Hearts?

29. What fraction of all arrangements of WISCONSIN without any pair of consecutive vowels have W adjacent to I?

30. How many arrangements are there with n 0's and m 1's, with k *runs* of 0's? (A *run* is a consecutive set (1 or more) of the same digit; e.g., 0001110100 has three (underlined) runs of 0's).

31. How many arrangements are there of seven a's, eight b's, five c's, and four d's with
    (a) No occurrence of the consecutive pair ca?
    (b) No occurrence of either ca or db?

32. How many arrangements are there of eight α's and β's in which each letter is beside (on at least one side) a similar letter? of n α's and m β's?

33. How many arrangements are there of three a's, four b's, and five c's with no pair of consecutive a's and no pair of consecutive b's? Arrangements of k_1 a's, k_2 b's, and k_3 c's?

34. (a) Use induction to give a rigorous proof of Theorem 1 (Proof 1).
    (b) Use induction to prove that:

$$\binom{n}{r_1}\binom{n-r_1}{r_2}\binom{n-r_1-r_2}{r_m} \cdots \left(\frac{n-r_1-r_2\cdots r_{m-1}}{r_m}\right) = \frac{n!}{r_1!r_2!\cdots r_m!}$$

5.4 DISTRIBUTIONS

Generally a distribution problem is equivalent to an arrangement or selection problem with repetition. Specialized distribution problems must be broken up into subcases that can be counted in terms of simple permutations and combinations (with and without repetition). A general guideline for modeling distribution problems is, *distributions of distinct objects correspond to arrangements* and *distributions of identical objects correspond to selections.*

Basic Models for Distributions

Distinct Objects   The process of distributing r distinct objects into n different boxes is equivalent to lining up the distinct objects in a row and stamping one of the n different box names on each object. Thus there are

$$\underbrace{n \cdot n \cdot n \cdots \cdots n}_{r \text{ times}} = n^r$$

distributions. If r_i objects must go in box i, there are $P(n;r_1,r_2, \ldots ,r_n)$ distributions.

Identical Objects   The process of distributing r identical objects into n different boxes is equivalent to choosing an (unordered) subset of r box names with repetition from among the n choices of boxes. Thus there are $C(r + n - 1,r) = (r + n - 1)!/r!(n - 1)!$ distributions.

Example 1

How many ways are there to assign 100 different diplomats to five different continents? How many ways if 20 diplomats must be assigned to each continent?

According to the model for distributions of distinct objects, this assignment can be done in 5^{100} ways, the number of ways to line the diplomats up in a row with one of the continent names stamped on each diplomat's attache case. The constraint that 20 diplomats go to each continent means that each continent name should appear on 20 attache cases in the row. This can be done $P(100;20,20,20,20,20) = 100!/(20!)^5$ ways.   ∎

Example 2

In bridge, the 52 cards of a standard card deck are randomly dealt 13 apiece to players North, East, South, and West. What is the probability that West has all 13 spades? That each player has one Ace?

There are $P(52;13,13,13,13)$ distributions of the 52 cards into four different 13-card hands, using the same reasoning as in the preceding example. Distributions in which West gets all the spades may be counted as the ways to distribute the remaining 39 nonspade cards among the 3 other hands—$P(39;13,13,13)$ ways (there is only 1 way to give West all 13 spades). So the probability that West has all the spades is

$$\frac{39!}{(13!)^3} \Big/ \frac{52!}{(13!)^4} = 1 \Big/ \frac{52!}{13!39!} = 1 \Big/ \binom{52}{13}$$

The simple form of this answer can be directly obtained by considering West's possible hands alone (ignoring the other three hands). A random deal gives West one of the $C(52,13)$ possible 13-card hands. Thus the unique hand of 13 spades has probability $1/C(52,13)$.

To count the ways to deal in which each player gets one Ace, we divide the distribution up into an Ace part—4! ways to arrange the 4 Aces among the four players—and a non-Ace part—$P(48;12,12,12,12)$ ways to distribute the remaining 48 non-Ace cards, 12 to each player. So the probability that each player gets an Ace is

$$\frac{4!48!}{(12!)^4} \Big/ \frac{52!}{(13!)^4} = 13^4 \Big/ \frac{52!}{4!48!} = 13^4 \Big/ \binom{52}{4} = 0.105$$

■

Example 3

How many ways are there to distribute 20 (identical) sticks of red licorice and
15 (identical) sticks of black licorice among five children?
   Using the identical objects model for distributions, we see that the ways to
distribute 20 identical sticks of red licorice among five children is equal to the
ways to select a collection of 20 names, one for each stick, from a set of 5
different names with repetition. This can be done $C(20 + 5 - 1,20) = 10{,}626$
ways. The 15 identical sticks of black licorice can be distributed in $C(15 + 5 -
1,15) = 3876$ ways, by the same modeling argument. The distributions of red
and of black licorice are disjoint procedures. So the number of ways to distribute red and black licorice is $10{,}626 \cdot 3876 = 41{,}186{,}376$. ■

Example 4

Show that the number of ways to distribute r identical balls into n distinct
boxes with at least one ball in each box is $C(r - 1, n - 1)$. With at least r_1 balls
in the first box, at least r_2 balls in the second box, . . . , and at least r_n balls in
the nth box, the number is $C(r - r_1 - r_2 - \cdots r_n + n - 1, n - 1)$.
   The requirement of at least one ball in each box can be incorporated into
the selection-with-repetition model (as in the doughnut selection problem in
the previous section). However, we will handle this constraint in terms of the
original distribution problem. We first put one ball in each box and then
"forget" that the balls are there (or think of putting a false bottom in the boxes
to conceal the ball in each box). Now it remains to count the ways to distribute
without restriction the remaining $r - n$ balls into the n boxes. By the selection-
with-repetition model, this can be done in

$$C((r - n) + n - 1,(r - n)) = \frac{(r - n + n - 1)!}{(r - n)!(n - 1)!} = C(r - 1, n - 1) \text{ ways}$$

In the case where at least r_i balls must be in the ith box, we first put r_i balls in
the ith box, and then distribute the remaining $r - r_1 - r_2 - \cdots - r_n$ in any
way into the n boxes. This can be done in

$$C((r - r_1 - r_2 - \cdots - r_n) + n - 1,(r - r_1 - r_2 - \cdots - r_n))$$

$$= C((r - r_1 - r_2 - \cdots - r_n) + n - 1, n - 1)$$

ways. ■

Example 5

How many arrangements of the letters a, e, i, o, u, x, x, x, x, x, x, x, x (eight x's) are there if no two vowels can be consecutive?

First we place a, e, i, o, u in any of the $5! = 120$ possible orders. Now we fill in the x's before, between, and after the five vowels with the constraint that at least one x is between each consecutive pair of vowels. The situation can be modeled as putting the eight x's into six boxes; before the first vowel, be-tween each consecutive pair of vowels, and after the last vowel. The following pattern depicts the situation.

$(v = \text{vowel})$

$$\underline{\qquad}\; v \;\underline{\qquad}\; v \;\underline{\qquad}\; v \;\underline{\qquad}\; v \;\underline{\qquad}\; v \;\underline{\qquad}$$
$$\text{box 1} \qquad \text{box 2} \qquad \text{box 3} \qquad \text{box 4} \qquad \text{box 5} \qquad \text{box 6}$$

Boxes 2, 3, 4, 5 must each have at least one x. So put one x in each of these boxes and then count the ways to distribute the remaining four x's into the six boxes—$C(4 + 6 - 1,4) = 126$ ways. In total, there are $120 \cdot 126 = 15{,}120$ ar-rangements. ∎

This model of non-consecutivity—of separating objects in arrangements—arises often. Remember it.

Example 6

How many integer solutions are there to the equation $x_1 + x_2 + x_3 + x_4 = 12$, with $x_i \geqslant 0$. How many solutions with $x_i > 0$? How many solutions with $x_1 \geqslant 2$, $x_2 \geqslant 2$, $x_3 \geqslant 4$, $x_4 \geqslant 0$?

By an integer solution to this equation, we mean an ordered set of integer values for the x_i's summing to 12, such as $x_1 = 2$, $x_2 = 3$, $x_3 = 3$, $x_4 = 4$. We can model this problem as a distribution-of-identical-objects problem or as a selec-tion-with-repetition problem. Let x_i represent the number of (identical) ob-jects in box i or the number of objects of type i chosen. Using either of these models, we see that the number of integer solutions is $C(12 + 4 - 1,12) = 455$.

Solutions with $x_i > 0$ correspond in these models to putting at least one object in each box or choosing at least one object of each type. Solutions with $x_1 \geqslant 2$, $x_2 \geqslant 2$, $x_3 \geqslant 4$, $x_4 \geqslant 0$ correspond to putting at least two objects in the first box, at least two in the second, at least four in the third, and any number in the fourth (or equivalently in the selection-with-repetition model). Formu-las for these two types of distribution problems were given in Example 4. The respective answers are $C(12 - 1,4 - 1) = 165$ and $C((12 - 2 - 2 - 4) + 4 - 1, 4 - 1) = 35$. ∎

Let us summarize the three equivalent forms we have seen for selection with repetition problems.

Equivalent Forms for
Selection with Repetition

1. The number of ways to select r objects with repetition from n different types of objects.

2. The number of ways to distribute r identical objects into n distinct boxes.

3. The number of nonnegative integer solutions to $x_1 + x_2 + \cdots + x_n = r$.

It is important that the reader be able to restate a problem given in one of the above settings in the other two. Most students find version 2 the most convenient way to look at such problems because a distribution is easiest to picture on paper (or in one's head). Indeed, the original argument with order forms for hot dogs used to derive our formula for selection with repetition was really a distribution model. Version 3 is the most general (abstract) form of the problem. It is the form needed in Chapter 6 to build generating functions.

Example 7

How many ways are there to distribute four identical oranges and six distinct apples (each a different variety) into five distinct boxes? In what fraction of these distributions does each box get a total of two objects?

There are $C(4 + 5 - 1,4) = 70$ ways to put four identical oranges in five distinct boxes and $5^6 = 15,625$ ways to put six distinct apples in five distinct boxes. These two processes are disjoint, and so there are $70 \cdot 15625 = 1,093,750$ ways to distribute the four identical and six distinct fruits.

The additional constraint of two objects in each box substantially complicates matters. To count constrained distributions of distinct objects, we must know exactly how many of the distinct objects should go in each box. But in this problem, the number of distinct objects that can go in a box depends on how many identical objects are in the box. So we deal with the (identical) oranges first. There are three possible categories of distributions of the four oranges (without exceeding two in any box).

Case 1

Two (identical) oranges in each of two boxes and no oranges in the other three boxes. The two boxes to get the pair of oranges can be chosen in $C(5,2) = 10$ ways, and the six (distinct) apples can then be distributed in those three other boxes (two to a box) in $P(6;2,2,2) = 90$ ways. So Case 1 has $10 \cdot 90 = 900$ possible distributions.

Case 2

Two (identical) oranges in one box, one orange in each of two other boxes, and the remaining two boxes empty. The box for two oranges can be chosen five ways, the two boxes with one orange can be chosen in $C(4,2) = 6$ ways (or combining these two steps, we can think of arranging the numbers 2, 1, 1, 0, 0, among the five boxes in $P(5;1,2,2) = 30$ ways), and the six (distinct) apples can then be distributed in $P(6;2,2,1,1) = 180$ ways, for a total of $5 \cdot 6 \cdot 180 = 5400$ ways.

Case 3

One orange in each of four boxes. This can be done in $C(5,4) = 5$ ways, and then the apples can be distributed in $P(6;2,1,1,1,1) = 360$ ways, for a total of $5 \cdot 360 = 1800$ ways.

Summing these cases, there are $900 + 5400 + 1800 + 8100$ distributions with two objects in each box. The fraction of such distributions among all ways to distribute the four oranges and six apples is thus $8100/1093750 = .0074$—a lot smaller than one might guess. (If all 10 fruits were distinct, the fraction with two in each box would be about 50 percent larger. Why?) ∎

Example 8

A warlock goes to a store to buy one dollar's worth of ingredients for his wife's Witch's Brew. The store sells bat tails for 5¢ apiece, lizard claws for 5¢ apiece, newt eyes for 5¢ apiece, and calf blood for 20¢ a pint bottle. How many different purchases (subsets) of ingredients will this dollar buy?

The first step is to make our unit of money 5¢ (a nickel). So the warlock has 20 units to spend with blood costing four units and the other three items 1 unit. An integer-solutions-of-equation model of this problem is

$$T + S + E + 4B = 20, \qquad T,S,E,B \geq 0$$

The simplest way to handle the fact that B has a coefficient of 4 is first to specify exactly how much blood is bought. If one pint is bought, $B = 1$, then we have $T + S + E = 16$, an equation with $C(16 + 3 - 1,16) = 153$ nonnegative integer solutions. In general, if $B = i$, then we have $T + S + E = 20 - 4i$, an equation with $C((20 - 4i) + 3 - 1,20 - 4i) = C(22 - 4i,2)$ solutions, for $i - 0, 1, 2, 3, 4, 5$. Summing these possibilities, we obtain

$$\binom{22}{2} + \binom{18}{2} + \binom{14}{2} + \binom{10}{2} + \binom{6}{2} + \binom{2}{2} = 536 \text{ different purchases}$$

∎

EXERCISES

SUMMARY OF EXERCISES   Be careful to distinguish whether a problem involves distinct or identical objects. Exercises 18–21 involve problem restatement (no actual numerical answers are sought). Related programming problems are given in Section 5.6.

1. How many ways are there to distribute 40 identical jelly beans among four children:
   (a) Without restrictions?
   (b) With each child getting 10 beans?
   (c) With each child getting at least 1 bean?

2. How many ways are there to distribute 20 different toys among five children?
   (a) Without restrictions?
   (b) If two children get 7 toys and three children get 2 toys?
   (c) With each child getting 4 toys?

3. In a bridge deal, what is the probability that:
   (a) West has four spades, three hearts, three diamonds, and three clubs?
   (b) North and South have five spades, West has two spades, and East has one spade?
   (c) One player has all the Aces?
   (d) All players have a (4,3,3,3) division of suits?

4. How many ways are there to distribute five (identical) apples, six oranges, six pears, and four pineapples among three people:
   (a) Without restriction?
   (b) With each person getting at least one pear?

5. How many ways are there to distribute 18 chocolate doughnuts, 12 cinnamon doughnuts, and 14 powdered sugar doughnuts among four school principals if each principal demands at least 2 doughnuts of each kind?

6. How many ways are there to arrange two a's, two b's, and eight c's so that there is a c on both sides of each a and each b?

7. How many ways are there to arrange the letters in VISITING with no pair of consecutive I's?

8. How many ways are there to arrange the 26 letters of the alphabet so that no pair of vowels appear consecutively?

9. How many integer solutions are there to $x_1 + x_2 + x_3 + x_4 + x_5 = 28$ with:
   (a) $x_i \geq 0$ (b) $x_i > 0$ (c) $x_i > i$ $(i = 1,2,3,4,5)$

10. How many integer solutions are there to $x_1 + x_2 + x_3 = 0$ with $x_i \geq -5$?

11. How many positive integer solutions are there to $x_1 + x_2 + x_3 + x_4 < 100$?

12. How many ways are there to distribute k balls into n distinct boxes $(k < n)$ with at most one ball in any box if:
     (a) The balls are distinct?
     (b) The balls are identical?

13. How many ways are there to distribute three different teddy bears and nine identical lollipops to four children:
     (a) Without restriction?
     (b) With no child getting two teddy bears?
     (c) With each child getting three "goodies"?

14. If n distinct objects are distributed randomly into n distinct boxes, what is the probability that:
     (a) No box is empty?
     (b) Exactly one box is empty?
     (c) Exactly two boxes are empty?

15. How many ways are there to distribute eight balls into six boxes with the first two boxes collectively having at most four balls if:
     (a) The balls are identical?
     (b) The balls are distinct?

16. How many ways are there to distribute 15 identical objects into four different boxes if the number of objects in box 4 must be a multiple of 3?

17. (a) How many ways are there to make 35 cents change in 1952 pennies, 1959 pennies, and 1964 nickels?
     (b) In 1952 pennies, 1959 pennies, 1964 nickels, and 1971 quarters?

18. State an equivalent distribution version of each of the following arrangement problems:
     (a) Arrangements of 8 letters chosen from piles of a's, b's, and c's.
     (b) Arrangements of 2 a's, 3 b's, 4 c's.
     (c) Arrangements of 10 letters chosen from piles of a's, b's, c's, and d's with the same number of a's and b's.
     (d) Arrangements of 4 letters chosen from 2 a's, 3 b's, and 4 c's.

19. State an equivalent arrangement version of each of the following distribution problems:
     (a) Distributions of n distinct objects into n distinct boxes.
     (b) Distributions of 15 distinct objects into five distinct boxes with 3 objects in each box.
     (c) Distributions in 15 distinct objects into five boxes with i objects in the ith box, $i = 1, 2, 3, 4, 5$.
     (d) Distributions of 12 distinct objects into three distinct boxes with at most 3 objects in box 1 and at most 5 objects in box 2.

20. State an equivalent distribution version and an equivalent integer-solution-of-an-equation version of the following selection problems.

    (a)   Selections of six ice cream cones from 31 flavors.

    (b)   Selections of five marbles from a group of five reds, four blues, and two pinks.

    (c)   Selections of 12 apples from four types of apples with at least 2 apples of each type.

    (d)   Selections of 20 jelly beans from four different types with an even number of each type and not more than 8 of any one type.

21. State an equivalent selection version and an equivalent integer-solution-of-an-equation version of the following distribution problems:

    (a)   Distributions of 30 black chips into five distinct boxes.

    (b)   Distributions of 18 red balls into six distinct boxes with at least 2 balls in each box.

    (c)   Distributions of 20 markers into four distinct boxes with the same number of markers in the first and second boxes.

22. How many election outcomes are possible (numbers of votes for different candidates) if there are three candidates and 30 voters? If, in addition, some candidate receives a majority of the votes?

23. How many election outcomes in the race for class president are there if there are five candidates and 40 students in the class and

    (a)   Every candidate receives at least two votes?

    (b)   One candidate receives at most one vote (and the others at least two votes)?

    (c)   No candidate receives a majority of the votes?

    (d)   Exactly three candidates tie for the most votes?

24. How many numbers between 0 and 10,000 have a sum of digits:

    (a) Equal to 7?   (b) Less than or equal to 7?   (c) Equal to 13?

25. How many ways are there to split four red, five blue, and seven black balls among:

    (a)   Two boxes without restriction?

    (b)   Four boxes with no box empty?

    (c)   Two boxes with eight balls in each box?

26.   (a)   How many arrangements are there of REVISITED with vowels not in increasing order—that is, an I before one (or both) of the E's.

    (b)   How many arrangements with no consecutive E's and no consecutive I's?

    (c)   How many arrangements with vowels not in increasing order and no consecutive E's and no consecutive I's?

27. If you flip a coin 20 times and get 14 heads and 6 tails, what is the probability that there is no pair of consecutive tails?

28. How many bridge deals are there in which North and South get all the spades?

29. What is the probability in a bridge deal that each player gets at least two honors (an honor is an Ace, or King, or Queen, or Jack)?

30. How many ways are there to place 14 different books into three identical parcels of 4 books each (with 2 books left unwrapped)?

31. How many ways can a deck of 52 cards be broken up into a collection of unordered piles of sizes:
(a) Four piles of 13 cards?
(b) Three piles of 8 cards and 4 piles of 7 cards?

32. How many ways are there for 10 people to have five simultaneous telephone conversations?

33. How many ways are there to distribute 15 oranges into five different boxes with at most 8 oranges in a box?

34. How many ways are there to distribute r identical balls into n distinct boxes with exactly m boxes empty?

35. How many ways are there to distribute 20 distinct flags onto 12 distinct flagpoles if:
(a) In arranging flags on a flagpole, the order of flags from the ground up makes a difference.
(b) No flagpole is empty and the order on each flagpole is counted?

36. How many ways are there to distribute r identical balls into n distinct boxes with the first m boxes collectively containing:
(a) At least s balls?   (b) At most s balls?

37. How many integer solutions are there to the equation $x_1 + x_2 + x_3 + x_4 \leqslant 15$ with $x_i \geqslant -10$?

38. How many nonnegative integer solutions are there to the equation $2x_1 + 2x_2 + x_3 + x_4 = 12$?

39. Show that the number of k subsets of the integers $1, 2, \ldots, 30$ with no pair of consecutive integers is equal to the number of integer solutions of $x_1 + x_2 + \cdots + x_{k+1} = 30$ with $x_1 \geqslant 1$, $x_{k+1} \geqslant 0$, and $x_i \geqslant 2$ for $2 \leqslant i \leqslant k$.

40. How many nonnegative integer solutions are there to the pair of equations $x_1 + x_2 + \cdots + x_6 = 20$ and $x_1 + x_2 + x_3 = 7$?

41. How many nonnegative integer solutions are there to the inequalities $x_1 + x_2 + \cdots + x_6 \leqslant 20$ and $x_1 + x_2 + x_3 \leqslant 7$?

42. How many positive integer solutions are there to $x_1 + x_2 + \cdots + x_5 = 20$:
(a) With $x_i \leqslant 10$?   (b) With $x_i \leqslant 8$?   (c) With $x_1 = 2x_2$?

43. How many ways are there to deal four cards to each of 13 different players so that exactly 10 players have a card of each suit?

44. How many distributions of 24 different objects into three different boxes are there with twice as many objects in one box as in the other two combined?

45. How many ways are there to distribute 20 toys to m children such that the first two children get the same number of toys if:

(a)   The toys are identical?   (b)   The toys are distinct?

46.   If 14 distinct objects are randomly distributed among four distinct boxes, what is the probability that exactly one box has exactly 4 objects in it?

47.   How many arrangements of 5 α's, 6 β's and 6 γ's are there with at least one β and at least one γ between each successive pair of α's?

48.   Among the set of all arrangements of 4 α's, 8 β's, and 10 γ's in which each α is beside a β and a γ, what is the probability that no β is beside a γ?

49.   How many ways are there to partition 18 different objects into an unordered set of four piles with a different (positive) number of objects in each pile?

50.   How many ways are there to distribute 25 different presents to four people (including the boss) at an office party so that the boss receives exactly twice as many presents as the second most popular person?

51.   How many ways are there to distribute six white, five red, and seven blue balls into six different boxes with exactly three objects in each box?

52.   How many ways are there to partition n distinct objects into an unordered collection of m piles with k_i piles of i objects?

5.5   BINOMIAL COEFFICIENTS

In this section we show why the numbers $C(n,r)$ are called binomial coefficients. Then we present some basic identities involving binomial coefficients. We use three techniques to verify the identities: combinatorial selection models, "block walking," and the binomial expansion. The study of binomial identities is itself a major subfield of combinatorial mathematics and we will necessarily just scratch the surface of this topic. The interested reader is referred to Riordan [5] for a more thorough introduction to combinatorial identities.

Consider the polynomial expression $(a + x)^3$. Instead of multiplying $(a + x)$ by $(a + x)$ and the product by $(a + x)$ again, let us formally multiply the three factors term by term:

$$(a + x)(a + x)(a + x) = aaa + aax + axa + axx + xaa + xax + xxa + xxx$$

Collecting similar terms, we reduce the right-hand side of this expansion to

$$a^3 + 3a^2x + 3ax^2 + x^3$$

The formal expansion of $(a + x)^3$ was obtained by systematically forming all products of a term in the first factor, a or x, times a term in the second factor times a term in the third factor. There are two choices for each term in such a product, and so there are $2 \cdot 2 \cdot 2 = 8$ formal products (8 products were obtained above). If we were expanding $(a + x)^{10}$, we would obtain $2^{10} = 1024$ different formal products.

Now we ask the question, how many of the formal products in the expansion of $(a + x)^3$ contain k x's and $3 - k$ a's? This question is equivalent to asking for the coefficient of $a^{3-k}x^k$ in the reduced expansion. Since all possible formal products of a's and x's are formed, and since formal products are just three-letter sequences of a's and x's, we are simply asking for the number of all three-letter sequences with k x's and $3 - k$ a's. The answer is thus $C(3,k)$ and so the reduced expansion for $(a + x)^3$ can be written as

$$\binom{3}{0} a^3 + \binom{3}{1}a^2x + \binom{3}{2}ax^2 + \binom{3}{3} x^3$$

By the same argument, we see that the coefficient of $a^{n-k}x^k$ in $(a + x)^n$ will be equal to the number of n-letter sequences formed by k x's and $n - k$ a's, that is, $C(n,k)$. If we set $a = 1$, we have the following theorem.

Binomial Theorem

$$(1 + x)^n = \binom{n}{0} + \binom{n}{1} x + \binom{n}{2} x^2 + \cdots + \binom{n}{k} x^k + \cdots + \binom{n}{n} x^n$$

Another proof of this expansion, using induction, is left as an exercise. Just as important as the Binomial Theorem is the equivalence we have established between the number of k-subsets of n objects and the coefficient of x^k in $(1 + x)^n$. We exploit this equivalence extensively in the next chapter.

Let us now consider some basic properties of binomial coefficients. The most important identity for binomial coefficients is the symmetry identity

$$\binom{n}{k} = \frac{n!}{k!(n - k)!} = \binom{n}{n - k} \tag{1}$$

In words, this identity says that the number of ways to select a subset of k objects out of a set of n objects is equal to the number of ways to select a group of $n - k$ of the objects to set aside (not to be in the subset).

The other fundamental identity is

$$\binom{n}{k} = \binom{n - 1}{k} + \binom{n - 1}{k - 1} \tag{2}$$

This identity can be verified algebraically. We will give a combinatorial selection argument instead. We classify the $C(n,k)$ committees of k people chosen from a set of n people into two categories, depending on whether or not the committee contains a given person p. There are $C(n - 1, k)$ ways to form a committee from the other $n - 1$ people, not including p. On the other hand, if p is in the subset, there are only $k - 1$ remaining members of the committee to be chosen from the other $n - 1$ people. This can be done $C(n - 1, k - 1)$ ways. Thus $C(n, k) = C(n - 1, k) + C(n - 1, k - 1)$.

The following example presents another binomial identity that can be verified algebraically or by a combinatorial argument.

Example

Show that

$$\binom{n}{k}\binom{k}{m} = \binom{n}{m}\binom{n - m}{k - m} \tag{3}$$

The left-hand side of (3) counts the ways to select a group of k people chosen from a set of n people and then to select a subset of m leaders of this group. Equivalently, as counted on the right side, we could first select the subset of m leaders from the set of n people and then select the remaining $k - m$ members of the group from the remaining $n - m$ people. Note the special form of (3) when $m = 1$:

$$k\binom{n}{k} = n\binom{n - 1}{k - 1}, \quad \text{or} \quad \binom{n}{k} = \frac{n}{k}\binom{n - 1}{k - 1} \tag{4}$$

∎

Using (2) and the fact that $C(n, 0) = C(n, n) = 1$ for all nonnegative n, we can recursively build successive rows in the following table of binomial coefficients, called **Pascal's triangle**

Table of binomial coefficients: kth number in row n is $\binom{n}{k}$

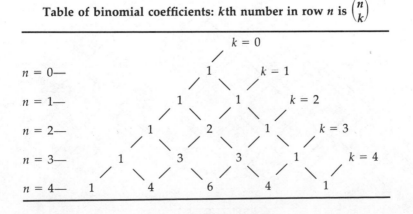

Each number in this table, except the first and last numbers in a row, is the sum of the two neighboring numbers in the preceding row.

Pascal's triangle has the following nice combinatorial interpretation. Consider the ways a person can traverse the blocks in the network of streets shown in Figure 5.1. The person begins at the top of the network, at the spot marked (0,0), and moves down the network (down the page) making a choice at each intersection to go right or left (for simplicity, let "right" be your right as you look at this page, not the right of the person moving down the network). We label each street corner in the network with a pair of numbers (n,k), where n indicates the distance from (0,0) (the number of blocks traversed) and k the number of times the person choose the right branch at intersections. Figure 5.1 shows a possible route from the start (0,0) to the corner (6,3).

Any route to corner (n,k) can be written as a list of the branches (left or right) chosen at the successive corners on the path from (0,0) to (n,k). Such a list is just a sequence of k R's (right branches) and $n - k$ L's (left branches).

Let $s(n,k)$ be the number of possible routes from the start (0,0) to corner (n,k) (moving downwards in the network), that is, the number of sequences of k R's and $n - k$ L's. Thus $s(n,k) = C(n,k)$. This block-walking correspondence with binomial coefficients is due to George Polya. Another useful correspondence is the committee selection model, used to verify (2) and (3).

Let us show how our "block-walking" model for binomial coefficients can be used to get an alternate proof of identity (2). At the end of a route from the start to corner (n,k), a block walker arrives at (n,k) from either corner $(n - 1,k)$ or corner $(n - 1, k - 1)$. For example, to get to corner (6,3) in Figure

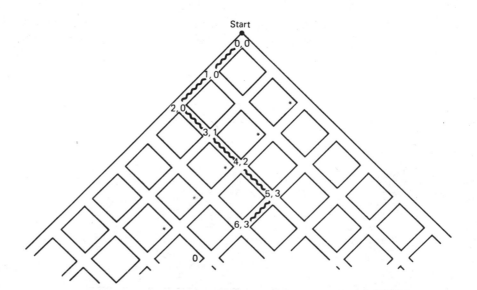

FIGURE 5.1

2.1, the person either goes to corner (5,3) and then branches left to (6,3), or goes to corner (5,2) and then branches right to (6,3). Then $s(n,k) = s(n-1,k) + s(n-1,k-1)$.

We now list seven well-known binomial identities and verify three of them (the others are exercises for the reader). These binomial identities are of much practical interest. Expressions involving sums and products of binomial coefficients arise frequently in complicated counting problems. These identities can be used to simplify such expressions.

$$\binom{n}{0} + \binom{n}{1} + \binom{n}{2} + \cdots + \binom{n}{n} = 2^n \tag{5}$$

$$\binom{n}{0} + \binom{n+1}{1} + \binom{n+2}{2} + \cdots + \binom{n+r}{r} = \binom{n+r+1}{r} \tag{6}$$

$$\binom{r}{r} + \binom{r+1}{r} + \binom{r+2}{r} + \cdots + \binom{n}{r} = \binom{n+1}{r+1} \tag{7}$$

$$\binom{n}{0}^2 + \binom{n}{1}^2 + \binom{n}{2}^2 + \cdots + \binom{n}{n}^2 = \binom{2n}{n} \tag{8}$$

$$\sum_{k=0}^{r} \binom{m}{k}\binom{n}{r-k} = \binom{m+n}{r} \tag{9}$$

$$\sum_{k=0}^{m} \binom{m}{k}\binom{n}{r+k} = \binom{m+n}{m+r} \tag{10}$$

$$\sum_{k=n}^{m} \binom{m-k}{r}\binom{n+k}{s} = \binom{m+n+1}{r+s+1} \tag{11}$$

Here $C(n,r) = 0$ if $0 \le n < r$. These identities can be explained by the type of combinatorial arguments used for (1), (2), and (3). We give such an argument for (5). Consider the following two ways of counting all subsets (of any size) of a set of n people: (a) summing the number of subsets of size 0, of size 1, of size 2, and so on—this yields the left-hand side of (5); and (b) counting all subsets by whether or not the first person is in the subset, whether or not the second person is in the subset, and so on—this yields $2 \cdot 2 \cdots$ (n times) $= 2^n$, the right-hand side of (5).

In general, a combinatorial argument for proving such identities consists of specifying a collection counted by the term on the right side, partitioning the collection into classes, and showing that the terms in the summation count the classes.

The advantage of proofs involving block walking is that we can draw pictures of the "proof." The picture shows that all the routes to a certain corner, this amount is the right-hand side of the identity, can be decomposed in terms of all routes to certain intermediate corners (or blocks), the sum on the left-hand side of the identity. (For some, the committee selection model also offers a helpful "picture," a mental picture.)

Example 2

Verify identity (7) by block-walking and committee selection arguments.

As an example of this identity, we consider the case where $r = 2$ and $n = 6$. The corners $(k,2)$, $k = 2, 3, 4, 5, 6$, are marked with a $*$ in Figure 2.1 and corner $(7,3)$ is marked with an o. Observe that the right branches at each starred corner are the set of last possible right branches on routes from the start $(0,0)$ to corner $(7,3)$, after traversing one of these right branches, there is just one way to continue on to corner $(7,3)$, by making all remaining branches left branches. In general, if we break all routes from $(0,0)$ to $(n + 1, r + 1)$ into subcases based on the corner where the last right branch is taken, we obtain identity (7).

We restate the block-walking model as a committee selection: if the kth turn is right this corresponds to selecting the kth person to be on the committee; if the kth turn is left, the kth person is not chosen. We break the ways to pick $r + 1$ members of a committee from $n + 1$ people into cases depending on who is the last person chosen: the $(r + 1)$st, the $(r + 2)$nd, $\ldots$, the $(n + 1)$st. If the $(r + k + 1)$st person is the last chosen, then there are $C(r + k, r)$ ways to pick the first r members of the committee. Identity (7) now follows. ∎

Example 3

Verify identity (8) by a block-walking argument.

The number of routes from (n,k) to $(2n,n)$ is equal to the number of routes from $(0,0)$ to $(n, n - k)$, since both trips go a total of n blocks with $n - k$ to the right (and k to the left). So the number of ways to go from $(0,0)$ to (n,k) and then on to $(2n,n)$ is $C(n,k) \cdot C(n, n - k)$. By (1), $C(n, n - k) = C(n,k)$, and thus the number of routes from $(0,0)$ to $(2n,n)$ via (n,k) is $C(n,k)^2$. Summing over all k—that is, all intermediate corners n blocks from the start—we count all $C(2n,n)$ routes from $(0,0)$ to $(2n,n)$. Identity (8) follows. ∎

Now we show how binomial identities can be used to evaluate sums whose terms are closely related to binomial coefficients.

Example 4

Evaluate the sum $1 \cdot 2 \cdot 3 + 2 \cdot 3 \cdot 4 + \cdots + (n - 2)(n - 1)n$.

The general term in this sum $(k - 2)(k - 1)k$ is equal to $P(k,3) = k!/(k - 3)!$. Recall that the number of r-permutations and of r-selections differ by a factor of $r!$. That is, $C(k,3) = k!/(k - 3)!3! = P(k,3)/3!$, or $P(k,3) = 3! \, C(k,3)$. So the given sum can be rewritten as

$$3! \binom{3}{3} + 3! \binom{4}{3} + \cdots + 3! \binom{n}{3} = 3! \left(\binom{3}{3} + \binom{4}{3} + \cdots + \binom{n}{3} \right)$$

By identity (7), this sum equals $3! \binom{n + 1}{4}$. ∎

Example 5

Evaluate the sum $1^2 + 2^2 + 3^2 + \cdots + n^2$.

A strategy for problems whose general term is not a multiple of $C(n,k)$ or $P(n,k)$ is to decompose the term algebraically into a sum of $P(n,k)$-type terms. In this case, the general term k^2 can be written as $k^2 = k(k - 1) + k$. So the given sum can be rewritten as

$$(1 \cdot 0 + 1) + (2 \cdot 1 + 2) + (3 \cdot 2 + 3) + \cdots + (n(n - 1) + n)$$

$$= (2 \cdot 1 + 3 \cdot 2 + \cdots + n(n - 1)) + (1 + 2 + 3 + \cdots + n)$$

$$= \left(2 \binom{2}{2} + 2 \binom{3}{2} + \cdots + 2 \binom{n}{2}\right) + \left(\binom{1}{1} + \binom{2}{1} + \cdots + \binom{n}{1}\right)$$

$$= 2 \binom{n + 1}{3} + \binom{n + 1}{2} \qquad \text{by identity (7)}.$$

Note that as part of Example 5, we showed that $1 + 2 + 3 + \cdots + n = C(n + 1,2) = \frac{1}{2}n(n + 1)$, a result we verify by induction in Example 1 of Appendix A.2. ∎

There is another simple way to verify identities with binomial coefficients. We start with the binomial expansion in the Binomial Theorem. Then we substitute appropriate values for x. The following identities can be obtained from the binomial expansion: (5) by setting $x = 1$, (12) by setting $x = -1$, and (13) by differentiating both sides of the binomial expansion and setting $x = 1$.

$$(1 + 1)^n = 2^n = \binom{n}{0} + \binom{n}{1} + \binom{n}{2} + \cdots + \binom{n}{n} \qquad (5)$$

$$(1 - 1)^n = 0 = \binom{n}{0} - \binom{n}{1} + \binom{n}{2} - \cdots - (-1)^n\binom{n}{n}, \qquad (12)$$

or

$$\binom{n}{0} + \binom{n}{2} + \cdots = \binom{n}{1} + \binom{n}{3} + \cdots = 2^{n-1} \qquad (12')$$

(since the sum of both sides is 2^n)

$$n(1 + x)^{n-1} = 1 \binom{n}{1} + 2 \binom{n}{2} x + 3 \binom{n}{3} x^2 + \cdots + n \binom{n}{n} x^{n-1} \qquad (13)$$

and so

$$n(1 + 1)^{n-1} = n2^{n-1} = \binom{n}{1} + 2 \binom{n}{2} + 3 \binom{n}{3} + \cdots + n \binom{n}{n} \qquad (13')$$

We close this section by extending our definition of binomial coefficients $C(n,k)$ to values of n that are not just integers $\geq k$. For any real number z and any positive integer k we define

$$\binom{z}{k} = \frac{z(z-1)(z-2)\cdots(z-(k-1))}{k!} \quad \text{and} \quad \binom{z}{0} = 1$$

Two basic identities involving generalized binomial coefficients are

$$\binom{-z}{k} = (-1)^k \binom{z+k-1}{k}, \qquad z \geq 0 \tag{14}$$

$$\binom{z}{0} - \binom{z}{1} + \binom{z}{2} - \cdots + (-1)^r \binom{z}{r} = (-1)^r \binom{z-1}{r} \tag{15}$$

Identities (2), (3), (6), (9), and (10) are true with n replaced by z. For z, these identities must be verified algebraically or by induction. The binomial expansion for an arbitrary power z is also valid when generalized binomial coefficients are used (the proof of this fact requires advanced calculus; see Buck [1]). However, when z is not a positive integer, the binomial expansion becomes an infinite series. For example,

$$(1 + x)^{-1/2} = \binom{-1/2}{0} + \binom{-1/2}{1} x + \binom{-1/2}{2} x^2 + \cdots$$

$$= 1 + \frac{-1/2}{1} x + \frac{(-1/2)(-3/2)}{2!} x^2 + \cdots$$

EXERCISES

SUMMARY OF EXERCISES   Combinatorial identities are a well-developed field whose surface was barely scratched in this section. The later problems in this exercise set go well beyond the level of examples worked in this section. Interested students should consult Riordan [5] for assistance with the harder problems.

1. (a) Verify identity (2) algebraically (writing out the binomial coefficients in factorials).
   (b) Verify identity (3) algebraically.
2. Verify the following identities by block walking.
   (a) (6)   (b) (9)   (c) (10)   (d) (11)
3. Verify the following identities by a committee selection model.
   (a) (6)   (e) (10)
   (b) (7)   (f) (11)
   (c) (8)   (g) (13)
   (d) (9)

4. Verify the following identities by mathematical induction [*Hint:* Use (2)].
   (a) (2) (b) (5) (c) (6) (d) (7) (e) (12′)

5. Prove the binomial theorem by mathematical induction.

6. Verify algebraically the identities for generalized binomial coefficients:
   (a) (14) (b) (2) (c) (3)

7. Show that identity (7) can be obtained as a special case of (11).

8. Use mathematical induction (on r) to verify these identities for generalized binomial coefficients.
   (a) (6) (b) (9) (c) (10) (d) (15)

9. Show that $C(2n,n) + C(2n,n - 1) = \frac{1}{2}C(2n + 2,n + 1)$.

10. If $C(n,3) + C(n + 3 - 1,3) = P(n,3)$, find n.

11. Show by a combinatorial argument that

   (a) $\binom{2n}{2} = 2\binom{n}{2} + n^2$

   (b) $(n - r)\binom{n + r - 1}{r}\binom{n}{r} = n\binom{n + r - 1}{2r}\binom{2r}{r}$

12. Show that $C(k + m + n,k)C(m + n,m) = (k + m + n)!/(k!m!n!)$.

13. (a) Show that $\binom{n}{1} + 6\binom{n}{2} + 6\binom{n}{3} = n^3$.

   (b) Evaluate $1^3 + 2^3 + 3^3 + \cdots + n^3$.

14. (a) Evaluate $\sum_{k=1}^{n} 12(k + 1)k(k - 1)$. (*Hint:* Use Example 4.)

   (b) Evaluate $\sum_{k=0}^{n} (2 + 3k)^2$.

   (c) Evaluate $\sum_{k=0}^{n} k(n - k)$.

15. (a) Evaluate the sum

$$1 + 2\binom{n}{1} + \cdots + (k + 1)\binom{n}{k} + \cdots + (n + 1)\binom{n}{n}$$

   by breaking this sum into two sums, each of which is an identity in this section.

   (b) Evaluate the sum

$$\binom{n}{0} + 2\binom{n}{1} + \binom{n}{2} + 2\binom{n}{3} + \cdots$$

16. By setting x equal to the appropriate values in the binomial expansion (or one of its derivatives, etc.) evaluate:

(a) $\displaystyle\sum_{k=1}^{n} (-1)^k k \binom{n}{k}$

(e) $\displaystyle\sum_{k=1}^{n} (-1)^k k^2 \binom{n}{k}$

(b) $\displaystyle\sum_{k=2}^{n} k(k-1) \binom{n}{k}$

(f) $\displaystyle\sum_{k=0}^{n} \frac{1}{k+1} \binom{n}{k}$

(c) $\displaystyle\sum_{k=0}^{n} 2^k \binom{n}{k}$

(g) $\displaystyle\sum_{k=0}^{n} (2k+1) \binom{n}{k}$

(d) $\displaystyle\sum_{k=1}^{n} k3^k \binom{n}{k}$

17. Show that $\displaystyle\sum_{k=m}^{n} \binom{k}{r} = \binom{n+1}{r+1} - \binom{m}{r+1}$.

18. Show that $\displaystyle\sum_{k=1}^{n} \binom{m+k-1}{k} = \sum_{k=1}^{m} \binom{n+k-1}{k}$

19. Show that $\displaystyle\sum_{k=0}^{n-1} P(m+k, m) = \frac{P(m+n, m+1)}{(m+1)}$

20. Consider a sequence of $2n$ distinct people in a line at a cashier. Suppose n of the people owe \$1 and n of the people are due a \$1 payment. Show that the number of arrangements in which the cashier never goes in debt (i.e., at every stage at least as many people have paid in \$1 as were paid out \$1) is equal to $\binom{2n}{n} - \binom{2n}{n+1}$. *Hint:* Use a symmetry argument in a blockwalking model to demonstrate a one-to-one correspondence between sequences where at some stage(s) the cashier goes (at least) \$1 in debt and all sequences of $2n$ people in which $n+1$ of the people are owed \$1.

21. Show that $\left(\binom{n}{0} + \binom{n}{1} + \cdots + \binom{n}{n} \right)^2 = \displaystyle\sum_{k=0}^{2n} \binom{2n}{k}$

22. Find the value of k that maximizes

(a) $\dbinom{n}{k}$ (b) $\dbinom{2n+k}{n}\dbinom{2n-k}{n}$

23. Evaluate $\displaystyle\sum_{k=0}^{n-1} \binom{n}{k}\binom{n}{k+1}$.

24. Prove that $\binom{n}{1} + 3\binom{n}{3} + 5\binom{n}{5} + \cdots = 2\binom{n}{2} + 4\binom{n}{4} + \cdots = n2^{n-2}$.

25. Evaluate $\binom{n}{0} - 2\binom{n}{1} + 3\binom{n}{2} + \cdots + (-1)^n(n+1)\binom{n}{n}$.

26. Show that $\displaystyle\sum_{k=0}^{n-1} (-1)^k \binom{n}{k+1} = 1$. (*Hint:* Rewrite as a binomial coefficient with k in the bottom.)

27. Show that $\displaystyle\sum_{k=0}^{n} \frac{(2n)!}{k!^2(n-k)!^2} = \binom{2n}{n}^2$.

28. Give a combinatorial argument to evaluate $\displaystyle\sum_{k=0}^{n} \binom{n}{k}\binom{m}{k}$, $n \le m$.

29. (a) For any given k, show that an integer n can be represented as:

$$n = \binom{m_1}{1} + \binom{m_2}{2} + \cdots + \binom{m_k}{k}, \quad \text{where} \quad 0 \le m_1 < m_2 < \cdots < m_k$$

   (b) Prove that for a given k, the representation in part a is unique.
   Hint: Use the fact that

$$\binom{n}{k} - 1 = \binom{n-1}{k} + \binom{n-2}{k-1} + \cdots + \binom{n-k}{1}$$

30. Give the first four terms in the generalized binomial expansions of
   (a) $(1+x)^{1/3}$ (b) $(1+x)^{-m}$ (c) $(1-4x)^{1/2}$.

31. Show that $(-1)^n \binom{-n}{k-1} = (-1)^k \binom{-k}{n-1}$.

32. Show that $(-1)^n \binom{-\frac{1}{2}}{n} 2^{2n} = \binom{2n}{n}$.

33. Show that $-\binom{2}{n}\binom{2n-2}{n-1} = (-1)^n \binom{\frac{1}{2}}{n} 2^{2n}$.

34. Show that $\displaystyle\sum_{j=0}^{k} \binom{n+k-j-1}{k-j}\binom{m+j-1}{j} = \binom{n+m+k-1}{k}$.

35. Show that $\displaystyle\sum_{k=0}^{m} \frac{P(m,k)}{P(n,k)} = \frac{1}{\binom{n}{m}}\sum_{k=0}^{m}\binom{n-k}{n-m} = \frac{n+1}{n-m+1}$, $m \le n$.

36. Show that $\displaystyle\sum_{k=0}^{m} \frac{m!(n-k)!}{n!(m-k)!} = \frac{n+1}{n-m+1}$, $m \le n$.

37. If $C_{2n} = \dfrac{1}{n+1}\binom{2n}{n}$, show that $C_{2m} = \displaystyle\sum_{k=1}^{m} C_{2k-2}C_{2m-2k}$.

38. Show that $\displaystyle\sum_{k=0}^{m}\binom{n}{k}\binom{n-k}{m-k} = 2^m\binom{n}{m}$, $m < n$.

39. Consider the problem of three-dimensional block-walking. Show by a combinatorial argument that $P(n;i-1,j,k) + P(n;i,j-1,k) + P(n;i,j,k-1) = P(n+1;i,j,k)$.

40. Give a combinatorial argument to show that

$$(x_1 + x_2 + \cdots + x_k)^n = \Sigma P(n;i_1,i_2,\ldots,i_k)x_1^{i_1}x_2^{i_2}\cdots x_k^{i_k}$$

where the sum is over all $i_1 + i_2 + \cdots + i_k = n$, $i_j \ge 0$.

41.   Show for sums over all $i_1 + i_2 + \cdots + i_k = n$, $i_j \geqslant 0$, that
   (a) $\Sigma P(n;i_1,i_2, \ldots , i_k) = k^n$
   (b) $\Sigma i_1 i_2 \cdots i_k P(n;i_1,i_2, \ldots ,i_k) = P(n,k)k^{n-k}$

5.6   GENERATING PERMUTATIONS AND COMBINATIONS AND PROGRAMMING PROJECTS

Having spent a whole chapter counting different types of arrangements and selections, it is only fair to say a little about actually enumerating, or listing, these permutations and combinations. In this section, we present algorithms for enumerating all permutations of an n-set, all combinations (of any size) of an n-set, and all r-combinations of an n-set. With these algorithms, one can write programs to enumerate all the elements in various combinatorial sets; we do this in the second half of the section.

The following examples illustrate how such algorithms are used in operations research. Suppose that we have eight different products to manufacture in a month at a factory. Given the cost of converting from product i to product j, we must find the arrangement of production that minimizes the sum of the conversion costs. The easiest solution to this problem is simply to generate all arrangements and see which one costs least (a "branch and bound" approach to this type of problem is given in Section 3.3).

Suppose that each of 10 workers knows a different subset of 30 skills needed on a project, and we must find all groups of 5 workers who collectively know all 30 skills. We need to check all five-subsets of the 10 workers to find the ones that satisfy this special condition (this also can be solved by "branch and bound" methods).

There are several different algorithms for enumerating permutations and combinations. We will present algorithms that enumerate these collections in **lexicographic order**, that is, we treat an arrangement of objects as a sequence of "letters" forming a "word" and list these "words" in alphabetical order. This approach requires that the objects be numbered (or otherwise ordered). We write $A < B$, if sequence A is lexicographically smaller than B. A subset is treated similarly, with the "letters" (objects) of each subset arranged in increasing alphabetical order. For example, Table 5.1 lists all permutations of a, b, c in lexicographic order, and Table 5.2 lists all three-combinations of 1, 2, 3, 4, 5 in lexicographic order.

To list all permutations of an n-set in lexicographic order, we need to be able to determine for any given permutation $A = a_1 a_2 a_3 \cdots a_n$, what is the next permutation $B = b_1 b_2 b_3 \cdots b_n$ in lexicographic order. If $a_{n-1} < a_n$, then interchanging the positions of a_{n-1} and a_n in A yields a lexicographically larger sequence, the desired sequence B. For example, 14325 is followed by 14352.

If $a_{n-1} > a_n$, then interchanging the positions of a_{n-1} and a_n in A yields a lexicographically smaller sequence. Instead, we must turn to a_{n-2}. If $a_{n-2} <$

TABLE 5.1		TABLE 5.2			
1.	*abc*	1.	123	6.	145
2.	*acb*	2.	124	7.	234
3.	*bac*	3.	125	8.	235
4.	*bca*	4.	134	9.	245
5.	*cab*	5.	135	10.	345
6.	*cba*				

a_{n-1}, then the last three digits of A can be rearranged to yield a lexicographically larger sequence. To increase the sequence as little as possible, position $n - 2$ should have the next larger "digit" (larger than a_{n-2}) chosen from a_{n-1} and a_{n-2}. The remaining two of the original last three "digits" go in positions $n - 1$ and n in increasing order. For example, 14352 would be followed by 14523, while 14253 would be followed by 14325. If $a_{n-2} > a_{n-1}$ (and $a_{n-1} > a_n$), then we must turn to a_{n-3}.

In general, we start from the right end of the sequence and moving left, look for the first pair a_i, a_{i+1} such that $a_i < a_{i+1}$ (and $a_{i+1} > a_{i+2} > a_{i+3} > \cdots > a_{n-1} > a_n$). Then the next larger sequence is obtained by placing a_k the minimum of the set $\{a_j | a_j > a_i \text{ and } j > i\}$ in position i and placing the remaining digits $a_i, a_{i+1}, \cdots, a_n$ (excluding a_k) in positions $i + 1$ through n in increasing order.

Algorithm for Next Sequence in Lexicographic Listing of Permutations

PROCEDURE NEXTPERMUT(A,n)
Comment—find lexicographically next permutation of *n*-element array A
Syntactic Comment—GOTO's are used instead of WHILE statement for simplicity
BEGIN

FOR $i \leftarrow n-1$ TO 1 STEP -1 DO IF $A(i) < A(i+1)$ THEN GOTO 1);
Comment—find min A(*h*) such that A(*h*)>A(*i*) and swap A(*h*), A(*i*)
1) FOR $h \leftarrow n$ TO $i+1$ STEP -1 DO IF $A(i) < A(h)$ THEN GOTO 2);
2) *TEMP←A(i); A(i)←A(h); A(h)←TEMP;*
Comment—complete next arrangement by reversing rest of sequence
FOR $h \leftarrow i+1$ TO n DO *TEM(h)←A(h)*;
FOR $h \leftarrow i+1$ TO n DO $A(h) \leftarrow TEM(n+i+1-h)$;
END *NEXTPERMUT.*

To list all combinations of any size of an *n*-set, we represent a subset as an *n*-digit binary sequence. Such a sequence is a membership vector: the *i*th digit is 1 if the *i*th object of the *n*-set is in the combination, and is 0 otherwise. Instead of listing all combinations, we present an algorithm for listing all *n*-

digit binary sequences. While our method lists these sequences lexicographically, (i.e., in ascending numerical order), the associated combinations will not be in lexicographic order (the necessary modification for combinations in this order is left an exercise). If we have access to the binary representation of numbers inside a computer, then we can simply list the numbers from 0 to $2^n - 1$ in binary form to produce all n-digit binary sequences. The following successor rule can be used otherwise to generate all n-digit binary sequences in increasing order. Given the binary sequence $a_1a_2 \cdot \cdot \cdot a_n$, we find the next larger sequence by searching from right to left until the first $a_i = 0$ is found. We set this ith digit to 1 and make all digits to its right 0. It is a straightforward matter to restate this rule in terms of combinations.

Finally, we show how to generate all r-combinations of an n-set in lexicographic order. For simplicity, let us assume that the n-set is the set of integers 1, 2, . . . , n. Recall that we list the elements in each r-combination in increasing order (see Table 5.2). The successor $b_1b_2 \cdot \cdot \cdot b_n$ to the current r-combination $a_1a_2 \cdot \cdot \cdot a_n$ is found by searching from right to left until the first $a_i \neq n - r + i$ is found. For this i, set $b_i = a_i + 1$ and $b_j = b_{j-1} + 1$, for $i < j \leq r$.

Algorithm for Next Subset in Lexicographic Listing of Combinations

PROCEDURE NEXTSELECT(B,n,r)
Comment—find lexicographically next *r-subset from n*-element array *B*
BEGIN
    FOR $i \leftarrow r$ TO 1 STEP -1 DO IF $B(i) < n - r + i$ THEN GOTO 1);
1) $B(i) \leftarrow B(i) + 1$;
    FOR $h \leftarrow i + 1$ TO r DO $B(h) \leftarrow B(h-1) + 1$;
END *NEXTSELECT*.

Programming Projects

Using the preceding algorithms we can enumerate most combinatorial sets. Recall from Section 5.3 that selection with repetition can be modeled as a selection without repetition problem, and that arrangement with repetition can be modeled as a succession of selection problems.

The following examples give computer programs to list all outcomes in two set enumeration problems from preceding sections.

Example 1

List all outcomes in Exercise 17 of Section 5.2. How many ways are there to pair off 10 women at a dance with 10 out of the 20 available men?

Let the men be named 1, 2, 3, . . . , 20. Let $A(i)$ be the man paired with the ith women, $i = 1, 2, . . . , 10$. $B(i)$ is the array with the current subset of 10 men being paired (in 10! ways) with the 10 women. Procedures *NEXTPERMUT* and *NEXTSELECT* are given earlier in this section.

```
BEGIN
    Comment—construct lexicographically first subset of 10 men
    FOR i←1 TO 10 DO B(i)←i
    Comment—loop to list all 10-subsets of the 20 men
    FOR j←1 TO (20
                 10) DO BEGIN
        FOR i←1 TO 10 DO A(i)←B(i);
        Comment—loop to list all arrangements of subset of 10 men
        FOR k←1 TO 10! DO BEGIN
            print out pairing (i,A(i),i=1,2, . . . ,10);
            NEXTPERMUT(A,10);
        END
        NEXTSELECT(B,20,10);
    END;
END.
```

Example 2

List all outcomes in Exercise 7 of Section 5.4. How many ways are there to arrange the letters in *VISITING* with no pair of consecutive I's?

Let $OUTCOME(k)$, $k = 1, 2, . . . , 8$, be the array where the possible outcomes will be constructed. Let $CONS(k)$, $k = 1, 2, . . . , 5$, be the permutation of the consonants V, S, T, N, G to be used in $OUTCOME(I)$.

Let $POS(j)$, $j = 1, 2, 3$, be the positions in $OUTCOME$ where I's will occur. We think of the I's as boundaries between four boxes and the consonants as x's being put in the boxes. The middle two boxes must each have one x (to avoid consecutive I's), and the remaining three x's can be distributed in $C(3 + 4 - 1,3) = C(6,3)$ ways.

Reversing this box model, we can find the positions for the I's by picking a 3-subset from 1, 2, 3, 4, 5, 6 and then adding 1 to the second number in the subset (for the extra consonant that must be between the first and second I) and adding 2 to the third number in the subset.

```
BEGIN
    Comment—construct lexicographically first 3-subset of 1,2,3,4,5,6
    FOR i←1 TO 3 DO P(i)←i;
    Comment—loop on all 3-subsets of 1,2,3,4,5,6
    FOR i←1 TO (6
                 3) DO BEGIN;
        Comment—convert 3-subset into positions for I's
        FOR k←1 TO 3 DO POS(k)←P(k)+k−1;
```

Comment—construct lexicographically first permutation of consonants
CONS(1)←'G'; *CONS*(2)←'N'; *CONS*(3)←'S'; *CONS*(4)←'T'; *CONS*(5)←'V';
Comment—loop on all permutations of the five consonants
FOR *j*←1 TO 5! DO BEGIN
    Comment—construct current *OUTCOME*
    i1←1; *j1*←1;
    FOR *k*←1 TO 8 DO
      IF *k*=*POS*(*i1*) THEN BEGIN
                OUTCOME(*K*)←'I';
                i1←*i1*+1;
          END;
        ELSE BEGIN
                OUTCOME(*k*)←*CONS*(*j1*);
                j1←*j1*+1;
          END;
    PRINT *OUTCOME*(*i*),*i*=1,2, . . . ,8;
    NEXTPERMUT(*CONS*,5);
  END;
  NEXTSELECT(*P*,6,3);
END;
END.

EXERCISES

SUMMARY OF EXERCISES The first 7 exercises use the three algorithms presented in this section. The next 11 exercises involve variants and related properties of these algorithms. The last 10 exercises are programming projects.

1. Enumerate all arrangements of 1, 2, 3, 4 in lexicographic order.
2. Enumerate all arrangements of *a, c, e, h* in lexicographic order.
3. Enumerate all subsets of 1, 2, 3, 4 as binary sequences in lexicographic order.
4. Enumerate all subsets of *a, f, g, h* as binary sequences in lexicographic order.
5. Enumerate all four-subsets of 1, 2, 3, 4, 5, 6 in lexicographic order.
6. Enumerate all three-subsets of *a, d, h, j* in lexicographic order.
7. What are the lexicographically first and last elements in:
    (a) The collection of all arrangements of the integers' 1, 2, . . . , 10?
    (b) The collection of all subsets of 1, 2, . . . , *n*?
    (c) The collection of all four-subsets of the 26 letters?

8. Write a simple recursive program to generate all n-digit binary sequences, where n is an input variable.

9. Give a method to list all subsets of an n-set in lexicographic order.

10. Modify the binary representation of a combination and the successor rule for binary sequences so that in listing all n-digit binary sequences, the associated combinations appear in lexicographic order (see Exercise 9).

11. Write a computer program to generate all subsets of 1, 2, . . . , n, where n is an input variable.

12. Prove by induction that the successor rule given in this section correctly finds the next lexicographic element in:
    (a) The collection of all arrangements of an n-set.
    (b) The collection of all n-digit binary sequences.
    (c) The collection of r-subsets of an n-set.

13. Devise another algorithm for enumerating all arrangements of an n-set.

14. Devise another algorithm for enumerating all subsets of an n-set.

15. Devise another algorithm for enumerating all r-subsets of an n-set.

16. (a) Develop a procedure to compute the location of a given arrangement in the lexicographic order of all arrangements of 1, 2, . . . , n.
    (b) What is the location of 314526 in the lexicographic order of 1, 2, . . . , 6?

17. (a) Develop a procedure to determine the location of a given subset in the lexicographic order of all subsets of 1, 2, . . . , n.
    (b) What is the location of $\{5,9\}$ in the subsets of 1, 3, 5, 7, 9?

18. (a) Develop a procedure to compute the location of a given r-subset in the lexicographic order of all r-subsets of 1, 2, . . . , n.
    (b) What is the location of $\{3,4,7\}$ in the three-subsets of 1, 2, . . . , 8?

PROGRAMMING PROJECTS

19. The following table gives the cost of converting from job i to job j. In what order should the four jobs be performed in order to minimize the total conversion costs?

From\To	1	2	3	4
1	—	4	5	2
2	6	—	4	8
3	3	5	—	7
4	2	7	6	—

20. The following table tells which people know each of a set of trades. Find all subsets of four people who know all 10 trades.

Person\Trade	1	2	3	4	5	6	7	8	9	10
1	1	1	0	1	0	0	0	1	0	0
2	1	0	1	0	1	0	1	0	0	1
3	0	1	0	0	0	1	0	0	1	1
4	0	0	1	1	1	0	1	1	0	0
5	0	1	0	1	0	0	1	0	1	0
6	1	0	1	0	1	0	0	0	1	0
7	1	0	0	1	0	1	0	0	0	1

21. A salesperson wants to make a tour visiting each of five cities, starting and ending at city 1. Use the intercity cost matrix on the right to find the cheapest tour.

From\To	1	2	3	4	5
1	—	4	5	3	6
2	7	—	3	4	7
3	3	4	—	3	6
4	2	5	6	—	5
5	3	3	7	4	—

22. Write a program to list all 5-subsets of 1, 2, 3, 4, 5, 6 with repetition.

23. Write a program to list all arrangements of three a's, three b's, and three c's.

24. Write a program to list all distributions of r identical objects into n different boxes with between three and six objects in each box.

25. Write a program to list all integer solutions of $x_1 + x_2 + \cdots + x_r = n$:
    (a) Where $x_i \geq k$.   (b) $0 \leq x_1 \leq x_2 < x_3 < \cdots < x_r$.

26. Write a program to list all distributions of r distinct objects into n different boxes with at least m objects in each box.

27. Write a program to list all r subsets of 1, 2, . . . , n with no pair of consecutive integers.

28. Write a program to list all (unordered) partitions of 1, 2, 3, 4, 5, 6.

5.7 SUMMARY AND REFERENCES

This chapter introduced the basic formulas and logical reasoning used in arrangement and selection problems. There were four formulas for arrangements and selections with and without repetition, and there were three principles for composing subproblems. But these formulas and principles were just the building blocks for constructing answers to dozens of examples and hundreds of exercises. These problems required a thorough logical analysis before one could begin to decompose them into tractable subproblems. Such logical analysis of possibilities arises often in computer science, probability, and operations research. It is the basic methodology of discrete mathematics and the most important skill for a student to develop in this course. Note that such logical reasoning is the very essence of mathematical model building.

The $n!$ formula for arrangements was known at least 2500 years ago (see David [2] for more details of the history of combinatorial mathematics). Problems involving binomial coefficients and the binomial expansion were mentioned in a primitive way in Chinese, Hindu, and Arab works 800 years ago. Pascal's triangle appears in a fourteenth-century work of Chu Shih-Chieh. The first appearance of the triangle in the West was 200 years later. However it was not until the end of the seventeenth century that Jacob Bernoulli gave a careful proof of the binomial theorem. The examination of probabilities related to gambling by Pascal and Fermat around 1650 was the beginning of modern combinatorial mathematics. The formula for selection with repetition was discovered soon afterward in the following context: the probability when flipping a coin that the nth head appears after exactly r tails is $(\frac{1}{2})^{r+n} C(r + n - 1, r)$ (there are n "boxes" before and between the occurrences of the n heads for placing the r tails). The formula for arrangements with repetition was discovered around 1700 by Leibnitz in connection with the multinomial theorem (see Exercise 41 of Section 5.5). Jacques Bernoulli's *Ars Conjectandi* (1713) was the first book presenting basic combinatorial methods. Around the same time, Newton found the generalized binomial coefficients in connection with his work on expanding $(1 + x)^z$, where z is any real number. The first comprehensive textbook on permutation and combination problems was written by Whitworth [6] in 1901.

The problem of systematically enumerating arrangements, selections, and other combinatorial collections was not really addressed until the advent of modern digital computers. One of the first people to work on such enumeration in the 1950s was Lehmer [3].

See General References (at back of text) for a list of other introductory texts on enumeration.

1. R. Buck, *Advanced Calculus*, 2nd ed., McGraw-Hill, New York, 1965.

2. F. David, *Games, Gods, and Gambling*, Hafner Press, New York, 1962.

3. D. Lehmer, "The Machine Tools of Combinatorics" in *Applied Combinatorial Mathematics*, E. Beckenbach (ed.), John Wiley & Sons, New York, 1964.

4. E. Reingold, J. Nievergelt, and N. Deo, *Combinatorial Algorithms*, Prentice-Hall, Englewood Cliffs, N.J., 1977.

5. J. Riordan, *Combinatorial Identities*, John Wiley & Sons, New York, 1968.

6. W. Whitworth, *Choice and Chance*, 5th ed. (1901), Hafner Press, New York, 1965.

Chapter Six
Generating Functions

6.1 GENERATING FUNCTION MODELS

In this chapter, we introduce the concept of a generating function. Generating functions are a convenient tool for handling special constraints in selection and arrangement problems with repetition. They are used in Chapters 7, 8, and 9 to solve other combinatorial problems. Generating functions are the most abstract problem-solving technique introduced in this text. But once understood, they are also the easiest way to model a broad spectrum of combinatorial problems.

Suppose a_r is the number of ways to select r objects in a certain procedure. Then $g(x)$ is a **generating function** for a_r if $g(x)$ has the polynomial expansion

$$g(x) = a_0 + a_1x + a_2x^2 + \cdots + a_rx^r + \cdots$$

If the function has an infinite number of terms, it is called a **power series**. In contrast to the exponential generating functions used in Section 6.4, the above $g(x)$ is sometimes called an **ordinary generating function**. In Section 5.5 we verified the well-known binomial expansion

$$(1 + x)^n = 1 + \binom{n}{1}x + \binom{n}{2}x^2 + \cdots + \binom{n}{r}x^r + \cdots + \binom{n}{n}x^n$$

Then $g(x) = (1 + x)^n$ is the generating function for $a_r = C(n,r)$, the number of ways to select an r-subset from an n set. Recall that we derived the expansion

222

of $(1 + x)^n$ by first considering the formal multiplication of $(a + x)^3$:

$$(a + x)(a + x)(a + x) = aaa + aax + axa + axx + xaa + xax + xxa + xxx$$

When $a = 1$, we had

$$(1 + x)(1 + x)(1 + x) = 111 + 11x + 1x1 + 1xx + x11$$

$$+ x1x + xx1 + xxx \tag{1}$$

Such a formal expansion lists all ways of multiplying a term in the first factor times a term in the second factor times a term in the third factor. The problem of determining the coefficient of x^r in $(1 + x)^3$, and more generally in $(1 + x)^n$, reduces to the problem of counting the number of different formal products with exactly r x's and $n - r$ 1's. So the coefficient of x^r in $(1 + x)^3$ is $C(3,r)$, and in $(1 + x)^n$ is $C(n,r)$.

It is very important that the multiplication of several polynomial factors be viewed as generating the collection of all formal products obtained by multiplying together a term from each polynomial factor. If the ith polynomial factor contains r_i different terms and there are n factors, then there will be $r_1 \cdot r_2 \cdot r_3 \cdot \ldots \cdot r_n$ different formal products. For example, there will be 2^n formal products in the expansion of $(1 + x)^n$. In the expansion of $(1 + x)^3 \cdot (1 + x + x^2)^2$, the set of all formal products will be sequences of the form:

$$\begin{Bmatrix} 1 \\ x \end{Bmatrix} \cdot \begin{Bmatrix} 1 \\ x \end{Bmatrix} \cdot \begin{Bmatrix} 1 \\ x \end{Bmatrix} \cdot \begin{Bmatrix} 1 \\ x \\ x^2 \end{Bmatrix} \cdot \begin{Bmatrix} 1 \\ x \\ x^2 \end{Bmatrix} \tag{2}$$

that is, a 1 or an x in each of the first three positions in the product and then a 1 or an x or an x^2 in each of the last two positions in the product, for example, $1x1x^2x$.

In this chapter we are primarily concerned with multiplying polynomial factors in which the powers of x in each factor have coefficient 1, factors such as $(1 + x + x^2 + x^3)$ or $(1 + x^2 + x^4 + x^6 + \cdot \cdot \cdot)$. These factors are completely specified by the set of different exponents of the x's. Note that $1 = x^0$. Thus expansion (1) can be rewritten

$$(x^0 + x^1)(x^0 + x^1)(x^0 + x^1) = x^0x^0x^0 + x^0x^0x^1 + x^0x^1x^0 + x^0x^1x^1 + x^1x^0x^0$$

$$+ x^1x^0x^1 + x^1x^1x^0 + x^1x^1x^1$$

And the formal products in (2) can be written as

$$\begin{Bmatrix} x^0 \\ x^1 \end{Bmatrix} \begin{Bmatrix} x^0 \\ x^1 \end{Bmatrix} \begin{Bmatrix} x^0 \\ x^1 \end{Bmatrix} \begin{Bmatrix} x^0 \\ x^1 \\ x^2 \end{Bmatrix} \begin{Bmatrix} x^0 \\ x^1 \\ x^2 \end{Bmatrix} \tag{3}$$

The problem of determining the coefficient of x^r when we multiply several polynomial factors together can be restated in terms of exponents. Consider the coefficient of x^4 in the expansion of $(1 + x)^3(1 + x + x^2)^2$. It is the number of different formal products of the form (3) whose sum of exponents is 4. Determining this coefficient can be modeled as an integer-solution-to-an-equation problem (Example 6 of Section 5.4). The coefficient of x^4 in $(1 + x)^3(1 + x + x^2)^2$ is the number of integer solutions to

$$e_1 + e_2 + e_3 + e_4 + e_5 = 4$$

where $0 \le e_1, e_2, e_3 \le 1$ and $0 \le e_4, e_5 \le 2$. The variable e_i stands for the value of the exponent of the ith term in the formal product (3).

By the equivalent-forms-of-selection-with-repetition principle in Section 5.4, the preceding integer-solution-to-an-equation problem is equivalent to the problem of selecting four objects from a collection of five types of objects, where there is one object of the first type, one of the second type, one of the third type, and two of each of the fourth and fifth types. It is also equivalent to the problem of distributing four identical objects into five distinct boxes with at most one object in each of the first three boxes and at most two objects in the last two boxes.

More generally, the coefficient of x^r in $(1 + x)^3(1 + x + x^2)^2$ will be the number of integer solutions to

$$e_1 + e_2 + e_3 + e_4 + e_5 = r, \qquad 0 \le e_1, e_2, e_3 \le 1, \qquad 0 \le e_4, e_5 \le 2$$

Or equivalently, the number of ways of selecting r objects from the collection mentioned above (or of distributing r identical objects into five boxes with the constraints mentioned above). Thus $(1 + x)^3(1 + x + x^2)^2$ is the generating function for a_r, the number of ways to select r objects from the given collection of five types (or to perform the equivalent distribution).

At this stage, we are only concerned with how to build generating function models for counting problems. In the next section we will see how various algebraic manipulations of generating functions permit us to evaluate desired coefficients.

We have shown how the coefficients of $(1 + x)^3(1 + x + x^2)^2$ can be interpreted as the solutions to a certain selection-with-repetition or distribution-of-identical-objects problem. This line of reasoning can be reversed: given a certain selection-with-repetition or equivalent distribution problem, we can build a generating function whose coefficients are the answers to this problem. We now give some examples of how to build such generating functions. Using an intermediate model of integer-solution-to-an-equation makes it easier to find the right generating function.

Example 1

Find a generating function for a_r, the number of ways to select r balls from a pile of three green, three white, three blue, and three gold balls.

This selection problem can be modeled as the number of integer solutions to

$$e_1 + e_2 + e_3 + e_4 = r, \qquad 0 \leqslant e_i \leqslant 3$$

Here e_1 represents the number of green balls chosen, e_2 the number of white, e_3 blue, and e_4 gold. We want to construct a multiplication of polynomial factors such that the formal products will contain four terms, each with an exponent between 0 and 3:

$$\begin{Bmatrix} x^0 \\ x^1 \\ x^2 \\ x^3 \end{Bmatrix} \cdot \begin{Bmatrix} x^0 \\ x^1 \\ x^2 \\ x^3 \end{Bmatrix} \cdot \begin{Bmatrix} x^0 \\ x^1 \\ x^2 \\ x^3 \end{Bmatrix} \cdot \begin{Bmatrix} x^0 \\ x^1 \\ x^2 \\ x^3 \end{Bmatrix}$$

Then the number of these formal products in which the four exponents sum to r will be the number of integer solutions to the above equation. The desired generating function is thus $(x^0 + x^1 + x^2 + x^3)^4 = (1 + x + x^2 + x^3)^4$. ∎

Example 2

Use a generating function to model the problem of counting all selections of six objects chosen from three types of objects with repetition of up to four objects of one type. Also model the problem with unlimited repetition.

This selection problem can be modeled as the number of integer solutions to

$$e_1 + e_2 + e_3 = 6, \qquad 0 \leqslant e_i \leqslant 4$$

This problem does not ask for the general solution of the ways to select r objects. However in building a generating function, we automatically model all values of r, not just $r = 6$. Wanting a solution to the problem for six objects means that we are only interested in the coefficient of x^6, that is, the ways the exponents can sum to 6. The desired generating function is $(1 + x + x^2 + x^3 + x^4)^3$ and the term in it we want is the coefficient of x^6.

Permitting unlimited repetition means that any number can be chosen of each type and so any exponent value is possible (although in this particular problem no exponent can exceed 6). From this point of view, the answer is the coefficient of x^6 in $(1 + x + x^2 + x^3 + \cdots)^3$, where "$+ \cdots$" means that the factor is an infinite series. We will use such infinite series extensively in the next section. They make for much easier computation. ∎

Example 3

Find a generating function for a_r, the number of ways to distribute r identical objects into five distinct boxes with an even number of objects not exceeding 10 in the first two boxes and between three and five in the other boxes.

While the constraints may be a bit contrived in this example, it is still easy to model the problem as an integer-solution-to-an-equation problem with appropriate constraints. We want to count all integer solutions to

$$e_1 + {}_2 + e_3 + e_4 + e_5 = r, \quad e_1, e_2 \text{ even}, \quad 0 \leqslant e_1, e_2 \leqslant 10, \quad 3 \leqslant e_3, e_4, e_5 \leqslant 5$$

The associated generating function is $g(x) = (1 + x^2 + x^4 + x^6 + x^8 + x^{10})^2 \cdot (x^3 + x^4 + x^5)^3$. The formal products in an expansion of this $g(x)$ will have an exponent of 0, 2, 4, 6, 8, or 10 in the first two terms and an exponent of 3, 4, or 5 in the other three terms. So the coefficient of x^r in $g(x)$ will be the number of integer solutions to the above equation. ∎

Once the original problem has been restated as an integer-solution-to-an-equation problem, the following conceptual aid makes it easier to write down a generating function. Think of each polynomial factor as an "inventory" of exponent values. That is, for each value k that e_i can assume, the ith polynomial factor must have a term x^k. The ith factor lists, or inventories, in its exponents the possible values for e_i.

With a little practice, generating function models become simple to build. However, the concept behind generating functions is far from simple. In the previous chapter, we solved similar selection and distribution problems with explicit formulas involving binomial coefficients. Now we are modeling these problems as some coefficient, which represents the number of formal products in the multiplication of certain polynomial factors. All we write down is the polynomial factors.

A generating function cannot be designed to model a selection or distribution problem for just one amount, say six objects. It must model the problem *for all possible numbers of objects*. Several of the harder problems in the previous chapter had to be broken into many simple subproblems. A generating function model implicitly stores all the necessary information about these subproblems in one function. In the next section, computing the value of a coefficient of such a function will be reduced to a series of rote algebraic operations. The fact that the function models a complex selection or distribution problem will not be needed in the actual computations.

EXERCISES

SUMMARY OF EXERCISES   The first 21 exercises involve simple generating function modeling. The remaining problems involve more challenging modeling; Exercises 25–29 use multinomial generating functions.

1. For each of the following expressions, list the set of all formal products in which the exponents sum to 4.
    (a) $(1 + x + x^2)^2(1 + x)^2$
    (b) $(1 + x + x^2 + x^3 + x^4)^3$
    (c) $(1 + x^2 + x^4)^2(1 + x + x^2)^2$
    (d) $(1 + x + x^2 + x^3 + \cdots)^3$

2. Build a generating function for a_r, the number of integer solutions to the equations:
    (a) $e_1 + e_2 + e_3 + e_4 + e_5 = r, 0 \leqslant e_i \leqslant 4$
    (b) $e_1 + e_2 + e_3 = r, 0 < e_i < 4$
    (c) $e_1 + e_2 + e_3 + e_4 = r, 2 \leqslant e_i \leqslant 8, e_1$ odd, e_2 even
    (d) $e_1 + e_2 + e_3 + e_4 = r, 0 \leqslant e_i$
    (e) $e_1 + e_2 + e_3 + e_4 = r, 0 < e_i, e_2, e_4$ odd, $e_4 \leqslant 3$

3. Build a generating function for a_r, the number of r selections from a pile of:
    (a) Three red, four black, and four white balls.
    (b) Five jelly beans, three licorice sticks, eight lollipops with at least one of each candy.
    (c) Unlimited amounts of pennies, nickels, dimes, and quarters.
    (d) Seven types of lightbulbs with an odd number of the first and second types.

4. Build a generating function for a_r, the number of distributions of r identical objects into:
    (a) Five different boxes with at most four objects in each box.
    (b) Four different boxes with between three and eight objects in each box.
    (c) Seven different boxes with at least one object in each box.
    (d) Three different boxes with at most five objects in the first box.

5. Use a generating function for modeling the number of 5-combinations of the letters M, A, T, H in which M and A can appear any number of times but T and H at most once. Which coefficient in this generating function do we want?

6. Use a generating function for modeling the number of different selections of r hot dogs when there are five types of hot dogs.

7. Use a generating function for modeling the number of distributions of

18 chocolate bunny rabbits into four Easter baskets with over two rabbits in each basket. Which coefficient do we want?

8. (a) Use a generating function for modeling the number of different election outcomes in an election for class president if 27 students are voting among four candidates. Which coefficient do we want?

   (b) Suppose each student who is a candidate votes for herself or himself. Now what is the generating function and the required coefficient?

   (c) Suppose no candidate receives a majority of the vote. Repeat part (a).

9. Find a generating function for a_r, the number of r-combinations of an n-set with repetition.

10. Given one of each of u types of candy, two of v types of candy, and three of w types of candy, find a generating function for the number of ways to select r candies.

11. Find a generating function for a_k, the number of k-combinations of n types of objects with an even number of the first type, an odd number of the second type, and any amount of the other types.

12. Find a generating function for a_r, the number of ways to distribute r identical objects into q distinct boxes with an odd number between r_1 and s_1 in the first box, an even number between r_2 and s_2 in the second box, and at most three in the other boxes.

13. Find a generating function for a_r, the number of ways n distinct dice can show a sum of r.

14. Find a generating function for a_r, the number of ways a roll of six distinct dice can show a sum of r if:

    (a) The first three dice are odd and the second three even.

    (b) The ith die does not show a value of i.

15. Build a generating function for a_r, the number of integer solutions to $e_1 + e_2 + e_3 + e_4 = r$, $-3 \leq e_i \leq 3$. (*Hint:* Use negative exponents.)

16. Find a generating function for the number of integers between 0 and 999,999 whose sum of digits is r.

17. Find a generating function for the number of selections of r sticks of chewing gum chosen from eight flavors if each flavor comes in packets of five sticks.

18. (a) Use a generating function for modeling the number of distributions of 20 identical balls into five distinct boxes if each box has between 2 and 7 balls.

    (b) Factor out an x^2 from each polynomial factor in part (a). Interpret this revised generating function combinatorially.

19. Use a generating function for modeling the number of ways to select

five integers from $1, 2, \ldots, n$, no two of which are consecutive. Which coefficient do we want for $n = 20$? For a general n?

20. Explain why $(1 + x + x^2 + \ldots + x^r)^4$ is not a proper generating function for a_r, the number of ways to select r objects from four types with repetition. What is the correct generating function?

21. Explain why $(1 + x + x^2 + x^3 + x^4)^r$ is not a proper generating function for the number of ways to distribute r jelly beans among r children with no child getting more than four jelly beans.

22. Show that the generating function for the number of integer solutions to $e_1 + e_2 + e_3 + e_4 = r,\ 0 \leqslant e_1 \leqslant e_2 \leqslant e_3 \leqslant e_4$, is

$$(1 + x + x^2 + \cdots)(1 + x^2 + x^4 + \cdots)(1 + x^3 + x^6 + \cdots) \cdot$$
$$(1 + x^4 + x^8 + \cdots)$$

23. Find a generating function for the number of ways to make r cents change in pennies, nickels, and dimes.

24. A national singing contest has 5 distinct entrants from each state. Use a generating function for modeling the number of ways to pick 20 semifinalists if:
    (a) There is at most 1 person from each state.
    (b) There are at most 3 people from each state.

25. Find a generating function $g(x, y)$ whose coefficient of $x^r y^s$ is the number of ways to distribute r chocolate bars and s lollipops among five children such that no child gets more than three lollipops.

26. (a) Find a generating function $g(x, y, z)$ whose coefficient $x^r y^s z^t$ is the number of ways to distribute r red balls, s blue balls, and t green balls to n people with between three and six balls of each type to each person.
    (b) Suppose also that each person gets at least as many red balls as blues.
    (c) Suppose also that the first three people equal numbers of reds and blues.
    (d) Suppose also that no one gets the same number of green and red balls.

27. Find a generating function $g(x, y, z)$ whose coefficient of $x^r y^s z^t$ is the number of ways 8 people can each pick two different fruits from a bowl of apples, oranges, and bananas for a total of r apples, s oranges, and t bananas.

28. Find a generating function $g(x_1, x_2, \ldots, x_m)$ whose coefficient of $x_1^{r_1} x_2^{r_2} \cdots x_m^{r_m}$ is the number of ways n people can pick a total of r_1 chairs of type 1, r_2 chairs of type 2, $\ldots$, r_m chairs of type m if:

(a) Each person picks one chair.
(b) Each person picks either two chairs of one type or no chairs at all.
(c) Person i picks up to i chairs of exactly one type.

29. If $g(x_1, x_2, \ldots, x_p) = (x_1 + x_2 + \ldots + x_p)^n$, how many terms of $g(x_1, \ldots, x_p)$'s expansion have no exponent of any x_i greater than 1. What is the coefficient of one of these terms?

6.2 CALCULATING COEFFICIENTS OF GENERATING FUNCTIONS

We now develop algebraic techniques for calculating the coefficients of generating functions. All these methods seek to reduce a given generating function to a simple binomial-type generating function or a product of binomial-type generating functions. For easy reference, we list first all the polynomial identities and expansions to be used in this section. They will be explained below.

Polynomial Expansions

(1) $\dfrac{1 - x^{n+1}}{1 - x} = 1 + x + x^2 + \ldots + x^n$

(2) $\dfrac{1}{1 - x} = 1 + x + x^2 + \ldots$

(3) $(1 + x)^n = 1 + \binom{n}{1} x + \binom{n}{2} x^2 + \ldots + \binom{n}{r} x^r + \ldots + \binom{n}{n} x^n$

(4) $(1 - x^m)^n = 1 - \binom{n}{1} x^m + \binom{n}{2} x^{2m} + \ldots (-1)^r \binom{n}{r} x^{rm} \ldots (-1)^n \binom{n}{n} x^{nm}$

(5) $\dfrac{1}{(1 - x)^n} = 1 + \binom{1 + n - 1}{1} x + \binom{2 + n - 1}{2} x^2 + \ldots$
$\qquad + \binom{r + n - 1}{r} x^r + \ldots$

(6) If $h(x) = f(x)g(x)$, where

$$f(x) = a_0 + a_1 x + a_2 x^2 + \cdots$$

and

$$g(x) = b_0 + b_1 x + b_2 x^2 + \cdots$$

then

$$h(x) = a_0 b_0 + (a_1 b_0 + a_0 b_1)x + (a_2 b_0 + a_1 b_1 + a_0 b_2)x^2 + \cdots$$
$$+ (a_r b_0 + a_{r-1} b_1 + a_{r-2} b_2 + \cdots + a_0 b_r)x^r + \cdots$$

The rule for multiplication of generating functions in (6) is simply the standard formula for polynomial multiplication. Identity (1) can be verified by polynomial "long division." We restate it, multiplying both sides of (1) by

$(1 - x)$, as $1 - x^{n+1} = (1 - x)(1 + x + x^2 + \cdots + x^n)$. We verify that the product of the right-hand side is $1 - x^{n+1}$ by "long multiplication."

(*)

$$
\begin{array}{l}
1 + x + x^2 + \qquad \cdots + x^n \\
1 - x \\
\hline
1 + x + x^2 + \qquad \cdots + x^n \\
\quad - x - x^2 - x^3 - \cdots - x^n - x^{n+1} \\
\hline
1 \qquad\qquad\qquad\qquad\qquad - x^{n+1}
\end{array}
$$

If n is made infinitely large, so that $1 + x + x^2 + \cdots + x^n$ becomes the infinite series $1 + x + x^2 + \cdots$, then the multiplication process (*) will yield a power series in which the coefficient of each x^k, $k > 0$, is zero [the reader can confirm this with (6)]. We conclude that $(1 - x)(1 + x + x^2 + \cdots) = 1$. [Numerically, this equation is valid for $|x| < 1$; the "remainder" term x^{n+1} in (*) goes to 0 as n becomes infinite.] Dividing both sides of this equation by $(1 - x)$ yields identity (2).

Expansion (3), the binomial expansion, was explained at the start of Section 6.1. Expansion (4) is obtained from (3) by expanding $(1 + y)^n$, where $y = -x^m$:

$$
[1 + (-x^m)]^n = 1 + \binom{n}{1}(-x^m) + \binom{n}{2}(-x^m)^2 + \cdots
$$
$$
+ \binom{n}{r}(-x^m)^r + \cdots + \binom{n}{n}(-x^m)^n
$$

Expansion (4) follows immediately.

By identity (2), $(1 - x)^{-n}$, or equivalently, $[1/(1 - x)]^n$, equals

(7) $(1 + x + x^2 + x^3 + \cdots)^n$

Let us determine the coefficient of x^r in (7) by counting the number of formal products whose sum of exponents is r. If e_i represents the exponent of the ith term in a formal product, then the number of formal products with exponents summing to r is the same as the number of integer solutions to the equation

$$
e_1 + e_2 + e_3 + \cdots + e_n = r \qquad e_i \geqslant 0
$$

In Example 6 of Section 5.4, we showed that the number of nonnegative integer solutions to this equation is $C(r + n - 1, r)$. Thus the coefficient of x^r in (7) is $C(r + n - 1, r)$. This verifies expansion (5). Expansion (5) can also be verified by using the generalized binomial expansion, introduced at the end of Section 5.5, and applying identity (14) of Section 5.5.

With formulas (1) to (6) we can determine the coefficients of a variety of generating functions: first, reduce a given generating function to one of the forms $(1 + x)^n$, $(1 - x^m)^n$, or $(1 - x)^{-n}$, or a product of two such expansions; next use expansions (3) to (5) and the product rule (6) to obtain any desired coefficient. We illustrate the most common reduction methods in the following examples.

Example 1

Find the coefficient of x^{16} in $(x^2 + x^3 + x^4 + \cdots)^5$. What is the coefficient of x^r?

To simplify the expression, we extract x^2 from each polynomial factor and then apply identity (2).

$$(x^2 + x^3 + x^4 + \cdots)^5 = [x^2(1 + x + x^2 + \cdots)]^5$$

$$= \left(\frac{x^2}{1 - x}\right)^5 = x^{10}\frac{1}{(1 - x)^5}$$

Thus the coefficient of x^{16} in the original generating function is the coefficient of x^{16} in $x^{10}(1 - x)^{-5}$. But the coefficient of x^{16} in this expression will be the coefficient of x^6 in $(1 - x)^{-5}$ [i.e., the x^6 term in $(1 - x)^{-5}$ is multiplied by x^{10} to become the x^{16} term in $x^{10}(1 - x)^{-5}$]. From expansion (5), we see that the coefficient of x^6 in $(1 - x)^{-5}$ is $C(6 + 5 - 1,6)$. More generally, the coefficient of x^r in $x^{10}(1 - x)^{-5}$ equals the coefficient of x^{r-10} in $(1 - x)^{-5}$, namely, $C((r - 10) + 5 - 1,(r - 10))$. ∎

Observe that $(x^2 + x^3 + x^4 + \cdots)^5$ is the generating function for the number of ways to distribute r identical objects into five distinct boxes with at least two objects in each box. In the last chapter, we solved such a problem by first putting two objects in each of the five boxes and then counting the ways of distributing the remaining $r - 10$ objects, $C((r - 10) + 5 - 1,(r - 10))$ ways. In the generating function analysis in Example 1, we algebraically picked out an x^2 from each factor for a total of x^{10} and then found the coefficient of x^{r-10} in $(1 + x + x^2 + \cdots)^5$, the generating function for (unrestricted) distributions of identical objects into five boxes.

The standard algebraic technique of extracting the highest common power of x from each factor corresponds to the technique used to solve the associated distribution problem. Such correspondences are a major reason for using generating functions: the algebraic techniques automatically do the combinatorial reasoning for us.

Example 2

Use generating functions to find the number of ways to collect $15 from 20 distinct people if each of the first 19 people can give a dollar (or nothing) and the twentieth person can give either $1 or $5 (or nothing).

The generating function for the number of ways to collect r dollars from these people is $(1 + x)^{19}(1 + x + x^5)$. We want the coefficient of x^{15}. The first part of this generating function has the binomial expansion

$$1 + \binom{19}{1} x + \binom{19}{2} x^2 + \cdots + \binom{19}{r} x^r + \cdots + \binom{19}{19} x^{19}$$

If we let $f(x)$ be this first polynomial and let $g(x) = 1 + x + x^5$, then we can use (6) to calculate the coefficient of x^{15} in $h(x) = f(x)g(x)$. Let $a_r = \binom{19}{r}$ be the coefficient of x^r in $f(x)$ and b_r the coefficient of x^r in $g(x)$. Then the coefficient of x^{15} in $h(x)$ is

$$a_{15}b_0 + a_{14}b_1 + a_{13}b_2 + \cdots + a_0 b_{15}$$

which reduces to

$$a_{15}b_0 + a_{14}b_1 + a_{10}b_5$$

since x^0, x^1, x^5 are the only non-zero terms in $g(x)$. Substituting for a_r and b_r, we have

$$\binom{19}{15} \cdot 1 + \binom{19}{14} \cdot 1 + \binom{19}{10} \cdot 1 = \binom{19}{15} + \binom{19}{14} + \binom{19}{10}$$

■

Example 3

How many ways are there to distribute 25 identical balls into seven distinct boxes if the first box can have no more than 10 balls but any amount can go in each of the other six boxes?

The generating function for the number of ways to distribute r balls into seven boxes with at most 10 balls in the first box is

$$(1 + x + x^2 + \cdots + x^{10})(1 + x + x^2 + \cdots)^6$$
$$= \left(\frac{1 - x^{11}}{1 - x}\right) \cdot \left(\frac{1}{1 - x}\right)^6 = (1 - x)^{-7} \cdot (1 - x^{11})$$

using identities (1) and (2). Let $f(x) = (1 - x)^{-7}$ and $g(x) = 1 - x^{11}$. Using expansion (5) we have

$$f(x) = (1 - x)^{-7} = 1 + \binom{1 + 7 - 1}{1} x + \binom{2 + 7 - 1}{2} x^2 + \cdots$$
$$+ \binom{r + 7 - 1}{r} x^r \cdots$$

We want the coefficient of x^{25} (25 balls distributed) in $h(x) = f(x)g(x)$. As in Example 2, we only need to consider the terms $a_{25-i}b_i$ in (6) in which the b_i's, coefficients of $g(x)$, are nonzero. So the coefficient of x^{25} is

$$
a_{25}b_0 + a_{14}b_{11}
$$
$$
= \binom{25 + 7 - 1}{25} \cdot 1 + \binom{14 + 7 - 1}{14} \cdot (-1) = \binom{31}{25} - \binom{20}{14}
$$

The combinatorial interpretation of the answer in Example 3 is that we count all the ways to distribute (without restriction) the 25 balls into the seven boxes, $C(25 + 7 - 1,25)$ ways and then subtract the distributions that violate the first box constraint, that is, distributions with at least 11 balls in the first box, $C((25 - 11) + 7 - 1,(25 - 11))$ (first put 11 balls in the first box and then distribute the remaining balls arbitrarily). Again, generating functions automatically performed this combinatorial reasoning.   ∎

The next example employs all the techniques used in the first three examples to solve a problem that cannot be solved by the combinatorial methods of the previous chapter.

Example 4

How many ways are there to select 25 toys from seven types of toys with between two and six of each type?

The generating function for a_r, the number of ways to select r toys from seven types with between 2 and 6 of each type, is

$$
(x^2 + x^3 + x^4 + x^5 + x^6)^7
$$

We want the coefficient of x^{25}. As in Example 1, we extract x^2 from each factor to get

$$
[x^2(1 + x + x^2 + x^3 + x^4)]^7 = x^{14}(1 + x + x^2 + x^3 + x^4)^7
$$

Now we reduce our problem to finding the coefficient of $x^{25-14} = x^{11}$ in $(1 + x + x^2 + x^3 + x^4)^7$. Using identity (1), we can rewrite this generating function as

$$
(1 + x + x^2 + x^3 + x^4)^7 = \left(\frac{1 - x^5}{1 - x}\right)^7 = (1 - x)^{-7} \cdot (1 - x^5)^7
$$

Let $f(x) = (1 - x)^{-7}$ and $g(x) = (1 - x^5)^7$. By expansions (5) and (4), respectively, we have

$$f(x) = (1 - x)^{-7} = 1 + \binom{1 + 7 - 1}{1} x + \binom{2 + 7 - 1}{2} x^2 + \cdots$$

$$+ \binom{r + 7 - 1}{r} x^r + \cdots$$

$$g(x) = (1 - x^5)^7 = 1 - \binom{7}{1} x^5 + \binom{7}{2} x^{10} - \cdots$$

$$- (-1)^r \binom{7}{r} x^{5r} - \cdots - \binom{7}{7} x^{35}$$

As in Examples 2 and 3, to find the coefficient of x^{11} in $h(x) = f(x)g(x)$ we only need to consider terms $a_{11-i}b_i$ in (6) in which the b_i's are nonzero. That is,

$$\begin{matrix} a_{11}b_0 & + & a_6 b_5 & + & a_1 b_{10} \end{matrix}$$

$$= \binom{11 + 7 - 1}{11} \cdot 1 + \binom{6 + 7 - 1}{6} \cdot \left(-\binom{7}{1}\right) + \binom{1 + 7 - 1}{1} \cdot \binom{7}{2}$$

∎

We close this section by showing how generating functions can be used to verify binomial identities. We express the right side of the identity as a particular coefficient in some generating function and the left side as the same coefficient in a product of generating functions, whose product turns out to be the original generating function.

Example 5

Verify the binomial identity

$$\binom{n}{0}^2 + \binom{n}{1}^2 + \cdots + \binom{n}{n}^2 = \binom{2n}{n}$$

The right-hand side of the identity is the coefficient of x^n in $(1 + x)^{2n}$. The product of generating functions we want is $(1 + x)^n (1 + x)^n$. So $f(x) = g(x) = (1 + x)^n$. Then $a_r = b_r = C(n, r)$. By the product rule (6), the coefficient of x^n in $f(x)g(x)$ is

$$\begin{matrix} a_0 b_n + a_1 b_{n-1} & + \cdots + a_n b_0 \end{matrix}$$

$$= \binom{n}{0}\binom{n}{n} + \binom{n}{1}\binom{n}{n - 1} + \cdots + \binom{n}{n}\binom{n}{0}$$

$$= \binom{n}{0}^2 + \binom{n}{1}^2 + \cdots + \binom{n}{n}^2, \quad \text{since } \binom{n}{r} = \binom{n}{n - r}$$

Equating coefficients of x^n is $(1 + x)^n(1 + x)^n = (1 + x)^{2n}$, we obtain the required identity. ∎

EXERCISES

SUMMARY OF EXERCISES   The first 25 exercises are similar to the examples of coefficient calculation in this section. Exercises 27–36 involve binomial identities (several of these problems are quite tricky). Exercises 38–44 introduce the topic of probability generating functions (some background in probability is helpful).

1.   Find the coefficient of x^8 in $(1 + x + x^2 + x^3 + \cdots)^n$.

2.   Find the coefficient of x^r in $(x^5 + x^6 + x^7 + \cdots)^8$.

3.   Find the coefficient of x^9 in $(1 + x^2 + x^4)(1 + x)^m$.

4.   Find the coefficient of x^{20} in $(x + x^2 + x^3 + x^4 + x^5)(x^2 + x^3 + x^4 + \cdots)^5$.

5.   Find the coefficient of x^{18} in $(x^2 + x^3 + x^4 + x^5 + x^6 + x^7)^4$.

6.   Find the coefficient of x^{50} in $(x^{10} + x^{11} + \cdots + x^{25})(x + x^2 + \cdots + x^{15})(x^{20} + \cdots + x^{45})$.

7.   Find the coefficient of x^{12} in:

   (a) $x^2(1 - x)^{-10}$ (d) $\dfrac{x + 3}{1 - 3x + x^2}$

   (b) $\dfrac{x^2 - 3x}{(1 - x)^4}$ (e) $\dfrac{b^m x^m}{(1 - bx)^{m+1}}$

   (c) $\dfrac{(1 - x)^5}{(1 - x)^5}$

8.   Give a formula similar to (1) for:
   (a) $1 + x^4 + x^8 + \cdots + x^{24}$ (b) $x^{20} + x^{40} + \cdots + x^{180}$

9.   Find the coefficient of x^9 in $(x^2 + x^3 + x^4 + x^5)^5$.

10.   Find the coefficient of x^{18} in $(1 + x^3 + x^6 + x^9 + \cdots)^7$.

11.   Find the coefficient of x^{12} in:
   (a) $(1 - x)^8$ (d) $(1 - 4x)^{-5}$
   (b) $(1 + x)^{-1}$ (e) $(1 + x^3)^{-4}$
   (c) $(1 + x)^{-8}$

12.   Find the coefficient of x^{25} in $(1 + x^3 + x^8)^{10}$.

13.   Use generating functions to find the number of ways to select 10 balls from a large pile of red, white, and blue balls if:
   (a)   The selection has at least two balls of each color.
   (b)   The selection has at most two red balls.
   (c)   The selection has an even number of blue balls.

14. Use generating functions to find the number of ways to distribute r jelly beans among eight children if:
    (a) Each child gets at least one jelly bean.
    (b) Each child gets an even number of beans.

15. How many ways are there to place an order for 12 chocolate sundaes if there are 5 types of sundaes, and at most 4 sundaes of one type are allowed?

16. How many ways are there to paint the 10 identical rooms in a hotel with five colors if at most three rooms can be painted green, at most three painted blue, at most three red, and no constraint on the other two colors, black and white?

17. How many ways are there to distribute 20 cents to n children and one parent if the parent receives either a nickel or a dime and:
    (a) The children receive any amounts?
    (b) Each child receives at most 1¢?
    (c) Each child receives at most 2¢?

18. How many ways are there to get a sum of 25 when 10 distinct dice are rolled?

19. How many ways are there to select 240 chocolate candies from seven types if each type comes in boxes of 20 and if at least one but not more than five boxes of each type are chosen? (*Hint:* Solve in terms of boxes of chocolate.)

20. How many different committees of 40 Senators can be formed if the two Senators from the same state (50 states in all) are considered identical?

21. How many ways are there to split 6 copies of one book, 7 copies of a second book, and 11 copies of a third book between two teachers if each teacher gets 12 books and each teacher gets at least 2 copies of each book?

22. How many ways are there to divide five pears, five apples, five doughnuts, five lollipops, five chocolate cats, and five candy rocks into two (unordered) piles of 15 objects each?

23. How many ways are there to collect 24 dollars from 4 children and r adults if each person gives at least \$1, but each child can give at most \$4 and each adult at most \$7?

24. If a coin is flipped 25 times with eight tails occurring, what is the probability that no run of six (or more) consecutive heads occurs.

25. If 10 steaks and 15 lobsters are distributed among four people. How many ways are there to give each person at most 3 steaks and at least 3 lobsters?

26. Show that $(1 - x - x^2 - x^3 - x^4 - x^5 - x^6)^{-1}$ is the generating function for the number of ways a sum of r can occur if a die is rolled any number of times.

27. Use generating functions to show that

$$\sum_{k=0}^{r} \binom{m}{k}\binom{n}{r-k} = \binom{m+n}{r}$$

28. Use the equation

$$\frac{(1-x^2)^n}{(1-x)^n} = (1+x)^n$$

to show that

$$\sum_{k=0}^{m/2} (-1)^k \binom{n}{k}\binom{n+m-2k-1}{n-1} = \binom{n}{m}, \quad m \leqslant n \text{ and } m \text{ even}$$

29. Use binomial expansions to evaluate:

(a) $\displaystyle\sum_{k=0}^{m} \binom{m}{k}\binom{n}{r+k}$

(d) $\displaystyle\sum_{k=0}^{n} 2^k \binom{n}{k}$

(b) $\displaystyle\sum_{k=0}^{n} k \binom{n}{k}^2$

(e) $\displaystyle\sum_{k=1}^{n} (-1)^{k-1} k \binom{n}{k} 3^{n-k}.$

(c) $\displaystyle\sum_{k=0}^{r} (-1)^k \binom{n}{k}\binom{n}{r-k}$

30. (a) Evaluate $\displaystyle\sum_{k=n_1}^{n_2} \binom{k+5}{k}.$
    (b) Evaluate $\displaystyle\sum_{k=0}^{m} \binom{n-k}{m-k}.$

31. (a) Evaluate $\displaystyle\sum_{k=0}^{n_1} (-1)^k \binom{2n-k}{k}.$
    (b) Evaluate $\displaystyle\sum_{k=0}^{n} \binom{2n-k}{k} 2^k.$

32. Use the generalized binomial expansion of Section 5.5 to reverify (5).

33. (a) Use the generalized binomial expansion of Section 5.5 to show that

$(1-4x)^{1/2}$ is the generating function for $a_r = -\dfrac{2}{r}\binom{2r-2}{r-1}.$

Note: This identity has an important use in Section 7.6.

(b) Show that $(1-4x)^{-1/2}$ is the generating function for $a_r = \binom{2r}{r}.$

(c) Evaluate $\displaystyle\sum_{k=0}^{n} \binom{2k}{k}\binom{2n-2k}{n-k}.$

(d) Evaluate $\displaystyle\sum_{k=0}^{n} Y_k Y_{n-k}$, where $Y_k = -\dfrac{2}{k}\binom{2k-2}{k-1}.$

34. (a) Show that $\displaystyle\sum_{k=0}^{\infty} \binom{n+k-1}{k} \tfrac{1}{2}^k = 2^n.$
    (b) Evaluate $\displaystyle\sum_{k=1}^{\infty} \tfrac{1}{2}^k k.$

35. Use Exercise 33b to evaluate $\sum_{k=0}^{\infty} \binom{2k}{k} m^{-k}$, $m > 4$.

36. Why cannot one set $x = -1$ in formula (5) to "prove" that

$$\tfrac{1}{2}^n = \sum_{k=0}^{\infty} (-1)^k \binom{k + n - 1}{k}?$$

37. If $g(x)$ is the generating function for a_r, then show that $g^{(k)}(0)/k! = a_r$.

38. A *probability generating function* $p_X(t)$ for a discrete random variable X has a polynomial expansion in which p_r, the coefficient of t^r, is equal to the probability that $X = r$.
     (a) If X is the number of heads that occur when a coin is flipped n times, show that $p_X(t) = \tfrac{1}{2}^n(1 + t)^n$.
     (b) If X is the number of heads that occur when a biased coin is flipped n times with probability p of heads (and $q = 1 - p$), show that $p_X(t) = (q + pt)^n$.
     (c) If X is the number of times a fair coin is flipped until the fifth head occurs, find $p_X(t)$.
     (d) Repeat part (c) until the mth head occurs and probability of a head is p.

39. (a) The *expected value* $E(X)$ of a discrete random variable X is defined to be $\Sigma p_r r$. Show that $E(X) = p_X'(1)$, that is $\frac{d}{dt} p_X(t)$, with t set equal to 1.
     (b) Find $E(X)$ for the random variables X in Exercise 38.

40. (a) The *second moment* $E_2(X)$ of a discrete random variable X is defined to be $\Sigma p_r r^2$. Show that $E_2(X) = p_X'(1) + p_X''(1)$.
     (b) Find $E_2(X)$ for the random variables X in Exercise 38.

41. Suppose a fair coin is flipped until the mth head occurs and suppose that no more than s tails in a row occur. If X is the number of flips, find $p_X(t)$, $E(X)$, and $E_2(X)$.

42. Experiments A' and A'' have probabilities p' and p'' of success in each trial and are performed n' and n'' times, respectively. Let X' and X'' be the number of success in the respective experiments, and let X be the total number of successes on both experiments. Verify the following.
     (a) $p_X(t) = p_{X'}(t) p_{X''}(t)$
     (b) $E(X) = E(X') + E(X'')$
     (c) $E_2(X) = E_2(X') + E_2(X'') + E(X'X'')$

43. Suppose a red die is rolled once and then a green die is rolled as many times as the value on the red die. If a_r is the number of ways that the (variable length) sequence of rolls of the green die can sum to r, show that the generating function for a_r is $f(f(x))$, $f(x) = (x + x^2 + x^3 + x^4 + x^5 + x^6)$.

44. Suppose X is the random variable of the number of minutes it takes to serve a person at a fast-food stand. Suppose Y is the random variable of the number of people who line up to be served at the stand in one minute. Let Z be the number of people who line up while a person is being served.

    (a) Show $p_Z(t) = p_X(p_Y(t))$.
    (b) Find $E(Z)$ and $E_2(Z)$.

6.3 PARTITIONS

In this section we discuss partitions and their generating functions. Unfortunately, there is no easy way to calculate the coefficients of most of these generating functions. A **partition** of a group of r identical objects divides the group into a collection of (unordered) subsets of various sizes. Analogously we define a partition of the integer r to be a collection of positive integers whose sum is r. Normally we write this collection as a sum and list the integers of the partition in increasing order. For example, the seven partitions of the integer 5 are

$$
\begin{array}{lll}
1 + 1 + 1 + 1 + 1 & 1 + 2 + 2 & 2 + 3 \\
1 + 1 + 1 + 2 & 1 + 1 + 3 & 1 + 4
\end{array} \quad 5
$$

Note that 5 is a "trivial" partition of itself.

Let us construct a generating function for a_r, the number of partitions of the integer r. A partition of an integer is described by specifying how many 1's, how many 2's, and so on, are in the sum. Let e_k denote the number of k's in a partition. Then

$$
1e_1 + 2e_2 + \cdots + ke_k + \cdots + re_r = r
$$

Intuitively, we can think of picking r objects from an unlimited number of piles where the first pile contains single objects, the second pile contains objects stuck together in pairs, the third pile contains objects stuck together in triples, and so on. To model this integer-solution-to-an-equation problem with a generating function, we need polynomial factors whose formal multiplication yields products of the form

$$
\left\{ \begin{array}{c} x^0 \\ x^1 \\ x^2 \\ \vdots \end{array} \right\}
\left\{ \begin{array}{c} (x^2)^0 \\ (x^2)^1 \\ (x^2)^2 \\ \vdots \end{array} \right\}
\left\{ \begin{array}{c} (x^3)^0 \\ (x^3)^1 \\ (x^3)^2 \\ \vdots \end{array} \right\}
\cdots
\left\{ \begin{array}{c} (x^k)^0 \\ (x^k)^1 \\ (x^k)^2 \\ \vdots \end{array} \right\}
\cdots
$$

Thus the generating function $g(x)$ must be

$$g(x) = (1 + x + x^2 + \cdots + x^n + \cdots)$$
$$\cdot (1 + x^2 + x^4 + \cdots + x^{2n} + \cdots)$$
$$\cdot (1 + x^3 + x^6 + \cdots + x^{3n} + \cdots)$$
$$\vdots$$
$$\cdot (1 + x^k + x^{2k} + \cdots + x^{kn} + \cdots)$$
$$\vdots$$

If we set $y = x^2$, then the second factor in $g(x)$ becomes $1 + y + y^2 \cdots =$ $(1 - y)^{-1}$. Thus

$$1 + x^2 + x^4 + \cdots + x^{2n} + \cdots = (1 - x^2)^{-1}$$

Similarly, the kth factor can be written as $(1 - x^k)^{-1}$. For partitions up to $r = m$, we need the first m polynomial factors. But for arbitrary values of r, we need an infinite number of polynomial factors. Then

$$g(x) = \frac{1}{(1 - x)(1 - x^2)(1 - x^3) \cdots (1 - x^k) \cdots}$$

We now consider some more specialized partition problems.

Example 1

Find the generating function for a_r, the number of ways to express r as a sum of distinct integers.

We must constrain the standard partition problem not to allow any repetition of an integer. For example, there are three ways to write 5 as a sum of distinct integers $2 + 3$, $1 + 4$, and 5. The appropriate modification of the generating function for unrestricted partitions is

$$g(x) = (1 + x)(1 + x^2)(1 + x^3)(1 + x^4) \cdots (1 + x^k) \cdots$$

■

Example 2

Find a generating function for a_r, the number of ways that we can choose 2¢, 3¢, and 5¢ stamps adding to a net value of r¢.

The problem is equivalent to the number of integer solutions to

$$2e_2 + 3e_3 + 5e_5 = r \qquad 0 \leqslant e_2, e_3, e_5$$

The appropriate generating function is

$$(1+x^2+x^4+x^6+\cdots)(1+x^3+x^6+x^9+\cdots)(1+x^5+x^{10}+x^{15}+\cdots)$$

$$= \frac{1}{(1-x^2)(1-x^3)(1-x^5)}$$

■

Example 3

Show with generating functions that every positive integer can be written as a unique sum of distinct powers of 2.

The generating function for a_r, the number of ways to write an integer r as a sum of distinct powers of 2, will be similar to the generating function for sums of distinct integers in Example 1, except that now only integers that are powers of 2 are used. The generating function is

$$g^*(x) = (1 + x)(1 + x^2)(1 + x^4)(1 + x^8) \cdots (1 + x^{2^k}) \cdots$$

To show that every integer can be written as a unique sum of distinct powers of 2, we must show that the coefficient of every power of x in $g^*(x)$ is 1. That is, show that

$$g^*(x) = 1 + x + x^2 + x^3 + \cdots = \frac{1}{1 - x}$$

or equivalently

$$(1 - x)g^*(x) = 1$$

We prove this identity with the following recursive technique.

$$(1 - x)g^*(x) = \underbrace{(1 - x)(1 + x)}(1 + x^2)(1 + x^4)(1 + x^8) \cdots$$
$$= \underbrace{(1 - x^2)(1 + x^2)}(1 + x^4)(1 + x^8) \cdots$$
$$= (1 - x^4)(1 + x^4)(1 + x^8) \cdots$$
$$\vdots$$
$$= 1$$

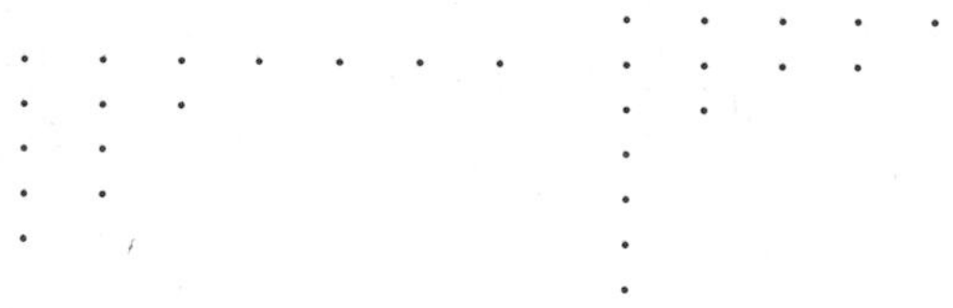

(a)   (b)

FIGURE 6.1

By successively making the replacement $(1 - x^{2^k})(1 + x^{2^k}) = 1 - x^{2^{k+1}}$, we eventually cancel all factors in $g^*(x)$, so that $(1 - x)g^*(x) = 1 - (?)$. Formally, the coefficient of each x^k must be 0. So $(1 - x)g^*(x) = 1$ and $g^*(x) = 1 + x + x^2 + \cdots$, as required. ∎

A convenient tool for studying partitions is a diagram known as Ferrers diagram. A **Ferrers diagram** displays a partition of r dots in a set of rows listed in order of decreasing size. The partition $1 + 2 + 2 + 3 + 7$ of 15 is shown in the Ferrers diagram in Figure 6.1a. If we transpose the rows and columns of a Ferrers diagram of a partition of r, we get a Ferrers diagram of another partition of r. This diagram is called the *conjugate* of the original Ferrers diagram. For example, Figure 6.1b shows the conjugate of the Ferrers diagram in Figure 6.1a. This new Ferrers diagram represents the partition of 15, $1 + 1 + 1 + 1 + 2 + 4 + 5$. Clearly, transposing is unique: two Ferrers diagrams have equal conjugates if and only if they are equal.

Example 4

Show that the number of partitions of an integer r as a sum of m positive integers is equal to the number of partitions of r as a sum of positive integers, the largest of which is m.

If we draw a Ferrers diagram of a partition of r into m parts, then the graph will have m rows. The transposition of such a diagram will have m columns, that is, the largest row will have m dots. Thus there is a one-to-one correspondence between these two classes of partitions. ∎

EXERCISES

SUMMARY OF EXERCISES   Exercises 1–12 involve generating function models for partitions. The next six exercises use Ferrers diagrams to verify partition identities. The last exercise calls for programming, and the preceding three are very challenging partition problems.

1.   List all partitions of the integer:
     (a) 4   (b) 6   (c) 8.

2. Find a generating function for a_r, the number of partitions of r into:
   (a) Even integers.   (b) Distinct odd integers.

3. Find a generating function for the number of ways to write the integer r as a sum of positive integers in which no integer appears more than three times.

4. Find a generating function for the number of integer solutions of $2x + 3y + 7z = r$ with:
   (a) $x, y, z \geqslant 0$ (b) $0 \leqslant z \leqslant 2 \leqslant y \leqslant 8 \leqslant x$

5. Find a generating function for the number of ways to make r cents change in pennies, nickels, dimes, and quarters.

6. Find the number of ways to give 13 cents change in 1954 pennies, 1974 pennies, 1959 nickels, 1979 nickels, and 1973 dimes.

7. Find a generating function for the number of ways to distribute r identical objects into:
   (a) Three indistinguishable boxes.
   (b) n indistinguishable boxes ($n \leqslant r$).
   (c) r indistinguishable boxes. (*Hint:* Remember r will be an index in the generating function, not a fixed parameter.)

8. (a) Show that the number of partitions of 10 into distinct parts (integers) is equal to the number of partitions of 10 into odd parts by listing all partitions of these two types.
   (b) Show algebraically that the generating function for partitions of r into distinct parts equals the generating function for partitions of into odd parts, hence the numbers of these two types of partitions are equal.

9. Show with generating functions that every positive integer has exactly one partition into distinct powers of 10; that is, it has a unique decimal representation.

10. (a) Prove the result in Example 3 by induction.
    (b) Prove Example 3's result directly by recursively substituting $(1 + x^k) = (1 - x^{2k})/(1 - x^k)$ in $g^*(x)$.

11. The equation in Example 3, $g^*(x)(1 - x) = 1$, can be rewritten $1 - x = 1/g^*(x)$. Use this latter equation to prove that among all partitions of an integer r, $r \geqslant 2$, into powers of 2, there are as many such partitions with an odd number of parts as with an even number of parts.

12. Let $R(r, k)$ denote the number of partitions of the integer r into k parts.
    (a) Show that $R(r, k) = R(r - 1, k - 1) + R(r - k, k)$.
    (b) Show that $\Sigma_{k=1}^r R(n - r, k) = R(n, r)$.

13. Use a Ferrers diagram to show that the number of partitions of an integer into at most m parts (a sum of m positive integers) is equal to the number of partitions of r into parts of size $\leqslant m$.

14. Use a Ferrers diagram to show that the number of partitions of an

integer into parts of even size is equal to the number of partitions into parts such that each part occurs an even number of times.

15. Interpret the integer multiplication mn, "m times n", to be the sum of m n's. Prove that $mn = nm$.

16. Show that the number of partitions of the integer n into three parts equals the number of partitions of $2n$ into three parts of size $<n$.

17. Show that the number of partitions of n is equal to the number of partitions of $2n$ into n parts.

18. Show that any number of partitions of $2r + k$ into $r + k$ parts is the same for any k.

19. Show that the number of partitions of $r + k$ into k parts is equal to:

   (a) The number of partitions of $r + \binom{k + 1}{2}$ into k distinct parts.

   (b) The number of partitions of r into parts of size $\leqslant k$.

20. (a) Find a generating function for a_n, the number of partitions that add up to at most n.

   (b) Find a generating function for a_n, the number of partitions of n into three parts in which no part is larger than the sum of the other two.

   (c) Find a generating function for a_n, the number of different (incongruent) triangles with integral sides and perimeter n.

21. Show that $\frac{1}{2}(1 - x)^{-3}[(1 - x)^{-3} + (1 + x)^{-3}]$ is the generating function for the number of ways to toss r identical dice and obtain an even sum.

22. (a) A partition of an integer r is *self-conjugate* if the Ferrers diagram of the partition is equal to its own transpose. Find a one-to-one correspondence between the self-conjugate partitions of r and the partitions of r into distinct odd parts.

   (b) The largest square of dots in the upper left-hand corner of a Ferrers diagram is called the *Durfee square* of the Ferrers diagram. Find a generating function for the number of self-conjugate partitions of r whose Durfee square is size k (a $k \times k$ array of dots). *Hint:* Use a 1-1 correspondence between these and the partitions of $r - k^2$ into even parts of size at most $2k$.

   (c) Show that

   $$(1 + x)(1 + x^3)(1 + x^5) \cdot \cdot \cdot$$

   $$= 1 + \sum_{k=1}^{\infty} \frac{x^{k^2}}{(1 - x^2)(1 - x^4)(1 - x^6) \cdot \cdot \cdot (1 - x^{2k})}$$

23. Write a computer program to determine the number of all partitions of an integer r:
   (a) Into k parts.

(b)   Into any number of parts.
(c)   Into distinct parts.

6.4 EXPONENTIAL GENERATING FUNCTIONS

In this section we discuss exponential generating functions. They are used to model and solve problems involving arrangements and distributions of distinct objects. Exponential generating functions involve a more complicated modeling process than do ordinary generating functions. Consider the problem of finding the number of different words (arrangements) of four letters when the letters are chosen from an unlimited supply of a's, b's and c's, and the word must contain at least two a's. The possible sets of four letters to form the word are $\{a,a,a,a\}$, $\{a,a,a,b\}$, $\{a,a,a,c\}$, $\{a,a,b,b\}$, $\{a,a,b,c\}$, and $\{a,a,c,c\}$. The number of arrangements possible with each of these six sets is

$$\frac{4!}{4!0!0!} \qquad \frac{4!}{3!1!0!} \qquad \frac{4!}{3!0!1!} \qquad \frac{4!}{2!2!0!} \qquad \frac{4!}{2!1!1!} \qquad \frac{4!}{2!0!2!}$$

respectively. So the total number of words will be the sum of these six terms.

The sets of letters used in such a word range over all sets of four letters chosen with repetition from a's, b's and c's with at least two a's. Equivalently, the number of such sets equals the number of integer solutions to the equation

$$e_1 + e_2 + e_3 = 4, \qquad 2 \leq e_1, \qquad 0 \leq e_2, e_3 \tag{1}$$

At first sight, the four-letter words problem seems similar to previous problems that could be modeled by (ordinary) generating functions. However, in this case we do not want each integer solution of (1) to contribute 1 to the total number of words. Instead it must contribute $4!/(e_1!e_2!e_3!)$ words. In terms of generating functions, the coefficient of x^4 will not count all formal products $x^{e_1}x^{e_2}x^{e_3}(e_1 \geq 2)$ whose exponents sum to 4 but instead will count all formal products with associated coefficients

$$\frac{(e_1 + e_2 + e_3)!}{e_1!e_2!e_3!} x^{e_1}x^{e_2}x^{e_3}, \qquad e_1 \geq 2$$

whose exponents sum to 4 (a much harder problem). Fortunately, exponential generating functions yield formal products of exactly this form.

An **exponential generating function** $g(x)$ for a_r, the number of arrangements (or distinct-object distributions) with r objects, is a function with the power series expansion

$$g(x) = a_0 + a_1 x + a_2 \frac{x^2}{2!} + a_3 \frac{x^3}{3!} + \cdots + a_r \frac{x^r}{r!} + \cdots$$

We build exponential generating functions in the same way that we build ordinary generating functions: one polynomial factor for each type of object (or each box); each factor has a collection of powers of x that are an inventory of the choices for the number of objects of that type. However now each power x^r is divided by $r!$

As an example, let us consider the four-letter word problem with at least two a's. We claim that the exponential generating function for the number of r-letter words formed from an unlimited number of a's, b's, and c's containing at least two a's is

$$g(x) = \left(\frac{x^2}{2!} + \frac{x^3}{3!} + \frac{x^4}{4!} + \cdots\right)\left(1 + x + \frac{x^2}{2!} + \frac{x^3}{3!} + \cdots\right)^2 \qquad (2)$$

The coefficient of x^r will be the sum of all such products where $e_1 + e_2 + e_3 = r, 2 \le e_1, 0 \le e_2, e_3$. More importantly, if we divide x^r by $r!$ and compensate by multiplying its coefficient by $r!$, then the x^r term in $g(x)$ becomes

$$\left(\sum_{\substack{e_1+e_2+e_3=r \\ 2 \le e_1, 0 \le e_2, e_3}} \frac{r!}{e_1! e_2! e_3!}\right) \frac{x^r}{r!}$$

The coefficient of x^r will be the sum of all such products where $e_1 + e_2 + e_3 = r$, $2 \le e_1, 0 \le e_2, e_3$. More importantly, if we divide x^r by $r!$ and compensate by multiplying its coefficient by $r!$, then the x^r term in $g(x)$ becomes

$$\left(\sum_{\substack{e_1+e_2+e_3=r \\ 2 \le e_1, 0 \le e_2, e_3}} \frac{r!}{e_1! e_2! e_3!}\right) \frac{x^r}{r!}$$

This coefficient of $x^r/r!$ is just what we wanted. So the exponential generating function for the number of such r-letter words is indeed the expression (2). What makes exponential generating functions work is the "trick" of dividing x^r by $r!$ and multiplying the coefficient of x^r by $r!$.

Before we show how to calculate coefficients of an exponential generating function, let us give a few other examples of exponential generating function models.

Example 1

Find the exponential generating function for a_r, the number of r arrangements (without repetition) of n objects.

The exponential generating function is $(1 + x)^n$—our old binomial friend! So the $x^r/r!$ term is

$$\binom{n}{r} x^r = \frac{n!}{(n - r)!} \cdot \frac{x^r}{r!}$$

and $a_r = n!/(n - r)! = P(n,r)$, as required. ∎

Example 2

Find the exponential generating function for a_r, the number of different arrangements of r objects chosen from four different types of objects with each type of object appearing at least two and no more than five times.

The number of ways to arrange the r objects with e_i objects of the ith type is $r!/e_1!e_2!e_3!e_4!$ To sum up all such terms with the given constraints, we want the coefficient of $x^r/r!$ in $(x^2/2! + x^3/3! + x^44! + x^5/5!)^4$. ∎

Example 3

Find the exponential generating function for the number of ways to place r (distinct) people into three different rooms with at least one person in each room. Repeat with an even number of people in each room.

The required exponential generating functions are

$$\left(x + \frac{x^2}{2!} + \frac{x^3}{3!} + \cdots\right)^3 \quad \text{and} \quad \left(1 + \frac{x^2}{2!} + \frac{x^4}{4!} + \cdots\right)^3$$

∎

There are few identities or expansions to use in conjunction with exponential generating functions. As a result, there are only a limited number of exponential generating functions whose coefficients can be easily evaluated. The fundamental expansion for exponential generating functions is

$$e^x = 1 + x + \frac{x^2}{2!} + \frac{x^3}{3!} + \cdots + \frac{x^r}{r!} + \cdots \tag{3}$$

Replacing x by nx in (3), we obtain

$$e^{nx} = 1 + nx + n^2\frac{x^2}{2!} + n^3\frac{x^3}{3!} + \cdots + n^r\frac{x^r}{r!} + \cdots \tag{4}$$

These expansions are valid for all values of x. Expansion (3) is proved in most calculus texts and will be simply accepted as fact in this text.

There is no way to factor out a common power of x in exponential generating functions (without destroying the $r!$ denominators). The best that can be done is to subtract the missing (lowest) powers of x from e^x. For example, the exponential factor representing two or more of a certain type of objects can be written

$$\frac{x^2}{2!} + \frac{x^3}{3!} + \frac{x^4}{4!} + \cdots = e^x - 1 - x$$

Two other special expansions we sometimes need are

$$\tfrac{1}{2}(e^x + e^{-x}) = 1 + \frac{x^2}{2!} + \frac{x^4}{4!} + \frac{x^6}{6!} + \cdots \qquad (5)$$

$$\tfrac{1}{2}(e^x - e^{-x}) = x + \frac{x^3}{3!} + \frac{x^5}{5!} + \frac{x^7}{7!} + \cdots \qquad (6)$$

Let us work some examples using (3) to (6).

Example 4

Find the number of different r arrangements of objects chosen from unlimited supplies of n types of objects.

In Chapter 5 we would have solved this problem by arguing that there are n choices for the type of object in each of the r positions for a total of n^r different arrangements. Now let us solve this problem with exponential generating functions. The exponential generating function for this problem is

$$\left(1 + x + \frac{x^2}{2!} + \frac{x^3}{3!} = \cdots\right)^n = (e^x)^n = e^{nx}$$

By (4), the coefficient of $x^r/r!$ in this generating function is n^r.   ■

Example 5

Find the number of ways to place 25 people into three rooms with at least 1 person in each room.

In Example 3, we found the exponential generating function for this problem

$$\left(x + \frac{x^2}{2!} + \frac{x^3}{3!} + \cdots\right)^3 = (e^x - 1)^3$$

To find the coefficient of $x^r/r!$ in $(e^x - 1)^3$, we first must expand this binomial expression in e^x

$$(e^x - 1)^3 = e^{3x} - 3e^{2x} + 3e^x - 1$$

From (4), we get

$$e^{3x} - 3e^{2x} + 3e^x - 1 = \sum_{r=0}^{\infty} 3^r \frac{x^r}{r!} - 3 \sum_{r=0}^{\infty} 2^r \frac{x^r}{r!} + 3 \sum_{r=0}^{\infty} \frac{x^r}{r!} - 1$$

$$= \sum_{r=0}^{\infty} (3^r - 3 \cdot 2^r + 3) \frac{x^r}{r!} - 1$$

So the coefficient of $x^{25}/25!$ is $3^{25} - 3 \cdot 2^{25} + 3$.   ■

Example 6

Find the number of r-digit quaternary sequences (whose digits are 0, 1, 2, and 3) with an even number of 0's and an odd number of 1's.
   The exponential generating function for this problem is

$$\left(1 + \frac{x^2}{2!} + \frac{x^4}{4!} + \frac{x^6}{6!} + \cdots\right)\left(x + \frac{x^3}{3!} + \frac{x^5}{5!} + \cdots\right)\left(1 + x + \frac{x^2}{2!} + \frac{x^3}{3!} + \cdots\right)^2$$

Using identities (5) and (6), we can write this expression as

$$\tfrac{1}{2}(e^x + e^{-x}) \cdot \tfrac{1}{2}(e^x - e^{-x}) \cdot e^x \cdot e^x = \tfrac{1}{4}(e^{2x} - e^0 + e^0 - e^{-2x}) \cdot e^x \cdot e^x$$

$$= \tfrac{1}{4}(e^{2x} - e^{-2x})e^{2x} = \tfrac{1}{4}(e^{4x} - 1)$$

Then for $r > 0$, the coefficient of $x^r/r!$ is $\tfrac{1}{4}4^r = 4^{r-1}$. The simple form of this answer suggests that there should be some combinatorial argument for obtaining this short answer directly.   ■

EXERCISES

SUMMARY OF EXERCISES   Most of the first 15 exercises are similar to the examples in this section. Exercises 21 and 22 are a continuation of the probability generating function exercises in Section 6.2.

1.   Find the exponential generating function for the number of arrangements of r objects chosen from five different types with at most five of each type.

2. Find the exponential generating function for the number of ways to distribute r people into six different rooms with between two and four in each room.

3. Find an exponential generating function for a_r, the number of r-letter words with no vowel used more than once (consonants can be repeated).

4. (a) Find the exponential generating function for $s_{n,r}$, the number of ways to distribute r distinct objects into n distinct boxes with no empty box. Consider n a fixed constant.
   (b) Determine $s_{n,r}$. The number $s_{n,r}/n!$ is called a *Stirling number of the second kind*.

5. Find the exponential generating function, and identify the appropriate coefficient, for the number of ways to deal a sequence of 13 cards (from a standard 52-card deck) if the suits are ignored and only the values of the cards are noted.

6. How many ways are there to distribute eight different toys among four children if the first child gets at least two toys?

7. How many r-digit ternary sequences are there with:
   (a) An even number of 0's?
   (b) An even number of 0's and even number of 1's?

8. How many ways are there to make an r-arrangement of pennies, nickels, dimes, and quarters with at least one penny and an odd number of quarters? (Coins of the same denomination are identical.)

9. How many 10-letter words are there in which each of the letters e, n, r, s occur:
   (a) At most once?   (b) At least once?

10. How many r-digit ternary sequences are there in which:
    (a) No digit occurs exactly twice?
    (b) 0 and 1 each appear a positive even number of times?

11. How many r-digit quaternary sequences are there in which the total number of 0's and 1's is even?

12. Find the exponential generating function for the number of ways to distribute r distinct objects into five different boxes when $b_1 < b_2 \leq 4$, where b_1, b_2 are the numbers of objects in boxes 1 and 2, respectively.

13. (a) Find the exponential generating functions for p_r, the probability that the first two boxes each have at least two objects when r distinct objects are randomly distributed into n distinct boxes. (*Hint:* See Exercise 4.)
    (b) Determine p_r.

14. Find an exponential generating function for the number of distributions of r distinct objects into n different boxes with exactly m nonempty boxes.

15. Find an exponential generating function with:
   (a) $a_r = 1/(r + 1)$ (b) $a_r = r!$

16. Show that if $g(x)$ is the exponential generating function for a_r, then $g^{(k)}(0) = a_k$, where $g^{(k)}(x)$ is the kth derivative of $g(x)$.

17. If

$$f(x) = \sum_{r=0}^{\infty} a_r \frac{x^r}{r!}, g(x) = \sum_{r=0}^{\infty} b_r \frac{x^r}{r!}$$

and

$$h(x) = f(x)g(x) = \sum_{r=0}^{\infty} c_r \frac{x^r}{r!}$$

then show that

$$c_r = \sum_{r=0}^{\infty} \binom{r}{k} a_k b_{r-k}$$

18. Show that $e^x e^y = e^{(x+y)}$ by formally multiplying the expansions of e^x and e^y together.

19. Find a combinatorial argument to show why the answer in Example 6 is 4^{r-1}.

20. Show that $e^x/(1 - x)^n$ is the exponential generating function for the number of ways to choose some subset (possibly empty) of r distinct objects and distribute them into n different boxes with the order in each box counted.

21. A Poisson random variable X has $p_r = \text{Prob}(x = r) = \dfrac{\mu^r}{r!} e^{-\mu}$. Find the probability generating function for X (see Exercise 38 in Section 6.2).

22. Let $P(x) = \sum_{k=0}^{\infty} p_k x^k$ be the probability generating function for the discrete random variable X, that is, p_k is the probability that X is k.
   (a)   Show that the exponential generating function for m_k, the kth moment of X, $m_k = \sum_{j=0}^{\infty} j^k p_j$, is $P(e^x)$.
   (b)   Show that the exponential generating function for m_k^*, the kth factorial moment of X, $m_k^* = \sum_{j=k}^{\infty} \dfrac{j!}{(j = k)!} p_j$, is $P(x + 1)$.
   (c)   If X is the number of heads when n coins are flipped, find m_1, m_2, m_2^*. (*Hint:*   Use Exercise 16.)
   (d)   If X is Poisson (see Exercise 21), find m_1 and m_1^*.

6.5 A SUMMATION METHOD

In this section we show how to construct an (ordinary) generating function $h(x)$ whose coefficient of x^r is some specified function $p(r)$ of r, such as r^2 or $C(r,3)$. Then we use $h(x)$ to calculate the sums $p(0) + p(1) + \cdots + p(n)$, for each positive n. The following four simple rules for constructing a new generating function from an old one will be used repeatedly in this section. Assume that $A(x) = \Sigma a_n x^n$, $B(x) = \Sigma b_n x^n$, and $C(x) = \Sigma c_n x^n$.

(i)   If $b_n = da_n$, then $B(x) = dA(x)$, for any constant d.

(ii)   If $c_n = a_n + b_n$, then $C(x) = A(x) + B(x)$.

(iii)   If $c_n = \Sigma_{i=0}^n a_i b_{n-i}$, then $C(x) = A(x)B(x)$.

(iv)   If $b_n = a_{n-k}$, except $b_i = 0$ for $i < k$, then $B(x) = x^k A(x)$.

Rule (iii) is simply expression (6) from Section 6.2. The other rules are immediate.

The other basic operation for our coefficient construction is to multiply each coefficient a_r in a generating function $g(x)$ by r. The new generating function $g^*(x)$ with $a_r^* = ra_r$ is obtained by differentiating $g(x)$ and then multiplying by x, that is, $g^*(x) = x\left(\dfrac{d}{dx} g(x)\right)$. For if

$$g(x) = a_0 + a_1 x + a_2 x^3 + \cdots + a_r x^r + \cdots \tag{1}$$

then differentiation of $g(x)$ yields

$$\frac{d}{dx} g(x) = a_1 + 2a_2 x + 3a_3 x^2 + \cdots + ra_r x^{r-1} + \cdots \tag{2}$$

and now multiplying by x (rule (iv)) gives

$$g^*(x) = x\left(\frac{d}{dx} g(x)\right) = a_1 x + 2a_2 x^2 + 3a_3 x^3 + \cdots + ra_r x^r + \cdots \tag{3}$$

Note that the a_0 term in $g(x)$ disappears in (2) because $0a_0 = 0$.

Combining this operation with rules (i) and (ii), we can repeatedly multiply the coefficient of x^r by r or a constant and can add such coefficients together. This permits us to form polynomials in r.

The natural question now is, to what coefficients do we apply these operations. The natural answer is, when in doubt start with the unit coefficients

$a_r = 1$ of the generating function

$$\frac{1}{1-x} = 1 + x + x^2 + \cdots + x^r + \cdots$$

Example 1

Build a generating function $h(x)$ with $a_r = 2r^2$.

Starting with $1/(1-x)$, we multiply coefficients by r using the generating function operations shown in (2) and (3) to obtain

$$x\left(\frac{d}{dx}\frac{1}{1-x}\right) = x\frac{1}{(1-x)^2} = 1x + 2x^2 + 3x^3 + \cdots + rx^r + \cdots$$

Now we repeat these operations on $x/(1-x)^2$ to obtain

$$x\left(\frac{d}{dx}\frac{x}{(1-x)^2}\right) = \frac{x(1+x)}{(1-x)^3} = 1^2x + 2^2x^2 + 3^2x^3 + \cdots + r^2x^r + \cdots$$

Finally we multiply by 2 to obtain the required generating function

$$h(x) = \frac{2x(1+x)}{(1-x)^3} = 2 \cdot 1^2x + 2 \cdot 2^2x^2 + 2 \cdot 3^2x^3 + \cdots + 2 \cdot r^2x^r + \cdots$$

■

Example 2

Build a generating function $h(x)$ with $a_r = (r+1)r(r-1)$.

We could multiply $(r+1)r(r-1)$ out getting $a_r = r^3 - r$, obtain generating functions for r^3 and r as an Example 1, and then subtract one generating function from the other. It is easier however to start with $3!(1-x)^{-4}$, whose coefficient a_r equals

$$a_r = 3!\binom{r+4-1}{r} = 3!\frac{(r+3)!}{r!3!} = \frac{(r+3)!}{r!} = (r+3)(r+2)(r+1)$$

Then the power series expansion of $3!(1-x)^{-4}$ is

$$\frac{3!}{(1-x)^4} = 3 \cdot 2 \cdot 1 + 4 \cdot 3 \cdot 2x + 5 \cdot 4 \cdot 3x^2 + \cdots$$

$$+ (r+3)(r+2)(r+1)x^r + \cdots \tag{4}$$

Compare (4) with the desired generating function

$$h(x) = 3 \cdot 2 \cdot 1 x^2 + 4 \cdot 3 \cdot 2 x^3 + 5 \cdot 4 \cdot 3 x^4 + \cdots$$
$$+ (r + 1)r(r - 1)x^r + \cdots$$

Clearly $h(x)$ is just the series in (4) multiplied by x^2. So $h(x) = 3!x^2(1 - x)^{-4}$.

■

Generalizing the construction in Example 2, we see that $(n - 1)!(1 - x)^{-n}$ has a coefficient

$$a_r = (n - 1)!C(r + n - 1, r) = (r + (n - 1))(r + (n - 2)) \cdots (r + 1).$$

Any coefficient involving a product of decreasing terms, such as $(r + 1)r\ (r - 1)$, can be built from $(n - 1)!(1 - x)^{-n}$ as in Example 2, where $n - 1$ is the number of terms in the product.

Thus far the construction of an $h(x)$ with specified coefficients has just been an exercise in algebraic manipulations of generating functions. The following easily verified theorem gives some purpose to this exercise.

Theorem 1

If $h(x)$ is a generating function with a_r the coefficient of x^r, then $h^*(x) = h(x)/(1 - x)$ is a generating function of the sums of the a_r's. That is,

$$h^*(x) = a_0 + (a_0 + a_1)x + (a_0 + a_1 + a_2)x^2 + \cdots + \left(\sum_{i=0}^{r} a_i\right)x^r + \cdots$$

This theorem follows from rule (iii) for the coefficients of the product $h^*(x) = f(x)h(x)$, where $f(x) = 1/(1 - x)$. Now we return to the previous examples.

Example 1 (continued)

Evaluate the sum $2 \cdot 1^2 + 2 \cdot 2^2 + 2 \cdot 3^2 + \cdots + 2n^2$.

The generating function $h(x)$ for $a_r = 2r^2$ was found to be $2x(1 + x)/(1 - x)^3$. Then by Theorem 1, the desired sum $a_1 + a_2 + \cdots + a_n$ is the coefficient of x^n in

$$h^*(x) = h(x)/(1 - x) = 2x(1 + x)/(1 - x)^4$$
$$= 2x(1 - x)^{-4} + 2x^2(1 - x)^{-4}$$

The coefficient of x^n in $2x(1 - x)^{-4}$ is the coefficient of x^{n-1} in $2(1 - x)^{-4}$, and the coefficient of x^n in $2x^2(1 - x)^{-4}$ is the coefficient of x^{n-2} in $2(1 - x)^{-4}$. Thus the given sum equals

$$2 \binom{(n-1)+4-1}{n-1} + 2 \binom{(n-2)+4-1}{n-2} = 2 \binom{n+2}{3} + 2 \binom{n+1}{3}$$

■

Example 2 (continued)

Evaluate the sum $3 \cdot 2 \cdot 1 + 4 \cdot 3 \cdot 2 + \cdots + (n+1)n(n-1)$.

The generating function $h(x)$ for $a_r = (r+1)r(r-1)$ was found to be $6x^2(1-x)^{-4}$. By Theorem 1 the desired sum is the coefficient of x^n in $h^*(x) = h(x)/(1-x) = 6x^2(1-x)^{-5}$. This coefficient is the x^{n-2} term in $6(1-x)^{-5}$, that is, $6C((n-2)+5-1, n-2) = 6C(n+2,4)$.   ■

Note that these two summation problems can also be solved with the summation method in Section 5.6 that was based on a binomial identity. Both methods have their advantages.

EXERCISES

1. Find ordinary generating functions whose coefficient a_r equals:
   (a) r (c) $3r^2$ (e) $r(r-1)(r-2)(r-3)$.
   (b) 13   (d) $3r+7$

2. Evaluate the sums (using generating functions):
   (a) $0 + 1 + 2 + \cdots + n$
   (b) $13 + 13 + \cdots + 13$
   (c) $0 + 3 + 12 + \cdots + 3n^2$
   (d) $7 + 10 + \cdots + 3n + 7$
   (e) $4 \cdot 3 \cdot 2 \cdot 1 + 5 \cdot 4 \cdot 3 \cdot 2 + \cdots + n(n-1)(n-2)(n-3)$.

3. Find a generating function with $a_r = r(r+2)$ (do not add together generating functions for r^2 and for $2r$).

4. (a) Show how r^2 and r^3 can be written as linear combinations of $P(r,3)$, $P(r,2)$, and $P(r,1)$.
   (b) Use part a to find a generating function for $3r^3 - 5r^2 + 4r$.

5. Find a generating function for
   (a) $a_r = (r-1)^2$ (b) $a_r = 1/r$.

6. If $h(x)$ is the ordinary generating function for a_r, what is the coefficient of x^r in $h(x)(1-x)$ (give your answer in terms of the a_r's).

7. Verify Theorem 1 in this section.

8. If $h(x)$ is the ordinary generating function for a_r, find the generating function for $s_r = \Sigma_{k=r+1}^{\infty} a_k$, assuming all s_r's are finite and $a_r \to 0$ as $n \to \infty$.

6.6 SUMMARY AND REFERENCES

Generating functions are at once a simpleminded and a sophisticated mathematical model for counting problems; simpleminded because polynomial multiplication is a familiar, seemingly well-understood part of high school algebra and sophisticated because with standard algebraic manipulations on generating functions, one can solve complicated counting problems (some of which cannot be solved by the combinatorial arguments of Chapter 5). These algebraic manipulations automatically perform the correct combinatorial reasoning for us! Note that generating functions are an elementary example of the algebraic approach that pervades the research frontiers of contemporary mathematics. Letting algebraic expressions, whether they model combinatorial, geometric, or functional information, do the work for us is what much of modern mathematics is all about.

In this chapter, generating functions were used to model selection and arrangement problems with constrained repetition. Partition problems were also modeled (but were not solved). Finally we showed how to construct generating functions whose coefficient for x^r was a given function of r and used these generating functions to evaluate related sums. In the next three chapters, generating functions will be used to model and solve other combinatorial problems. Exercise 38 of Section 6.2 introduces one type of generating function used in probability theory (for more, see Feller [2]). Laplace and Fourier transforms in analysis are also generating functions [the Fourier transform of a function $f(t)$ can be viewed as a generating function for $f(t)$'s Fourier coefficients].

The first use of combinatorial generating functions was by DeMoivre around 1720. He used them to derive a formula for Fibonacci numbers (this derivation is given in Section 7.6). In 1748, Euler used generating functions in his work on partition problems. The theory of combinatorial generating functions, developed in the late eighteenth century, was primarily motivated by parallel work on probability generating functions (see Exercises 38 to 44 in Section 6.2 and Exercises 21 and 22 in Section 6.4). Laplace made many contributions to both theories and presented the first complete treatment of both in his 1812 classic *Theorie Analytique des Probabilities*.

For a full discussion of the use of generating functions in combinatorial mathematics, see MacMahon [3] or Riordan [4]. For a nice presentation of partition problems, also see Berge [1].

1. C. Berge, *Principles of Combinatorics*, Academic Press, New York, 1971.

2. W. Feller, *An Introduction to Probability Theory and Its Applications*, vol. I, 2nd ed., John Wiley & Sons, New York, 1957.

3. P. MacMahon, *Combinatory Analysis*, vols. I and II (1915), reprinted in one volume, Chelsea Publishing, New York, 1960.

4. J. Riordan, *An Introduction to Combinatorial Analysis*, John Wiley & Sons, New York, 1958.

Chapter Seven
Recurrence Relations

7.1 RECURRENCE RELATION MODELS

In this chapter we show how a variety of counting problems can be modeled with recurrence relations. We then discuss methods of solving several common types of recurrence relations.

A **recurrence relation** is, in general, a recursive scheme for computing any sequence of numbers. In this chapter, we view a recurrence relation as a recursive formula that counts the number of ways to do a procedure involving n objects in terms of the number of ways to do it with fewer objects. That is, if a_k is the number of ways to do the procedure with k objects, for $k = 0, 1, 2, \ldots$, then a recurrence relation is an equation that expresses a_n as some function of preceding a_k's, $k < n$. The simplest recurrence relation is an equation such as $a_n = 2a_{n-1}$. The following equations display some of the forms of recurrence relations that we will derive to model counting problems in the chapter.

$$a_n = c_1 a_{n-1} + c_2 a_{n-2} + \cdots + c_r a_{n-r} \quad \text{where } c_i \text{ are constants,}$$

$$a_n = c a_{n-1} + f(n) \quad \text{where } f(n) \text{ is some function of } n,$$

$$a_n = a_0 a_{n-1} + a_1 a_{n-2} + \cdots + a_{n-1} a_0$$

$$a_{n,m} = a_{n-1,m} + a_{n-1,m-1}$$

Just as mathematical induction is a proof technique that verifies a formula or assertion by recursively checking its validity for sets of increasing size, so a recurrence relation is a counting technique that solves an enumeration problem by recursively computing the answer for successively larger values of n.

The observant reader will remember that mathematical induction also involves an initial step of verifying the formula or assertion for some starting (smallest) value of n. The same is true for a recurrence relation. We cannot recursively compute the next a_n unless we initially are given a starting a_0. If the right-hand side of a recurrence relation involves the r preceding a_k's, then we need to be given the first r values, $a_0, a_1, \ldots, a_{r-1}$. For example, in the relation $a_n = a_{n-1} + a_{n-2}$, knowing only that $a_0 = 2$ is insufficient. If given also that $a_1 = 3$, then we use the relation to obtain $a_2 = a_1 + a_0 = 3 + 2 = 5$; $a_3 = a_2 + a_1 = 5 + 3 = 8$; $a_4 = 8 + 5 = 13$; $a_5 = 13 + 8 = 21$; $a_6 = 34$; $a_7 = 55$; and so on. The information about starting values needed to compute with a recurrence relation is called the **initial conditions**.

If we can devise a recurrence relation to model a counting problem we are studying and also determine the initial conditions, then it is usually possible to solve the problem for moderate sizes of n, such as $n = 15$, in a few minutes by recursively computing successive values of a_n up to the desired n. For larger values of n, a programmable calculator or computer is needed.

It is not difficult to write recursively structured programs to enumerate all outcomes of a counting problem that has been modeled with a recurrence relation. Such programs are the topic of Section 7.3. For many common types of recurrence relations, there are explicit solutions for a_n. Sections 7.4–7.6 discuss some of these solutions.

Example 1

Find a recurrence relation for the number of ways to arrange n distinct objects in a row. Find the number of arrangements of 8 objects.

Let a_n denote the number of arrangements of n objects. There are n choices for the first object in the row. This choice can be followed by any arrangement of the remaining $n - 1$ objects; that is, by a_{n-1} arrangements of the remaining $n - 1$ objects. Thus $a_n = na_{n-1}$. Substituting recursively in this relation we see that

$$a_n = na_{n-1} = n[(n - 1)a_{n-2}] = \cdots = n(n - 1)(n - 2) \cdots 2 \cdot a_1 = n!a_1$$

There is just one way to arrange one object; so $a_1 = 1$. Then $a_n = n!$. The number of arrangements of 8 objects is $a_8 = 8! = 40{,}320$. ∎

Of course, we showed in Chapter 5 that the number of arrangements of n objects is $n!$. It is often useful to try out a new technique on an old problem to see how it works before using it on new problems. Now for some new problems.

Example 2

An elf has a staircase of n stairs to climb. Each step it takes can cover either one stair or two stairs. Find a recurrence relation for a_n, the number of different ways for the elf to ascend the n-stair staircase.

It is easy to see that $a_1 = 1$, $a_2 = 2$ (two one-steps or one two-steps) and $a_3 = 3$. For $n = 4$, Figure 7.1a depicts one way of climbing the stairs, taking successive steps of sizes 1, 2, 1. Other possibilities are a two-step either preceded or followed by two one-steps, or two two-steps, or four one-steps. In all, we count five ways to climb the four stairs; so $a_4 = 5$.

Is there some systematic way to enumerate the ways to climb four stairs that breaks the problem into parts involving the ways to climb three or fewer stairs? Clearly, once the first step is taken there are three or fewer stairs remaining to climb. Thus we see that after a first step of one stair, there are a_3 ways to continue the climb up the remaining three stairs. If the first step covers two stairs, then there are a_2 ways to continue up the remaining two stairs. So $a_4 = a_3 + a_2$ (the values for a_2, a_3, a_4 satisfy this relation: $5 = 3 + 2$). This argument applies to the first step when climbing any number of stairs, as is shown in Figures 7.1b and 7.1c. Thus $a_n = a_{n-1} + a_{n-2}$. ∎

In Section 7.4 we obtain an explicit solution to this recurrence relation. The relation $a_n = a_{n-1} + a_{n-2}$ is called the **Fibonacci relation**. The numbers a_n generated by the Fibonacci relation with the initial conditions $a_0 = a_1 = 1$ are called the **Fabonacci numbers**. They begin 1, 1, 2, 3, 5, 8, 13, 21, 34, 55, 89. Fibonacci numbers arise naturally in many areas of combinatorial mathematics. There is even a journal, *Fibonacci Quarterly*, devoted solely to research involving the Fibonacci relation and Fibonacci numbers. Fibonacci numbers have been applied to other fields of mathematics, such as numerical analysis. They occur in the natural world—for example, the arrangements of petals in some flowers. For more information about the instances of Fibonacci numbers in nature see [1].

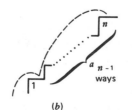

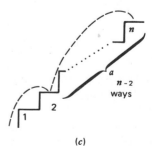

(a)   (b)   (c)

FIGURE 7.1

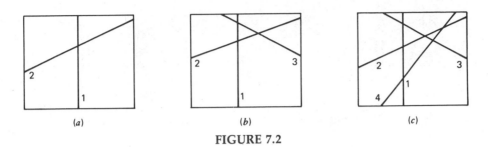

FIGURE 7.2

Example 3

Suppose we draw n straight lines on a piece of paper so that every pair of lines intersect (but no three lines intersect at a common point). Into how many regions do these n lines divide the plane?

Again we approach the problem by examining the situation for small values of n. With one line, the paper is divided into two regions. With two lines, four regions, that is, $a_2 = 4$. See Figure 7.2a. From Figure 7.2b, we see that $a_3 = 7$. The skeptical reader may ask, how do we know that three intersecting lines will always create seven regions? Let us go back one step then.

Clearly two intersecting lines will always yield four regions. Now let us examine the effect of drawing the third line (labeled "3" in Figure 7.2b). It must cross each of the other two lines (at different points). Before, between, and after these two intersection points, the third line cuts through three of the regions formed by the first two lines (this action of the third line does not depend on how it is drawn, just that it intersects the other two lines). So in severing three regions, the third line must form three new regions, actually create six new regions out of three old regions. Thus $a_3 = a_2 + 3 = 4 + 3 = 7$, independently of how the third line is drawn. Similarly, the fourth line severs four regions before, between, and after its three intersection points with the first 3 lines (see Figure 7.2), so that $a_4 = a_3 + 4 = 7 + 4 = 11$.

In general, the nth line must sever n regions before, between, and after its $n - 1$ intersection points with the first $n - 1$ lines. So $a_n = a_{n-1} + n$. ■

Example 4

The Tower of Hanoi is a game consisting of n circular rings of varying size and three pegs on which the rings fit. Initially the rings are placed on the first (leftmost) peg with the largest ring at the bottom covered by successively smaller rings. See Figure 7.3a. By transferring the rings among the pegs, one seeks to achieve a similarly tapered pile on the third (right-most) peg. The complication is that each time a ring is transferred to a new peg, the transferred ring must be smaller than any of the rings already piled on this new peg; equiva-

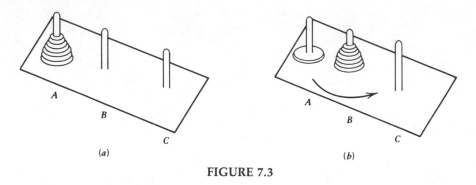

(a)   (b)

FIGURE 7.3

lently, at every stage in the game there must be a tapered pile (or no pile) on each peg.

Find a recurrence relation for the minimum number of moves required to play the Tower of Hanoi. How many moves are needed to play the six-ring game? Try playing this game with a dime, penny, nickel, and quarter (four rings) before reading our solution.

The key observation is that if, say, the four smallest rings are on peg A and we want to move them to peg C, we must first "play the three-ring Tower-of-Hanoi game" to get the three smallest rings from peg A to peg B, then move the fourth smallest ring from peg A to peg C, and then again "play the three-ring game" from peg B to peg C. (Of course, to move the three smallest rings we must move the third smallest from A to B, which means playing two-ring games from A to C and from C to B, etc.) Thus to move the nth ring from A to C, the $n - 1$ smaller rings must first be moved from A to B and later from B to C.

If a_n is the number of moves needed to transfer a tapered pile of n rings from peg A to peg C, then $a_n = a_{n-1} + 1 + a_{n-1} = 2a_{n-1} + 1$. The initial condition is $a_1 = 1$, and so $a_2 = 2a_1 + 1 = 3$; $a_3 = 2a_2 + 1 = 7$; $a_4 = 2a_3 + 1 = 15$; $a_5 = 2a_4 + 1 = 31$; and $a_6 = 2a_5 + 1 = 63$. So the six-ring game requires 63 moves. Note that the a_n's thus far fit the formula $a_n = 2^n - 1$. ∎

Example 5

A bank pays 8 percent interest each year on money in savings accounts. Find recurrence relations for the amounts of money a gnome would have after n years if it follows the investment strategy of:
(a) Investing \$1000 and leaving it in the bank for n years.
(b) Investing \$100 at the end of each year.

If an account has x dollars at the start of a year, then at the end of the year (i.e., start of the next year) it will have x dollars plus the interest on the x dollars, provided no money was added or removed during the year. Then for part a, the recurrence relation is $a_n = a_{n-1} + 0.08a_{n-1} = 1.08a_{n-1}$. The initial condition is $a_0 = 1000$. For part b, the relation must reflect the \$100 added (which earns no interest since it comes at the end of the year). So $a_n = 1.08a_{n-1} + 100$ with $a_0 = 0$. ∎

Example 6

Find a recurrence relation for the number of different ways to hand out a piece of chewing gum (worth 1¢) or a candy bar (worth 10¢) or a donut (worth 20¢) on successive days until n¢ worth of food has been given away.

This problem can be treated similarly to the stair climbing problem in Example 2. That is, if on the first day we hand out 1¢ worth of chewing gum, we are left with $(n - 1)$¢ worth of food to give away on following days; if the first day we hand out 10¢ worth of candy, we have $(n - 10)$¢ worth to dispense the next days; and if 20¢ then $(n - 20)$¢ the next days. So $a_n = a_{n-1} + a_{n-10} + a_{n-20}$ with $a_1 = 1$ and $a_0 = 1$ (one way to give nothing—by giving 0 pieces of each item). ∎

Example 7

Use recurrence relations to model the number of comparisons that must be made in the following algorithm for finding the largest number l and the smallest number s in a set S of n distinct integers.

Initially suppose that n is an even number. Assume that we have already found l_1 and s_1 the largest and smallest numbers, respectively, in the first half of S (the first $n/2$ numbers) and have found l_2 and s_2 the largest and smallest numbers in the second half of S. Then make two comparisons, one between l_1 and l_2 and the other between s_1 and s_2 to find the largest number l and smallest number s in S.

The associated recurrence relation for the number of comparisons in this procedure is $a_n = 2a_{n/2} + 2$, for $n \geqslant 4$ and even. If n is odd and we split S almost equally, the relation would be $a_n = a_{(n+1)/2} + a_{(n-1)/2}$. Observe that $a_1 = 0$, since the one number in both largest and smallest. And $a_2 = 1$, since we can determine the larger and smaller number in a two-element set with one comparison. With these two relations along with a_1 and a_2, we can recursively determine the number of comparisons needed for any n. ∎

The preceding examples developed recurrence relation models either by breaking a problem into a first (or last) step followed by the same problem for a smaller set (Examples 1, 2, 6, and 7) or by observing the *change* in going from the case of $n - 1$ to the case of n (Examples 3 and 5). Example 4 was a modified form of the former procedure in which the move of the nth smallest piece was seen to be preceded and followed by an $(n - 1)$-ring game.

There are two simple methods for solving some of the relations seen thus far. The first is recursive backward substitution: wherever a_{n-1} occurs in the relation for a_n, we replace a_{n-1} by the relation's formula for a_{n-1} (involving a_{n-2}) and then replace a_{n-2}, and so on. In Example 3, the relation $a_n =$

$a_{n-1} + n$ becomes $a_n = (a_{n-2} + n - 1) + n = \cdots = 1 + 2 + 3 + \cdots + n - 1 + n$. We used this method in Example 1 to obtain $a_n = n(n - 1)(n - 2) \cdots 2 \cdot 1$.

The second method is to guess the solution to the relation and then to verify it by mathematical induction. In Example 4, we noted that $a_n = 2^n - 1$ for the first six values of the recurrence relation $a_n = 2a_{n-1} + 1$, $a_1 = 1$. In particular, this was true for $n = 1$. Assuming this solution is true for a_{n-1}, we then have $a_n = 2a_{n-1} + 1 = 2(2^{n-1} - 1) + 1 = 2^n - 2 + 1 = 2^n - 1$. It follows by mathematical induction that the formula is true for all positive n.

Example 8

Use the following recurrence relation to determine the chromatic polynomial $P_k(G)$ of the graph G in Figure 7.4a. Recall that $P_k(G)$ gives the number of different k-colorings of G (see Example 9 in Section 5.2). For any two nonadjacent vertices x, y in G, define G_{xy}^+ and G_{xy}^c as follows: G_{xy}^+ is obtained from G by adding the edge (x,y); G_{xy}^c is obtained from G by coalescing vertices x and y into a single that is adjacent to all vertices formerly adjacent to x or y. Then

$$P_k(G) = P_k(G_{xy}^+) + P_k(G_{xy}^c)$$

since G_{xy}^+ corresponds to colorings of G in which x and y receive different colors, and G_{xy}^c corresponds to colorings of G in which x and y have the same color.

To apply this recurrence relation to the graph in Figure 7.4a, we need some "initial condition" graphs whose chromatic polynomial is known. The required building blocks are the collection of complete graphs on n vertices K_n (all vertices mutually adjacent). In Example 9 of Section 5.2 we showed that $P_k(K_n) = P(k,n)$.

Consider the non-adjacent pair b, d in G. Then $P_k(G) = P_k(G_{bd}^+) + P_k(G_{bd}^c)$. Figure 7.4b shows G_{bd}^+ and Figure 7.4c shows G_{bd}^c. To determine $P_k(G_{bd}^+)$ we use the non-adjacent pair c, e: $P_k(G_{bd}^+) = P_k((G_{bd}^+)_{ce}^+) + P_k((G_{bd}^+)_{ce}^c) = P_k(K_5) + P_k(K_4)$. So $P_k(G_{bd}^+) = P(k,5) + P(k,4)$. Similarly, $P_k(G_{bd}^c) = P_k((G_{bd}^c)_{ce}^+) + P_k((G_{bd}^c)_{ce}^c) = P_k(K_4) + P_k(K_3) = P(k,4) + P(k,3)$. In total, $P_k(G) = P(k,5) + 2P(k,4) + P(k,3)$. ∎

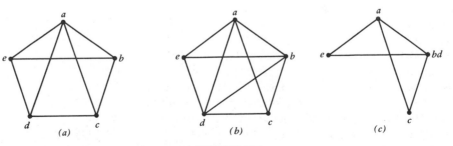

FIGURE 7.4

We now consider some more complex recurrence relations involving two-variable relations and simultaneous relations.

Example 9

Let $a_{n,k}$ denote the number of ways to select a subset of k objects from a set of n distinct objects. Find a recurrence relation for $a_{n,k}$.

Observe that $a_{n,k}$ is simply $C(n,k)$. We break the problem into two subcases based on whether or not the first object is used. There are $a_{n-1,k}$ k-subsets that do not use the first object, and there are $a_{n-1,k-1}$ k-subsets that use the first object. So we have $a_{n,k} = a_{n-1,k} + a_{n-1,k-1}$. This is identity (2) from Section 5.5. The initial conditions are $a_{n,0} = a_{n,n} = 1$ for all $n \geqslant 0$ (and $a_{n,k} = 0$, $k > n$). ∎

Example 10

Find a recurrence relation for the ways to distribute n identical balls into k distinct boxes with between two and four balls in each box. Repeat the problem with balls of three colors.

In the spirit of the stair-climbing problem (Example 2), we consider how many balls go in the first box. If there are two in the first box, then there are $a_{n-2,k-1}$ ways to put the remaining $n - 2$ identical balls in the remaining $k - 1$ boxes. Continuing this line of reasoning we see that $a_{n,k} = a_{n-2,k-1} + a_{n-3,k-1} + a_{n-4,k-1}$. The initial conditions are $a_{2,1} = a_{3,1} = a_{4,1} = 1$ and $a_{k,1} = 0$ otherwise.

If now three colors are allowed, there are $C(2 + 3 - 1,2) = 6$ ways to pick a subset of two balls for the first box from three types of colors with repetition. Similarly, there are $C(3 + 3 - 1,3) = 10$ ways to pick three balls and $C(4 + 3 - 1,4) = 15$ ways to pick four balls. Then $a_{n,k} = 6a_{n-2,k-1} + 10a_{n-3,k-1} + 15a_{n-4,k-1}$ with initial conditions $a_{2,1} = 6$, $a_{3,1} = 10$, $a_{4,1} = 15$, and $a_{k,1} = 0$ otherwise.

The problem with all balls identical can be solved by generating functions (see Example 3 of Section 6.1), but recurrence relations are the only practical approach with the extra constraint of different types of balls. ∎

Example 11

Find a system of recurrence relations from which you can compute the number of n-digit binary sequences with an even number of 0's and for the number of n-digit ternary sequences with an even number of 0's and an even number of 1's.

Again we use the stair-climbing first-step analysis. If the sequence starts with a 1, then we require an even number of 0's in the remaining $(n - 1)$-digit sequence—a_{n-1} such sequences. If the sequence starts with a 0, we require an *odd* number of 0's in the remaining $n - 1$ digits. Since odd $(n - 1)$-digit sequences plus even $(n - 1)$-digit sequences are all 2^{n-1} $(n - 1)$-digit sequences, there are $2^{n-1} - a_{n-1}$ odd $(n - 1)$-digit sequences. Thus we get $a_n = a_{n-1} +$

$(2^{n-1} - a_{n-1}) = 2^{n-1}$. Our recurrence relation reduces into a formula for a_n. In words, every $(n - 1)$ digit yields an even n-digit sequence by adding a 1 in front if the sequence is already even and by adding a 0 if the sequence is odd.

Next we consider the case of even 0's and even 1's in a ternary (0,1,2) sequence. We will need simultaneous recurrence relations for a_n, the number of n-digit (ternary) sequences with even 0's and even 1's; b_n, n-digit sequences with even 0's and odd 1's; and c_n, n-digit sequences with odd 0's and even 1's. Observe that $3^n - a_n - b_n - c_n$ is the number of n-digit ternary sequences with odd 0's and odd 1's. Now to get an n-digit sequence with even 0's and even 1's, we either have a 1 followed by a $(n - 1)$-digit sequence with even 0's and odd 1's, or a 0 followed by an $(n - 1)$-digit sequence with odd 0's and even 1's, or a 2 followed by an $(n - 1)$ sequence with even 0's and even 1's. Thus $a_n = b_{n-1} + c_{n-1} + a_{n-1}$. Similar analyses yield $b_n = a_{n-1} + (3^{n-1} - a_{n-1} - b_{n-1} - c_{n-1}) + b_{n-1} = 3^{n-1} - c_{n-1}$ and $c_n = a_{n-1} + (3^{n-1} - a_{n-1} - b_{n-1} - c_{n-1}) + c_{n-1} = 3^{n-1} - b_{n-1}$. The initial conditions are $a_1 = b_1 = c_1 = 1$. To recursively compute values for a_n, we must simultaneously compute b_n and c_n. ∎

We close this section with a few words about difference equations. The **first (backward) difference** Δa_n of the sequence $(a_0, a_1, a_2, \ldots)$ is defined to be $a_n - a_{n-1}$. The second difference is $\Delta^2 a_n = \Delta a_n - \Delta a_{n-1} = a_n - 2a_{n-1} + a_{n-2}$, and so on. A difference equation is an equation involving a_n and its differences, for example, $2\Delta^2 a_n - 3\Delta a_n + a_n = 0$. Observe that

$$a_{n-1} = a_n - (a_n - a_{n-1}) = a_n - \Delta a_n$$

and

$$a_{n-2} = a_{n-1} - \Delta a_{n-1} = (a_n - \Delta a_n) - \Delta(a_n - \Delta a_n)$$
$$= a_n - 2\Delta a_n + \Delta^2 a_n$$

Similar equations can express a_{n-k} in terms of $a_n, \Delta a_n, \ldots, \Delta^k a_n$. Thus any recurrence relation can be rewritten as a difference equation, by expressing the a_{n-k}'s on the right-hand side in terms of a_n and its differences. Conversely, by writing Δa_n as $a_n - a_{n-1}$, and so on, any difference equation can be written as a recurrence relation. Difference equations are commonly used to approximate differential equations when solving differential equations on a computer. Difference equations have wide use in their own right as models for dynamical systems for which differential equations (which require continuous functions) are inappropriate. They are used in economics in models for predicting the gross national product in successive years. They are used in ecology to model the numbers of various species in successive years. As noted above, any difference equation model can also be formulated as a recurrence relation; however the behavior of a dynamical system is easier to analyze and explain in terms of differences. Refer to Goldberg [2] for further information about difference equations and their applications.

Example 12

Assume that if undisturbed by foxes, the number of rabbits increases each year by an amount αr_n, where r_n is the number of rabbits, but when foxes are present, each rabbit has probability βf_n of being eaten by a fox (f_n is the number of foxes). Foxes alone decrease by an amount γf_n each year, but when rabbits are present, each fox has probability δr_n of feeding and raising up a new young fox (death of foxes is included in the γf_n term). Give a pair of simultaneous difference equations describing the number of rabbits and foxes in successive years.

The information given about yearly changes in the two populations yields the difference equations:

$$\Delta r_n = \alpha r_n - \beta r_n f_n$$
$$\Delta f_n = -\gamma f_n + \delta r_n f_n$$

■

EXERCISES

SUMMARY OF EXERCISES   The first 34 exercises call for recurrence relation modeling similar to that in the Examples, with multiple indices and equations required in Exercises 22–30 and difference equations in Exercises 31–34. The remaining exercises are more advanced problems.

1. Find a recurrence relation for the number of ways to distribute n distinct objects into five boxes. What is the initial condition?

2. (a) Find a recurrence relation for the number of ways the elf in Example 2 can climb n stairs if each step covers either one or two or three stairs?
   (b) How many ways are there for the elf to climb five stairs?

3. Find a recurrence relation for the number of ways to arrange cars in a row with n spaces if we can use Cadillacs or Continentals or Fords. A Cadillac or Continental requires two spaces, while a Ford requires just one space.

4. (a) Find a recurrence relation for numbers of ways to go n miles by foot walking at 3 miles per hour or jogging at 6 miles per hour or running at 10 miles per hour; at the end of each hour a choice is made of how to go the next hour.
   (b) How many ways are there to go 20 miles?

5. Find a recurrence relation for the number of ways to distribute a total of n cents on successive days using 1947 pennies, 1958 pennies, 1971 pennies, 1951 nickels, 1967 nickels, 1959 dimes, and 1975 quarters.

6.   (a)   Find a recurrence relation for the number of n-digit binary sequences with no pair of consecutive 1's.
     (b)   Repeat for n-digit ternary sequences.
     (c)   Repeat for n-digit ternary sequences with no consecutive 1's or consecutive 2's.

7.   Find a recurrence relation for the number of pairs of rabbits after n months if (1) initially there is one pair of rabbits who were just born; and (2) every month each pair of rabbits that are over one month old have a pair of offspring (a male and a female).

8.   Show that the binomial sum

$$S_n = \binom{n+1}{0} + \binom{n}{1} + \binom{n-1}{2} + \cdots$$

satisfies the Fibonacci relation.

9.   Find a recurrence relation for a_n for the number of bees in the nth previous generation of a male bee, if a male bee is born asexually from a single female and a female bee has the normal male and female parents. The ancestoral chart at the right shows that $a_1 = 1, a_2 = 2, a_3 = 3$.

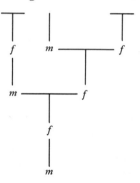

10.   Find a recurrence relation for a_n for the number of ways for an image to be reflected n times by internal faces of two adjacent panes of glass. The diagram below shows that $a_0 = 1, a_1 = 2,$ and $a_2 = 3$.

O reflections   1 reflection   2 reflections

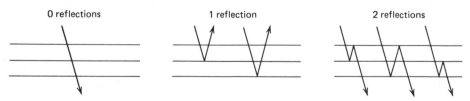

11.   Show that each of the following rules for playing the Tower of Hanoi works.
     (a)   On odd-numbered moves, move the smallest ring clockwise one peg (think of the pegs being at the corners of a triangle); and on

even-numbered moves, make the only legal move not using the smallest ring.

(b) Number the rings from 1 to n in order of increasing size. Never move the same ring twice in a row. Always put even-numbered rings on top of odd-numbered rings (or on an empty peg) and put odd-numbered rings on top of even-numbered rings (or on an empty peg).

12. Find a recurrence relation for the number of regions created by n mutually overlapping circles on a piece of paper (no three circles have a common intersection point).

13. (a) Find a recurrence relation for the number of regions created by n lines on a piece of paper if k of the lines are parallel and the other $n - k$ lines intersect all other lines (no three lines intersect at one point).
    (b) If $n = 9$ and $k = 3$, find the number of regions.

14. Find a recurrence relation for the amount of money in a savings account after n years if the interest rate is 6 percent and $50 are added at the start of each year.

15. (a) Find a recurrence relation for the amount of money outstanding on a $30,000 mortgage after n years if the interest rate is 8 percent and the yearly payment (paid at the end of each year after interest is paid) is $3000.
    (b) Use a calculator to determine how many years it will take to pay off the mortgage.

16. Suppose a coin is flipped until two heads appear (the two heads need not be consecutive), then the experiment stops. Find a recurrence relation for the number of experiments that end on the nth flip or sooner.

17. Find a recurrence relation for the number of n-digit quaternary (0,1,2,3) sequences with at least one 1 and the first 1 occurring before the first 0 (possibly no 0's).

18. A switching game has n switches, all initially in the OFF position. In order to be able to flip the ith switch, the $(i - 1)$st switch must be ON and all earlier switches OFF. The first switch can always be flipped. Find a recurrence relation for the total number of times the n switches must be flipped to get the nth switch ON and all others OFF.

19. (a) Let T be a binary tree with all leaves at level n. Find a recurrence relation for the number of leaves in T. Repeat for T an m-ary tree.
    (b) Let T be a binary tree with all leaves at level n. Find a recurrence relation for the sum of the lengths of all paths from the root to each leaf (one path for each leaf). Repeat for T an m-ary tree.

20. Find a recurrence relation for the number of ways to pair off $2n$ people for tennis matches.

21. Determine the chromatic polynomial for the following graphs:

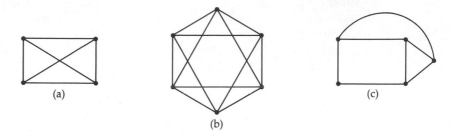

(a)

(b)

(c)

22. Find a recurrence relation for the number of ways to pick k objects with repetition from n types.

23. Find a recurrence relation for the number of ways to distribute n distinct objects into k indistinguishable boxes with no box empty.

24. Find a recurrence relation for the number of partitions of the integer n into k parts.

25. Find a recurrence relation for the number of ways to select n objects from k types with at most three of any one type.

26. Find a system of recurrence relations for the number of n-digit binary sequences with k adjacent pairs of 1's and no adjacent pairs of 0's.

27. Find a recurrence relation for the number $a_{n,m,k}$ of distributions of n identical objects into k distinct boxes with at most four objects in a box and with exactly m boxes having four objects.

28. Find a system of recurrence relations for computing the number of n-digit binary sequences with an even number of 0's and an even number of 1's.

29. Find a system of recurrence relations for computing the number of n-digit quaternary sequences with:
    (a) An even number of 0's.
    (b) An even total number of 0's and 1's.
    (c) An even number of 0's and an even number of 1's.

30. Find a system of recurrence relations for computing the number of ways to hand out a penny or a nickel or a dime on successive days until n cents are given such that the same amount of money is not handed on two consecutive days.

31. Find Δa_n and $\Delta^2 a_n$ if:
    (a) $a_n = 3n + 2$ (b) n^2 (c) n^3

32. If the half-life of the element combinatorium is one year (it gives off half as much radiation each successive year), give a difference equation for c_n, the radiation in the nth year.

33. Let f_n be the amount of food that can be bought with n dollars. Let p_n be the "perceived" value of the $\$n$ of food. Suppose the increase in per-

ceived value with one dollar more of food equals the relative, or percentage, increase in the actual amount of food. Find a difference equation relating Δp_n and Δf_n.

34. In Example 12, find equilibrium values for f_n and r_n (i.e., $\Delta f_n = \Delta r_n = 0$).

35. (a) Find a recurrence relation for the number of permutations of the first n integers such that each integer differs by one (except for the first digit) from some integer to the left of it in the permutation. What is the initial condition?
    (b) Solve the relation in part a by guessing and verifying the guess by induction.
    (c) Give a direct combinatorial answer to this problem.

36. Find a recurrence relation for the number of ways to pair off with nonintersecting lines $2n$ points on a circle. (*Hint:* The recurrence involves products of a_k's, i.e., $\ldots a_2a_{n-4} + \cdots$.)

37. Find a recurrence relation for the number of n-digit ternary sequences that have the pattern "012" occurring:
    (a) For the first time at the end of the sequence.
    (b) For the second time at the end of the sequence.

38. (a) Find a recurrence relation for $f(n,k)$, the number of k subsets of the integers 1 through n with no pair of consecutive integers.
    (b) Show that $\sum_{k=0}^{n/2} f(n,k) = F_{n+1}$, the $(n + 1)$st Fibonacci number (n even), where $F_0 = F_1 = 1$.
    (c) Repeat part a with the additional condition that n and 1 are considered consecutive.

39. Find a system of recurrence relations for computing the number of n-digit binary sequences with exactly one pair of consecutive 0's.

40. Find a system of recurrence relations for computing the number of ways to seat n couples around an (unoriented) round table with no pair of spouses adjacent.

41. Person A has r dollars and person B has s. They flip a coin and A wins \$1 from B if the coin comes up heads and vice versa for tails. This continues until one player is broke. Find a recurrence relation for the probability that A goes broke on the nth round.

42. Find a system of recurrence relations for computing the number of spanning trees in the "ladder" graph with $2n$ vertices.

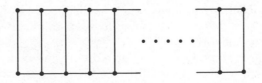

43. Find a system of recurrence relations for computing a_n, the number of

(unordered) collections of (identical) pennies, (identical) nickels, (identical) dimes, and (identical) quarters whose value is n cents.

44. Find a recurrence relation for the number of ways a coin can be flipped $2n$ times with:
    (a) The number of heads at any time never being less than the number of tails.
    (b) The number of heads never less than twice the number of tails.
    (c) The number of heads equaling the number of tails only after all $2n$ flips.

45. Find a recurrence relation for the number of incongruent integral-sided triangles whose perimeter is n (the relation is different for n odd and n even).

46. Verify the following identities for Fibonacci numbers (F_i is the ith Fibonacci number) by induction or combinatorial argument. Here $F_0 = F_1 = 1$.

    (a) $\sum_{i=0}^{n} F_i = F_{n+2} - 1$ (b) $\sum_{i=0}^{n} F_i^2 = F_n F_{n+1}$

    (c) $\sum_{k=0}^{n} F_{2k} = F_{2n+1}$ (d) $F_n F_{n+2} = F_{n+1}^2 + (-1)^n$

    (e) $F_1 - F_2 + F_3 - \cdots - F_{2n} = -F_{2n-1}$

47. (a) Show that $F_{n+m} = F_m F_n + F_{m-1} F_{n-1}$.
    (b) From part a conclude that F_{n-1} divides F_{kn-1}.

48. A set of cards is numbered from 1 to n. If the top card is k, we invert the order of the top k cards in the pile and again look at the new top card and invert again, and so on. Show that the maximum number of rounds until 1 is at the top is at most F_{n+1}. (*Hint:* Induct on the number k of top cards.)

7.2 DIVIDE-AND-CONQUER RELATIONS

In this section we present a special class of recurrence relations that arise frequently in the analysis of recursive computer algorithms. These are algorithms that use a "divide-and-conquer" approach to recursively split a problem into two subproblems of half the size. Many computer algorithms use a "divide-and-conquer" strategy, such as dictionary searches (see Example 3 in Section 3.1). A binary tree is explicitly or implicitly associated with "divide-and-conquer" algorithms.

The total number of steps a_n required by a divide-and-conquer algorithm to process an n-element problem frequently satisfies a recurrence relation of the form:

$$a_n = ca_{n/2} + f(n) \tag{1}$$

The following table indicates the form of the solution of (1) for some common values of c, d, and $f(n)$ ($[r]$ denotes the smallest integer m with $m \geq r$):

c	$f(n)$	a_n
$c = 1$	d	$d[\log_2 n] + A$
$c = 2$	d	$An + d$
$c \neq 2$	dn	$An^{\log_2 c} + \left(\dfrac{2d}{2 - c}\right) n$
$c = 2$	dn	$An\log_2 n + 2dn$

The constant A is to chosen to fit the initial condition.

If a problem is recursively split into k parts instead of two parts, then one should replace 2 by k everywhere in the above table. For example, the recurrence relation

$$a_n = ca_{n/k} + dn$$

has the solution

$$a_n = An^{\log_2 c} + \left(\frac{kd}{k - c}\right) n$$

The solutions for a_n given in the above table are easily verified by substitution. Consider the case: $a_n = ca_{n/2} + dn$, $c \neq d$. Substituting the solution in the table into the right-hand side of the relation, we have

$$ca_n + dn = c\left[A \left(\frac{n}{2}\right)^{\log_2 c} + \left(\frac{2d}{2 - c}\right)\frac{n}{2}\right] + dn$$

$$= \frac{cAn^{\log_2 c}}{2^{\log_2 c}} + \frac{cdn + (2 - c)\, dn}{2 - c}$$

$$= \frac{cAn^{\log_2 c}}{c} + \left(\frac{2d}{2 - c}\right) n$$

But this last expression is just a_n.

The following examples illustrate such "divide-and-conquer" recurrences and their solution.

Example 1

In a tennis tournament, each entrant plays a match the first round. Next all winners from the first round play a second-round match. Winners continue to move on to the next round, until finally only one player is left—the tournament winner. Assuming that tournaments always involve $n = 2^k$ players, for some k, find and solve a recurrence relation for the number of rounds in a tournament of n players.

In terms of binary trees, a_n is the height of a balanced binary tree with $n = 2^k$ leaves. The derived recurrence relation for a_n, the number of rounds, is

$$a_n = an_{n/2} + 1$$

since to get to the final round each finalist must win a subtournament involving $n/2$ players. The last round, one match, picks the tournament winner from the two finalists.

From the above table, the solution of this recurrence relation is $a_n = \log_2 n$. (If n is not a power of 2, we can "pad" the tournament by adding dummy players who play, and lose to, real players in the first round.) ∎

Example 2

Solve the recurrence relation $a_n = 2a_{n/2} + 2$, $n \geq 4$, with $a_2 = 1$, obtained in Example 7 of Section 7.1, for the number of comparisons to find the largest and smallest numbers in a set S_n of n distinct numbers when n is a power of 2.

The above table tells us that the solution will be of the form $a_n = A_1 n + A_2$. Substituting this form in the relation, we have

$$A_1 n + A_2 = a_n = 2a_{n/2} + 2 = 2(A_1 n/2 + A_2) + 2$$
$$= A_1 n + 2A_2 + 2$$

The $A_1 n$ terms on both sides of the relation cancel to yield $A_2 = 2A_2 + 2$ or $A_2 = -2$. Now we can use the initial condition $a_2 = 1$ to determine A_1:

$$1 = a_2 = A_1 2 - 2 \qquad \text{or} \qquad A_1 = \tfrac{3}{2}$$

So $a_n = (\tfrac{3}{2})n - 2$ is the number of comparisons needed to find the largest and smallest number according to the recursive procedure given in Example 7 of Section 7.1. (It is possible to prove that one cannot do better than $(\tfrac{3}{2})n - 2$ comparisons.) ∎

Example 3

Normally one must do n^2 digit-times-digit multiplications to multiply two n-digit numbers. Use a divide-and-conquer approach to develop a faster algorithm.

Let us initially assume that n is a power of 2. Let the two n-digit numbers be g and h. We split each of these numbers into two $n/2$-digit parts:

$$g = g_1 10^{n/2} + g_2, \qquad h = h_1 10^{n/2} + h_2$$

Then

$$g \cdot h = g_1 \cdot h_1 10^n + (g_1 \cdot h_2 + g_2 \cdot h_1) 10^{n/2} + g_2 \cdot h_2 \qquad (2)$$

Observe that

$$g_1 \cdot h_2 + g_2 \cdot h_1 = (g_1 + g_2) \cdot (h_1 + h_2) - g_1 \cdot h_1 - g_2 \cdot h_2$$

and so we only need to make three $n/2$-digit multiplications, $g_1 \cdot h_1$, $g_2 \cdot h_2$, and $(g_1 + g_2) \cdot (h_1 + h_2)$ to determine $g \cdot h$ in (2) (actually $(g_1 + g_2)$ or $(h_1 + h_2)$ may be $(n/2 + 1)$-digit numbers, but this slight variation does not affect the general magnitude of our solution). If a_n represents the number of digit-times-digit multiplications needed to multiply two n-digit numbers by the above procedure, then the above procedure yields the recurrence $a_n = 3a_{n/2}$.

By the table at the start of this section, a_n is proportional to $n^{\log_2 3} = n^{1.6}$—a substantial improvement over n^2. ■

EXERCISES

1. Solve the following recurrence relations:
    (a) $a_n = 3a_{n/3} + 5$ (b) $a_n = 2a_{n/4} + n$ (c) $a_n = a_{n/2} + 2n - 1$

2. Find and solve a recurrence relation for the number of matches played in a tournament with n players, where n is a power of 2.

3. In a large corporation with n salespeople, every 10 salespeople report to a local manager, every 10 local managers report to a district manager, and so forth until finally 10 vice-presidents report to the firm's president. If the firm has n salespeople, where n is a power of 10, find and solve recurrence relations for:
    (a) The number of different managerial levels in the firm.
    (b) The number of managers (up through president) in the firm.

4. In a tennis tournament, each player wins k hundreds of dollars where k is the number of people in the subtournament won by the player (the subsection of the tournament including the player, the player's victims, and their victims, and so forth; a player who loses in the first round gets $100). If the tournament has n contestants, where n is a power of 2, find and solve a recurrence relation for the total prize money in the tournament.

5. Consider the following method for rearranging the n distinct numbers x_1, x_2, . . . , x_n in order of increasing size (n is a power of 2). Pair the integers off $\{x_1, x_2\}$, $\{x_3, x_4\}$, and so forth. Compare each pair and put the smaller number first. Next pair off the pairs into sets of four numbers and merge the ordered pairs to get ordered four-tuples. Continue this process until the whole set is ordered. Find and solve a recurrence relation for the

total number of comparisons required to rearrange n distinct numbers. (*Hint:* First find the number of comparisons needed to merge two ordered k-tuples into an ordered $2k$-tuple.)

6. Verify by substitution the form of solution given in the text to the recurrence relations
   (a) $a_n = a_{n/2} + d$ (b) $a_n = 2a_{n/2} + d$ (c) $a_n = 2a_{n/2} + en$

7. (a) Use a divide-and-conquer approach to devise a procedure to find the largest number in a set of n distinct integers.
   (b) Give a recurrence relation for the number of comparisons performed by your procedure.
   (c) Solve the recurrence relation obtained in (b).

8. (a) Use a divide-and-conquer approach to devise a procedure to find the largest and next-to-largest numbers in a set of n distinct integers.
   (b) Give a recurrence relation for the number of comparisons performed by your procedure.
   (c) Solve the recurrence relation obtained in (b).

9. In a round-robin tournament, each of n players plays one match with all of the other $n - 1$ players. Assume n is a power of 2. We seek a schedule for $n - 1$ days in which each player has (exactly) one match each day. We represent a schedule as a table with n rows and $n - 1$ columns, where entry (i, j) is the number of the person that player i plays on the jth day; see the following example:

1
2

(with row labels 1, 2)

1	2	3
2	3	4
1	4	3
4	1	2
3	2	1

(with row labels 1, 2, 3, 4)

Use a divide-and-conquer approach to devise a procedure for constructing a round-robin tournament schedule (when n is a power of 2). Find a recurrence relation for the number of comparisons performed by your procedure.

7.3   RECURSIVE PROGRAMMING

The recursive reasoning involved in building a recurrence relation model of a counting problem is exactly the same logic used in designing recursive computer subroutines (that call themselves). Conversely, the number of steps in a recursive program is naturally modeled by a recurrence relation. Example 7 in

Section 7.1 about finding the largest and smallest numbers in a set illustrates this correspondence. The recursive reasoning in counting the number of comparisons implicitly specifies the following recursive subroutine LARGESMALL(S, l, s). Here S is the given set of numbers, and l and s are the largest and smallest, respectively, numbers in S; l and s are found and returned by a call of this subroutine.

```
LARGESMALL(S,l,s);
IF |S|=1 THEN BEGIN l=s=the one number in S; GO TO 1); END
let S₁ be the first half of S and S₂ the second half of S;
LARGESMALL(S₁,l₁,s₁);
LARGESMALL(S₂,l₂,s₂);
s←min(s₁,s₂);
l←max(l₁,l₂);
1) RETURN,
END LARGESMALL
```

It is very convenient to use linked lists in recursive programs that generate all outcomes to some counting problem. Linked lists were introduced in Supplement I of Chapter 1 for adjacency lists of graphs; Figure S4 shows such a set of linked lists. With linked lists, it is easy to delete and replace elements in the middle of lists.

Recall that a linked list is a pair of arrays, or for simplicity a two-dimensional array $A(i, j)$, where $j = 1$ or 2. $A(i,1)$ is the array of elements being stored and $A(i,2)$ is an array of "pointers" to the next element in the array. We assume that the pointer to the first element in the array is always found in location $A(0,2)$. The last element in the array can be marked by making its pointer zero.

The following examples illustrate how a recurrence relation for a counting problem can be used to design a recursive procedure to list all outcomes of the problem.

Example 1

Design a program to list all ways for an elf to climb n stairs if the elf takes steps of one stair or two stairs.

In Example 2 of Section 8.1 we derived the recurrence relation $a_n = a_{n-1} + a_{n-2}$, with $a_1 = 1$, $a_2 = 2$, for this problem. Our outcomes will be a sequence of 1's and 2's (1 for a one-stair step and 2 for a two-stair step) totaling to n steps.

The procedure is activated by the call $CLIMB(n,1)$.

```
PROCEDURE CLIMB(m,k)
BEGIN
    Comment—if m=0, climb is complete, print out sequence of 1's &
    IF m=0 THEN BEGIN                                              2's
```

```
      PRINT OUTCOME(i), i=1,2, . . . ,k=1;
      GOTO 1);
   END;
   Comment—use 1-stair step
   OUTCOME(k)←1;
   CLIMB(m−1,k+1);
   Comment—use 2-stair step (if 2 or more stairs left)
   IF m>2 THEN BEGIN
         OUTCOME(k)←2;
         CLIMB(m−2,k+1);
   END;
1) END CLIMB                                                    ■
```

Example 2

Design a program to list ways to arrange n distinct objects in a row.

In Example 1 of Section 8.1 we derived the recurrence relation $a_n = na_{n-1}$, with $a_1 = 1$, for counting all arrangements of n distinct objects. Following the logic of the recurrence relation, we will generate all arrangements by successively placing each of the objects in the first position in the arrangement, and then recursively handling the problem of listing all ways to arrange the remaining objects in the remaining positions.

Assume that $A(i,j)$ is a linked list, as described above, of the n distinct objects [with $A(0,2)$ pointing to the index of the first object]. Let m be the length of the remaining arrangement and let k be the next position in the arrangement to be filled. Assume that all variables in the procedure *ARRANGE* are local variables; that is, each time *ARRANGE* is called, a new (different) set of variables is created.

The procedure is activated by the call $ARRANGE(A,n,1)$.

```
   PROCEDURE ARRANGE(A,m,k)
   BEGIN
       POINTER←0;
       Comment—if m=0, arrangement is complete
       IF m=0 THEN BEGIN
           PRINT OUTCOME(l), l=1,2,. . .,n;
           GOTO 2)
       END
       Comment—loop to place each of the m remaining elements in po-
       sition k
1)   T←A(POINTER,2);
       Comment—put next object in position k & delete it from linked
       list
       OUTCOME(k)←A(T,1);
```

$A(POINTER,2) \leftarrow A(T,2);$
Comment—recursive call to make arrangements in remaining positions
$ARRANGE \ (A,m-1,k+1);$
Comment—put back in linked list object just used in position k
$A(POINTER,2) \leftarrow T;$
Comment—advance pointer to next object in list
$POINTER \leftarrow T;$
$IF \ A(POINTER,2) > 0 \ GOTO \ 1);$
2) END *ARRANGE*   ∎

Example 3

Design a program to list all ways to select a subset of k objects from a set of n distinct objects.

In Example 9 of Section 8.1 we derived the recurrence relation $a_{n,k} = a_{n-1,k} + a_{n-1,k-1}$, with $a_{n,0} = a_{n,n} = 1$ for $n > 0$ (and $a_{n,k} = 0$ for $k > n$).

The procedure is activated by the call $SUBSET(A,n,k,1)$. The last parameter in the procedure call is a pointer to the first location in the list of remaining objects. This time we only need an ordinary one-dimensional list (array A) of the objects.

```
PROCEDURE SUBSET(A,m,j,P)
  BEGIN
    Comment—if j=0, subset is complete
    IF j=0 THEN BEGIN
      PRINT OUTCOME(i),i=1,2, . . . ,k;
      GOTO 1);
    END;
    Comment—first object in remaining list not used
    IF m>j THEN SUBSET(A,m−1,j,P+1);
    Comment—first object in remaining list used
    OUTCOME(k+1−j)←A(P);
    SUBSET(A,m−1,j−1,P+1);
1) END SUBSET                                                         ∎
```

Compare the algorithms in Examples 2 and 3 with the nonrecursive algorithms to perform the same arrangement and subset enumerations in Section 5.6. Do the algorithms in this section list the outcomes in lexicographic order?

We close this section with an important warning. NEVER USE A RECURSIVE PROGRAM TO COMPUTE VALUES IN A RECURRENCE RELATION. Recurrence relations are meant for forward, direct tabulation of values, not for backward, recursive computation. Or they can be used to enumerate outcomes, as in the previous examples.

Consider the following two programs to compute the nth Fibonacci number F_n, when $F_0 = F_1 = 1$:

```
F(0)←1;  F(1)←1;              PROCEDURE F(n)
FOR i←2 TO n DO              IF n<2 THEN F←1
   F(i)←F(i−1) + F(i−2);        ELSE F←F(n−1)+F(n−2);
                             END F
```

The program on the left-hand side tabulates the Fibonacci numbers in an array F and requires $n-1$ additions to obtain $F(n)$. The program on the right-hand side uses a recursive procedure $F(n)$ and requires many, many recursive calls of the procedure; with a little thought, one can see than F_n (the nth Fibonacci number) calls are required to compute $F(n)$. So for $n=15$, the left program requires 14 iterations of its simple FOR statement, while the right program requires 610 ($=F_{15}$) recursive calls. The Fibonacci numbers grow exponentially (see Example 2 of Section 7.4), and so for large n, the recursive program will never terminate.

EXERCISES

There are no exercises for this section. Rather, the reader should write recursive programs to enumerate all outcomes for problems in other sections of this chapter. Note that some problems do not involve enumeration—for example, amount of money in a savings account after n years—and for some problems there is no simple way to describe outcomes, as in counting the number of regions into which the plane is divided by n mutually intersecting lines.

7.4   SOLUTION OF LINEAR RECURRENCE RELATIONS

In this section we sketch the theory for solving recurrence relations of the form

$$a_n = c_1 a_{n-1} + c_2 a_{n-2} + \cdots + c_r a_{n-r} \tag{1}$$

where the c_i's are given constants. There is a simple technique for solving such relations. Readers who have studied the theory of linear differential equations with constant coefficients will see a great similarity between that theory and the one we present here. The general solution to (1) will involve a sum of individual solutions of the form $a_n = \alpha^n$. To determine what α is, we simply substitute α^k for a_k in (1), yielding

$$\alpha^n = c_1 \alpha^{n-1} + c_2 \alpha^{n-2} + \cdots + c_r \alpha^{n-r} \tag{2}$$

We can reduce the power of α in all terms in (2) by dividing both sides by α^{n-r}:

$$\alpha^r = c_1\alpha^{r-1} + c_2\alpha^{r-2} + \cdots + c_r \tag{3}$$

or equivalently

$$\alpha^r = c_1\alpha^{r-1} - c_2\alpha^{r-2} - \cdots - c_r = 0 \tag{4}$$

Equation (4) is called the **characteristic equation** of the recurrence relation (1). It has r roots, some of which may be complex (but we shall initially assume that there are no multiple roots).

If $\alpha_1, \alpha_2, \ldots, \alpha_r$ are the r roots of (4), then $a_n = \alpha_i^n$ is a solution to the recurrence relation (1). It is easy to check that any linear combination of such solutions is also a solution (Exercise 8). That is,

$$a_n = A_1\alpha_1^n + A_2\alpha_2^n + \cdots + A_r\alpha_r^n \tag{5}$$

for any choice of constants A_i, $1 \le i \le r$, is a solution to (1).

Recall that for a recurrence relation involving $a_{n-1}, a_{n-2}, \ldots, a_{n-r}$, we need to be given the initial conditions of the first r values $a_0, a_1, a_2, \ldots, a_{r-1}$. Let us denote such a set of initial values by $a_0', a_1', a_2', \ldots, a_{r-1}'$. Then for each of these a_k' we require

$$a_k' = A_1\alpha_1^k + A_2\alpha_2^k + \cdots + A_r\alpha_r^k, \qquad 0 \le k \le r - 1 \tag{6}$$

The r equations (6) can be solved to determine the r constants A_i (remember that at this stage the α_i's are known). With the A_i's determined, we will have the desired solution for a_n, a solution that satisfies (4), and hence the recurrence relation (1), and satisfies the initial conditions $a_0 = a_0'$, $a_1 = a_1'$, \ldots, $a_{r-1} = a_{r-1}'$. If the solution of the characteristic equation (4) has a root α_* of multiplicity m, then $\alpha_*^n, n\alpha_*^n, n^2\alpha_*^n, \ldots, n^{(m-1)}\alpha_*^n$ are the m associated individual solutions to be used in (5) and (6).

Example 1

Every year Dr. Finch's rabbit population doubles. He started with 6 rabbits. How many rabbits does he have after 8 years? After n years?

If a_n is the number of rabbits, then a_n satisfies the relation $a_n = 2a_{n-1}$. We are also given $a_0 = 6$. Substituting $a_n = \alpha^n$, we obtain $\alpha^n = 2\alpha^{n-1}$ or, dividing by α^{n-1}, $\alpha = 2$. So $a_n = 2^n$ is the one individual solution, and $a_n = A_1 2^n$ is the general solution. The initial condition is $6 = a_0 = A_1 2^0$, or $A_1 = 6$. The desired solution is then $a_n = 6 \cdot 2^n$. After 8 years, we have $a_8 = 6 \cdot 2^8 = 6 \cdot 256 = 1536$ rabbits. ∎

Example 2

Solve the recurrence relation $a_n = 2a_{n-1} + 3a_{n-2}$ with $a_0 = a_1 = 1$.
Setting $a_n = \alpha^n$, we get the characteristic equation:

$$\alpha^n = 2\alpha^{n-1} + 3\alpha^{n-2}, \text{ which yields } \alpha^2 = 2\alpha + 3$$

This may be written $\alpha^2 - 2\alpha - 3 = 0$ or $(\alpha - 3)(\alpha + 1) = 0$; that is, the roots are $+3$ and -1. So the basic solutions of the recurrence relation are $a_n = 3^n$ and $a_n = (-1)^n$, and the general solution is

$$a_n = A_1 3^n + A_2(-1)^n$$

Now we determine A_1 and A_2 by using the initial conditions:

$$1 = a_0 = A_1 3^0 + A_2(-1)^0 = A_1 + A_2$$
$$1 = a_1 = A_1 3^1 + A_2(-1)^1 = 3A_1 - A_2$$

We solve these two simultaneous equations to obtain $A_1 = \frac{1}{2}$, $A_2 = \frac{1}{2}$ (add the two equations together to eliminate A_2 yielding $2 = 4A_1$ or $A_1 = \frac{1}{2}$). The required solution to the recurrence relation with the given initial conditions is

$$a_n = \tfrac{1}{2}3^n + \tfrac{1}{2}(-1)^n$$

■

Example 3

Find a formula for the number of ways for the elf in Example 2 of Section 7.1 to climb the n stairs.

The recurrence relation obtained in Example 2 of Section 7.1 was $a_n = a_{n-1} + a_{n-2}$, which the initial conditions $a_1 = 1$, $a_2 = 2$. The associated characteristic equation is obtained by setting $a_n = \alpha^n$:

$$\alpha^n = \alpha^{n-1} + \alpha^{n-2}, \text{ which reduces to } \alpha^2 = \alpha + 1, \text{ or } \alpha^2 - \alpha - 1 = 0$$

Using the quadratic formula, we get

$$\alpha = \frac{1}{2(1)}[-(-1) \pm \sqrt{(-1)^2 - 4(1)(-1)}] = \tfrac{1}{2}(1 \pm \sqrt{5})$$

That is, we have roots $\frac{1}{2} + \frac{1}{2}\sqrt{5}$ and $\frac{1}{2} - \frac{1}{2}\sqrt{5}$, and the general solution is

$$a_n = A_1(\tfrac{1}{2} + \tfrac{1}{2}\sqrt{5})^n + A_2(\tfrac{1}{2} - \tfrac{1}{2}\sqrt{5})^n$$

The determination of A_1 and A_2 are left as an exercise. We note the surprising fact that to generate the sequence of Fibonacci numbers 1, 1, 2, 3, 5, 8, 13, . . ., we need powers of $\frac{1}{2} + \frac{1}{2}\sqrt{5}$ and $\frac{1}{2} - \frac{1}{2}\sqrt{5}$. ∎

The following example illustrates a solution of a recurrence relation that has complex and multiple roots.

Example 4

Find a formula for a_n satisfying the relation $a_n = -2a_{n-2} - a_{n-4}$ with $a_0 = 0$, $a_1 = 1$, $a_2 = 2$, and $a_3 = 3$.

Substituting $a_n = \alpha^n$, we obtain $\alpha^n = -2\alpha^{n-2} - \alpha^{n-4}$, which yields the characteristic equation $\alpha^4 + 2\alpha^2 + 1 = (\alpha^2 + 1)^2 = 0$. The roots of this equation are $\alpha = +i$ and $\alpha = -i$ ($i = \sqrt{-1}$) and each root has multiplicity 2. So the general solution is

$$a_n = A_1 i^n + A_2 n i^n + A_3(-i)^n + A_4 n(-i)^n$$

The initial conditions yield the equations:

$$
\begin{aligned}
0 = a_0 &= A_1 i^0 + A_2 0 i^0 + A_3(-i)^0 + A_4 0(-i)^0 \\
&= A_1 \quad + \quad\quad A_3 \\
1 = a_1 &= A_1 i^1 + A_2 1 i^1 + A_3(-i)^1 + A_4 1(-i)^1 \\
&= i(A_1 + A_2 - A_3 - A_4) \\
2 = a_2 &= A_1 i^2 + A_2 2 i^2 + A_3(-i)^2 + A_4 2(-i)^2 \\
&= -A_1 - 2A_2 - A_3 - 2A_4 \\
3 = a_3 &= A_1 i^3 + A_2 3 i^3 + A_3(-i)^3 + A_4 3(-i)^3 \\
&= i(-A_1 - 3A_2 + A_3 + 3A_4)
\end{aligned}
$$

Solving these four simultaneous equations in four unknown A_i's, we obtain

$$A_1 = -\tfrac{3}{2}i, \quad A_2 = -\tfrac{1}{2} + i, \quad A_3 = \tfrac{3}{2}i, \quad A_4 = -\tfrac{1}{2} - i$$

Then the solution of the recurrence relation is

$$a_n = -\tfrac{3}{2}i^{n+1} + (-\tfrac{1}{2} + i)n i^n + \tfrac{3}{2}i(-i)^n + (-\tfrac{1}{2} - i)n(-i)^n$$

∎

We remind the reader that for specific values of n, such as $n = 12$, it is easier to determine a_{12} in the two preceding examples by recursively calculating a_3, a_4, a_5 up to a_{12} from the recurrence relation than to solve the initial-condition equations.

EXERCISES

1. If \$500 is invested in a savings account earning 8 percent a year, how much money is in the account after n years?

2. Find and solve a recurrence relation for the number of n-digit ternary sequences with no consecutive digits being equal.

3. Solve the following recurrence relations
    (a) $a_n = 3a_{n-1} + 4a_{n-2}, \quad a_0 = a_1 = 1$
    (b) $a_n = 2a_{n-1} - a_{n-2}, \quad a_0 = a_1 = 2$
    (c) $a_n = a_{n-2}, \quad a_0 = a_1 = 0$
    (d) $a_n = 3a_{n-1} + 3a_{n-2} - a_{n-3}, \quad a_0 = a_1 = 1, a_2 = 2$

4. Determine the constants A_1 and A_2 in Example 3. First show that the initial conditions $a_1 = 1$, $a_2 = 2$ are equivalent to the initial conditions $a_0 = 1$, $a_1 = 1$.

5. Find and solve a recurrence relation for the number of ways to arrange flags on an n-foot flagpole using three types of flags: red flags two feet high, yellow flags one foot high, and blue flags one foot high.

6. Find and solve a recurrence relation for the number of ways to make a pile of n chips using red, white, and blue chips and such that no two red chips are together.

7. Find and solve a recurrence relation for p_n, the value of a stock market indicator that obeys the rule, the change this year (from the previous year) equals twice last year's change. Suppose $p_0 = 1$, $p_1 = 4$.

8. Show that any linear combination of solutions to Equation (1) is itself a solution to (1).

9. Show that if the characteristic equation (4) has a root α_* of multiplicity 3, then $n^j \alpha_*^n$ for $j = 0, 1, 2$ are solutions of (1).

10. Show that

$$\lim_{n \to \infty} \frac{F_{n+1}}{F_n} = \frac{1 + \sqrt{5}}{2}$$

11. Suppose that interest is compounded k times a year, so that instead of i percent interest paid once at the end of each year, interest of (i/k) percent is paid k times a year (as if the year were k periods with (i/k) percent per period).
    (a) Show that if interest is compounded continuously ($k \to \infty$), then each dollar is worth $e^{i/100}$ dollars at the end of a year.
    (b) Use the continuous approximation from part a to estimate the value of \$100 after 50 years at 3 percent interest per year.

12. If the recurrence relation $a_n = c_1 a_{n-1} + c_2 a_{n-2}$ has a general solution $a_n = A_1 3^n + A_2 6^n$, find c_1 and c_2.

7.5 SOLUTION OF INHOMOGENEOUS RECURRENCE RELATIONS

A recurrence relation is called **homogeneous** if all the terms of the relation involve some a_k. In the preceding section, we discussed homogeneous, constant-coefficient, linear recurrence relations. When an additional term involving a constant or function of n appears in the recurrence relation, such as

$$a_n = ca_{n-1} + f(n) \qquad (1)$$

c a constant, $f(n)$ a function of n, the recurrence relation is said to be **inhomogeneous.**

In this section, we discuss methods for solving inhomogeneous recurrence relations of the form (1). The key idea in solving these relations is that a general solution for an inhomogeneous relation is made up of a general solution to the associated homogeneous relation (obtained by deleting the $f(n)$ term) plus *any* one particular solution to the inhomogeneous relation.

For (1), the homogeneous relation is $a_n = ca_{n-1}$, whose general solution is $a_n = Ac^n$. Suppose a_n^* is some particular solution of (1); that is, $a_n^* = ca_{n-1}^* + f(n)$. Then we see that $a_n = Ac^n + a_n^*$ satisfies (1):

$$a_n = Ac^n + a_n^*$$
$$= (c \cdot Ac^{n-1}) + (ca_{n-1}^* + f(n))$$
$$= c(Ac^{n-1} + a_{n-1}^*) + f(n) = ca_{n-1} + f(n)$$

The constant A in the general solution is chosen to satisfy the initial condition, as in the previous section (A cannot be determined until a_n^* is found).

There is one special case of (1) we can restate as an enumeration problem treated in previous chapters. If $c = 1$ so that (1) becomes

$$a_n = a_{n-1} + f(n) \qquad (2)$$

then we can iterate to get

$$a_1 = a_0 + f(1)$$
$$a_2 = a_1 + f(2) = (a_0 + f(1)) + f(2)$$
$$a_3 = a_2 + f(3) = (a_0 + f(1) + f(2)) + f(3)$$
$$\vdots$$
$$a_n = a_0 + f(1) + f(2) + f(3) + \cdots + f(n) = a_0 + \sum_{k=1}^{n} f(k)$$

So a_n is just the sum of the $f(k)$'s plus a_0. In Sections 5.5 and 6.5, we presented methods for summing functions of n. Either method can be used to solve (2).

Example 1

Solve the recurrence relation $a_n = a_{n-1} + n$ with initial condition $a_1 = 2$, obtained in Example 3 of Section 7.1, for the number of regions created by n mutually intersecting lines.

The initial condition of $a_1 = 2$ can be replaced by the initial condition $a_0 = 1$ (no lines means that the plane is one big region). By the above discussion, we see that $a_n = 1 + (1 + 2 + 3 + \cdots + n)$. The expression to be summed can be written as

$$\binom{1}{1} + \binom{2}{1} + \binom{3}{1} + \cdots + \binom{n}{1}. \tag{3}$$

By identity (7) of Section 5.5, this sum equals $C(n + 1, 2) = \frac{1}{2}n(n + 1)$. Then $a_n = 1 + \frac{1}{2}n(n + 1)$. ∎

When $c \neq 1$ in (1), there are known solutions to (1) to use for various $f(n)$. We present a table for the simplest $f(n)$'s. These solutions can be derived by generating function methods introduced in the next section.

$f(n)$	Particular Solution $p(n)$
d, a constant	B
dn	$B_1 n + B_0$
dn^2	$B_2 n^2 + B_1 n + B_0$
d^n	Bd^n

The B's are constants to be determined. If $f(n)$ were a sum of several different terms, we would separately solve the relation for each separate $f(n)$ term and then add these solutions together to get a particular solution for the composite $f(n)$.

There is one case in which the particular solutions will not work. This is when the appropriate particular solution is itself a solution to the homogeneous recurrence relation. Then one must try as the particular solution a polynomial in n of a higher degree, or if $f(n) = d^n$, try Bnd^n.

Example 2

Solve the recurrence relation $a_n = 2a_{n-1} + 1$ with $a_1 = 1$ obtained in Example 4 of Section 7.1 for the number of moves required to play the n-ring Tower of Hanoi puzzle.

The general solution to the homogeneous equation $a_n = 2a_{n-1}$ is $a_n = A2^n$.

We find a particular solution to the inhomogeneous relation by setting $a_n^* = B$ (this is the form of a particular solution given in the above table when $f(n)$ is a constant). Substituting in the relation, we have

$$B = a_n^* = 2a_{n-1}^* + 1 = 2B + 1 \quad \text{or} \quad B = -1$$

So $a_n^* = -1$ is the particular solution, and the general inhomogeneous solution is $a_n = A2^n + a_n^* = A2^n - 1$. We now can determine A from the initial condition: $1 = a_1 = A2^1 - 1$, or $2 = 2A$. Hence $A = 1$. The desired solution is then $a_n = 2^n - 1$. ∎

Example 3

Solve the recurrence relation $a_n = 3a_{n-1} - 4n + 3 \cdot 2^n$ to find its general solution. Also find the solution when $a_1 = 8$.

The general solution to the homogeneous equation $a_n = 3a_{n-1}$ is $a_n = A3^n$. We solve for a particular solution of the relation separately for each inhomogeneous term. For $a_n = 3a_{n-1} - 4n$, we try the form $a_n^* = B_1 n + B_0$, obtaining

$$B_1 n + B_0 = a_n^* = 3a_{n-1}^* - 4n = 3(B_1(n - 1) + B_0) - 4n. \tag{4}$$

We now equate the constant terms and the coefficients of n on each side of (4):

$$\text{Constant terms:} \quad B_0 = -3B_1 + 3B_0 \tag{5}$$

$$n \text{ terms:} \quad B_1 n = 3B_1 n - 4n \quad \text{or} \quad B_1 = 3B_1 - 4 \tag{6}$$

Solving for B_1 in (6), we obtain $B_1 = 2$. Substituting $B_1 = 2$ in (5), we obtain $B_0 = 3$. So $a_n^* = 2n + 3$ is a particular solution of $a_n = 3a_{n-1} - 4n$.
Next for $a_n = 3a_{n-1} + 3 \cdot 2^n$, we try $a_n^{**} = B2^n$, obtaining

$$B2^n = a_n^{**} = 3a_{n-1}^{**} + 3 \cdot 2^n = 3(B2^{n-1}) + 3 \cdot 2^n \tag{7}$$

Dividing both sides of (7) by 2^{n-1}, we get $2B = 3B + 6$, or $B = -6$. So $a_n^{**} = -6 \cdot 2^n$ is a particular solution of $a_n = 3a_{n-1} + 3 \cdot 2^n$. Combining our particular solutions with the general homogeneous solution, we obtain the general inhomogeneous solution

$$a_n = A3^n + 2n + 3 - 6 \cdot 2^n$$

When $a_1 = 8$, we can determine A:

$$8 = a_1 = A3^1 + 2 \cdot 1 + 3 - 6 \cdot 2^1 = 3A - 7 \quad \text{or} \quad A = 5$$

and the solution is $a_n = 5 \cdot 3^n + 2^n + 3 - 6 \cdot 2^n$. ∎

EXERCISES

1. Solve the following recurrence relations:
    (a) $a_n = a_{n-1} + 3(n - 1)$, $a_0 = 1$
    (b) $a_n = a_{n-1} + n(n + 1)$, $a_0 = 3$
    (c) $a_n = a_{n-1} + 3n^2$, $a_0 = 10$

2. Find and solve a recurrence relation for the number of infinite regions formed by n infinite lines drawn in the plane so that each pair of lines intersect at a different point.

3. (a) Find and solve a recurrence relation for the number of different square subboards of any size that can be drawn on an $n \times n$ chessboard.
    (b) Repeat part a for rectangular subboards of any size.

4. Find and solve a recurrence relation for the number of different regions formed when n mutually intersecting planes are drawn in three-dimensional space such that no four planes intersect at a common point and no two planes have parallel intersection lines in a third plane. [*Hint:* Reduce to a two-dimensional problem (Example 1).]

5. Find and solve a recurrence relation for the number of regions into which a convex n-gon is divided by all its diagonals, assuming no three diagonals intersect at a common point. *Hint:*   Sum the inhomogeneous term using a special case of an identity from Section 5.5.)

6. If the average number of two successive years' production $\frac{1}{2}(a_n + a_{n-1})$ is $2n + 5$ and $a_0 = 3$, find a_n.

7. Solve the recurrence relation $a_n = 1.08a_{n-1} + 100$, $a_0 = 0$, from part b of Example 5 in Section 7.1.

8. Suppose a savings account earns 5 percent a year. Initially there is $1000 in the account and in year k, $10k$ are withdrawn. How much money is in the account at the end of n years if:
    (a) Annual withdrawal is at year's end?
    (b) Withdrawal is at start of year?

9. Solve the following recurrence relations
    (a) $a_n = 3a_{n-1} - 2$, $a_0 = 0$ (b) $a_n = 2a_{n-1} + n$, $a_0 = 1$
    (c) $a_n = 2a_{n-1} + (-1)^n$, $a_0 = 2$ (d) $a_n = 2a_{n-1} + \frac{1}{2}n^2$, $a_0 = 3$

10. Solve the recurrence relation $a_n = 3a_{n-1} + n^2 - 3$ with $a_0 = 1$.

11. Solve the recurrence relation $a_n = 3a_{n-1} - 2a_{n-2} + 3$, $a_0 = a_1 = 1$.

12. Find and solve a recurrence relation for the number of n-digit ternary sequences in which no 1 appears to the right of any 2.

13. Find and solve a recurrence relation for the earnings of a company whose profits increase each year but the increase is $10 \cdot 2^k$ more in the kth year, where $a_0 = 20$ and $a_1 = 1020$.

14. Show that the general solution to any inhomogeneous linear recurrence relation is the general solution to the associated homogeneous relation plus one particular inhomogeneous solution.

15. Show that the form of the particular solution of Equation (1) given in the table in this section is correct for:
    (a) $f(n) = d$ (b) $f(n) = dn$ (c) $f(n) = dn^2$ (d) $f(n) = d^n$.

16. Show that if $f(n)$ in (1) is the sum of several different terms, a particular solution for this $f(n)$ may be obtained by summing particular solutions for the individual terms.

17. Find a general solution to $a_n - 5a_{n-1} + 6a_{n-2} = 2 + 3n$.

18. If the recurrence relation $a_n + c_1 a_{n-1} + c_2 a_{n-2} = c_3 n + c_4$ has a general solution $a_n = A_1 2^n + A_2 5^n + 3n - 5$, find c_1, c_2, c_3, c_4.

19. Solve the following recurrence relations when $a_0 = 1$:
    (a) $a_n^2 = 2a_{n-1}^2 + 1$ (*Hint:* Let $b_n = a_n^2$.)
    (b) $a_n = -na_{n-1} + n!$ (*Hint:* Define an appropriate b_n as in part a.)

7.6 SOLUTIONS WITH GENERATING FUNCTIONS

Most recurrence relations for a_n can be converted into an equation involving the generating function $g(x) = a_0 + a_1 x + \cdots + a_n x^n + \cdots$. This associated functional equation for $g(x)$ can often be solved algebraically and the resulting expression for $g(x)$ expanded in a power series to obtain a_n, the coefficient of x^n. Some of the algebraic manipulations of the functional equations may be new to the reader.

We will treat $g(x)$ as if it were a standard single variable, such as y, and treat other functions of x as constants. For example, the functional equation $g(x) = x^2 g(x) - 2x$ can be solved by rewriting the equation as $g(x)(1 - x^2) = -2x$ and hence $g(x) = -2x(1 - x^2)^{-1}$. Similarly, the functional equation

$$(1 - x^2)[g(x)]^2 - 4xg(x) + 4x^2 = 0 \tag{1}$$

can be solved by the quadratic formula that we normally apply to equations such as $ay^2 + by + c = 0$. Now $a = (1 - x^2)$, $b = -4x$, and $c = 4x^2$. Intuitively, for each particular value of x, $g(x)$ is the solution of (1). Thus by the quadratic formula,

$$g(x) = \frac{1}{2(1 - x^2)}\left[4x \pm \sqrt{16x^2 - 16x^2(1 - x^2)}\right] = \frac{1}{2(1 - x^2)}(4x \pm 4x^2)$$

So $g(x) = 2(x + x^2)/(1 - x^2)$ or $2(x - x^2)/(1 - x^2)$. If there are two (or more) possible solutions, only one will normally make sense as a generating function for a_n (e.g., have a power series expansion with the correct value for the initial condition a_0').

Now let us show by example how recurrence relations for a_n can be converted into functional equations for an associated generating function.

Example 1

Find a functional equation for $g(x) = a_0 + a_1 x + \cdots + a_n x^n$, where a_n satisfies the recurrence relation $a_n = a_{n-1} + n$, with $a_0 = 1$, obtained in Example 3 of Section 7.1. Solve the functional equation and expand $g(x)$ to find a_n.

Using this recurrence relation for every term in $g(x)$ except a_0, we have $a_n x^n = a_{n-1} x^n + n x^n$, $n \geq 1$. Summing the terms, we can write

$$g(x) - a_0 = \sum_{n=1}^{\infty} a_n x^n = \sum_{n=1}^{\infty} (a_{n-1} x^n + n x^n) \tag{2}$$

$$= x \sum_{n=1}^{\infty} a_{n-1} x^{n-1} + \sum_{n=1}^{\infty} n x^n \tag{3}$$

$$= x \sum_{m=0}^{\infty} a_m x^m + \sum_{n=0}^{\infty} \binom{n}{1} x^n \tag{4}$$

$$= x g(x) + x/(1 - x)^2 \tag{5}$$

Line (3) is obtained from (2) by breaking up the sum of the two x^n terms into sums of each term, and by rewriting $a_{n-1} x^n$ as $x a_{n-1} x^{n-1}$ (in order to make the power of x correspond with the subscript of a_{n-1}). Line (4) is obtained from line (3) by re-indexing the first sum with $m = n - 1$, and by adding the "phantom" (zero) term $0 x^0$ to the second sum and rewriting n as $C(n,1)$. The first series is the generating function $g(x)$ multiplied by x. The second series has a generating function obtained by the construction presented in Section 6.5. Equating line (5) with $g(x) - a_0$ [the left side of line (2)] and setting $a_0 = 1$, we have the required functional equation for $g(x)$:

$$g(x) - 1 = x g(x) + x/(1 - x)^2 \tag{6}$$

Solving for $g(x)$, we rewrite (6) as

$$g(x) - x g(x) = 1 + x/(1 - x)^2$$

$$g(x) = 1/(1 - x) + x/(1 - x)^3$$

The coefficient of x^n in $(1 - x)^{-1}$ is just 1 and in $x(1 - x)^{-3}$ is $C((n - 1) + 3 - 1, n - 1) = C(n + 1, n - 1) = C(n + 1, 2)$. Then

$$g(x) = \sum_{n=0}^{\infty} 1x^n + \sum_{n=0}^{\infty} \binom{n+1}{2} x^n = \sum_{n=0}^{\infty} \left(1 + \binom{n+1}{2}\right) x^n,$$

and so $a_n = 1 + C(n + 1, 2)$—the same answer as obtained for this recurrence relation in Example 1 of Section 7.5. ∎

Example 2

Use generating functions to solve the recurrence relation $a_n + a_{n-1} + a_{n-2}$, with $a_1 = 1$, $a_2 = 2$ obtained in Example 2 of Section 7.1.

The initial conditions, $a_1 = 1$, $a_2 = 2$ are equivalent to $a_0 = 1$, $a_1 = 1$. Then using the same power series summation approach as in the previous example, we obtain

$$g(x) - a_0 - a_1 x = \sum_{n=2}^{\infty} a_n x^n$$

$$= \sum_{n=2}^{\infty} (a_{n-1} x^n + a_{n-2} x^n)$$

$$= x \sum_{n=2}^{\infty} a_{n-1} x^{n-1} + x^2 \sum_{n=2}^{\infty} a_{n-2} x^{n-2}$$

$$= x \sum_{m=1}^{\infty} a_m x^m + x^2 \sum_{k=0}^{\infty} a_k x^k$$

$$= x(g(x) - a_0) + x^2 g(x)$$

Setting $a_0 = 1$ and $a_1 = 1$, we have the functional equation

$$g(x) - 1 - x = x(g(x) - 1) + x^2 g(x)$$

and so

$$g(x)(1 - x - x^2) = 1 \quad \text{or} \quad g(x) = 1/(1 - x - x^2)$$

This denominator is closely related to the characteristic equation of this recurrence relation, given in Example 3 of Section 7.4. *In general, $g(x)$ will have a denominator $1 + c_1 x + c_2 x^2 + \cdots + c_r x^r$ if and only if $x^r + c_1 x^{r-1} + c_2 x^{r-2} + \cdots + c_r$ is the characteristic equation of the associated recurrence relation, and so $1 - \alpha x$ is a factor of the denominator of $g(x)$ if and only if α is a root of the characteristic equation.*

By the quadratic formula, we can factor

$$1 - x - x^2 = (1 - \tfrac{1}{2}(1 + \sqrt{5})x)(1 - \tfrac{1}{2}(1 - \sqrt{5})x)$$

For simplicity, let us define $\alpha_1 = \frac{1}{2}(1 + \sqrt{5})$ and $\alpha_2 = \frac{1}{2}(1 - \sqrt{5})$. Then we have

$$g(x) = \frac{1}{(1 - \alpha_1 x)(1 - \alpha_2 x)} = \frac{\alpha_1/\sqrt{5}}{1 - \alpha_1 x} - \frac{\alpha_2/\sqrt{5}}{1 - \alpha_2 x}$$

The decomposition of $g(x)$ into two fractions is the "partial fraction" method of calculus, the reverse process of combining two fractions into a common fraction. This particular fraction decomposition depends on the numerical values of α_1 and α_2. Setting $y = \alpha_1 x$, we see that $(\alpha_1/\sqrt{5})(1 - y)^{-1}$ expands as

$$\frac{\alpha_1}{\sqrt{5}} \sum_{n=0}^{\infty} \alpha_1^n x^n$$

The same type of expansion exists for $y = \alpha_2 x$. Then a_n, the coefficient of x^n in the power series expansion of $g(x)$, is

$$a_n = \frac{1}{\sqrt{5}} \alpha_1^{n+1} - \frac{1}{\sqrt{5}} \alpha_2^{n+1}$$

$$= \frac{1}{\sqrt{5}} \left(\frac{1}{2}(1 + \sqrt{5})\right)^{n+1} - \frac{1}{\sqrt{5}} \left(\frac{1}{2}(1 - \sqrt{5})\right)^{n+1}$$

■

Example 3

Let $g_n(x)$ be a family of generating functions $g_n(x) = a_{n,0} + a_{n,1}x + \cdots + a_{n,k}x^k + \cdots + a_{n,n}x^n$ satisfying the relation $a_{n,k} = a_{n-1,k} + a_{n-1,k-1}$, with $a_{n,0} = a_{n,n} = 1$ (and $a_{n,k} = 0$, $k > n$) obtained in Example 9 of Section 7.1. Find a functional relation among the $g_n(x)$'s and solve it to obtain a formula for $a_{n,k}$.

Using the polynomial summation method, we obtain

$$g_n(x) = \sum_{k=0}^{n} a_{n,k}x^k = \sum_{k=0}^{n} (a_{n-1,k}x^k + a_{n-1,k-1}x^k)$$

$$= \sum_{k=0}^{n} a_{n-1,k}x^k + x \sum_{k=0}^{n} a_{n-1,k-1}x^{k-1}$$

$$= g_{n-1}(x) + xg_{n-1}(x) = (1 + x)g_{n-1}(x)$$

Note that in the second line, the x^n coefficient in the first sum $a_{n-1,n}$ and the x^0 coefficient in the second sum $a_{n-1,0-1}$ are both 0. The resulting recurrence among the $g_n(x)$'s, $g_n(x) = (1 + x)g_{n-1}(x)$, is solved just like the recurrence $a_n =$

ca_{n-1}. The solution is

$$g_n(x) = (1 + x)^n g_0(x) = (1 + x)^n$$

since $g_0(x) = a_{0,0} = 1$. Now by the binomial theorem, we have $a_{n,k} = C(n,k)$.

■

The rest of this section involves more complicated computations. The reader may skip this material.

Now we consider a nonlinear recurrence relation and solve it using generating functions.

Example 4

Find a recurrence relation for a_n, the number of ways to place parentheses to multiply the n numbers $k_1 \cdot k_2 \cdot k_3 \cdot k_4 \cdot \cdots \cdot k_n$ on a calculator. Solve for a_n using generating functions.

To clarify the problem, observe that there is one way to multiply $(k_1 \cdot k_2)$; so $a_2 = 1$. There are two ways to multiply $k_1 \cdot k_2 \cdot k_3$, namely, $((k_1 \cdot k_2) \cdot k_3)$ and $(k_1 \cdot (k_2 \cdot k_3))$; so $a_3 = 2$. It is not clear what a_0 and a_1 should be, but to make the eventual recurrence relation have a simple form we let $a_0 = 0$ and $a_1 = 1$. To find a recurrence relation for a_n we look at the last multiplication (the outermost parenthesis) in the product of the n numbers. This last multiplication involves the products of two multiplication subproblems:

$$((k_1 \cdot k_2 \cdot \cdots \cdot k_i) \cdot (k_{i+1} \cdot k_{i+2} \cdot \cdots \cdot k_n))$$

where i can range from 1 to $n - 1$. The numbers of ways to parenthesize the two respective subproblems are a_i and a_{n-i}, and so there are $a_i a_{n-1}$ ways to parenthesize both subproblems. Summing over all i, we obtain the recurrence relation (for $n \geq 2$)

$$a_n = a_1 a_{n-1} + a_2 a_{n-2} + \cdots + a_i a_{n-i} + \cdots + a_{n-1} a_1$$

Observe that the right-hand side of this equation is simply the coefficient of x^n in the product $g(x)g(x) = (0 + a_1 x + \cdots + a_n x^n + \cdots)^2$. Using the power series summation method, we have (recall that $a_0 = 0$ and $a_1 = 1$)

$$g(x) - 1x = \sum_{n=2}^{\infty} a_n x^n = \sum_{n=2}^{\infty} (a_1 a_{n-1} + a_2 a_{n-2} \cdots + a_{n-1} a_1) x^n = (g(x))^2$$

Solving this quadratic equation in $g(x)$ as described at the start of this section, we obtain $g(x) = \frac{1}{2}(1 \pm \sqrt{1 - 4x})$. To make $a_0 = 0$ (i.e., $g(0) = 0$), we want the solution $g(x) = \frac{1}{2} - \frac{1}{2}\sqrt{1 - 4x}$. This $g(x)$ requires a new type of generating function expansion. We use the generalized binomial theorem for

$(1 + y)^{1/2}$, where $y = -4x$, introduced at the end of Section 5.5. This generalized binomial expansion uses the generalized binomial coefficients

$$\binom{\frac{1}{2}}{k} = \frac{\frac{1}{2}(\frac{1}{2} - 1)(\frac{1}{2} - 2) \cdots (\frac{1}{2} - (k - 1))}{k!}$$

The coefficient of x^n in $\sqrt{1 - 4x}$ is

$$\binom{\frac{1}{2}}{n}(-4)^n = \frac{\frac{1}{2}(-\frac{1}{2})(-\frac{3}{2}) \cdots (-\frac{1}{2}(2n - 3))}{n!}(-4)^n$$

$$= \frac{-1 \cdot 1 \cdot 3 \cdot 5 \cdots \cdots (2n - 3)}{n!} 2^n$$

$$= -\frac{2}{n}\binom{2n - 2}{n - 1}$$

The last step is a fairly well-known identity obtained by multiplying certain numbers in the numerator by appropriately selected powers of 2 (see Exercise 33 in Section 6.2). Then a_n $(n \geq 1)$, the coefficient of x^n in $-\frac{1}{2}\sqrt{1 - 4x}$, called a *Catalan number*, is

$$a_n = \frac{1}{n}\binom{2n - 2}{n - 1}$$

■

Next let us consider generating functions for simultaneous recurrence relations.

Example 5

Use generating functions to solve the set of simultaneous recurrence relations obtained in Example 11 of Section 7.1.

$$a_n = a_{n-1} + b_{n-1} + c_{n-1}, \qquad b_n = 3^{n-1} - c_{n-1},$$

$$c_n = 3^{n-1} - b_{n-1}, \qquad a_1 = b_1 = c_1 = 1$$

Let $A(x)$, $B(x)$, and $C(x)$ be the generating functions for a_n, b_n, and c_n, respectively. We use the polynomial summation method to obtain

$$A(x) - a_0 = \sum_{n=1}^{\infty} a_n x^n = \sum_{n=1}^{\infty} a_{n-1} x^n + \sum_{n=1}^{\infty} b_{n-1} x^n + \sum_{n=1}^{\infty} c_{n-1} x^n$$

$$= xA(x) + xB(x) + xC(x)$$

$$B(x) - b_0 = \sum_{n=1}^{\infty} b_n x^n = \sum_{n=1}^{\infty} 3^{n-1} x^n - \sum_{n=1}^{\infty} c_{n-1} x^n$$

$$= x(1 - 3x)^{-1} - xC(x)$$

$$C(x) - c_0 = \sum_{n=1}^{\infty} c_n x^n = \sum_{n=1}^{\infty} 3^{n-1} x^n - \sum_{n=1}^{\infty} b_{n-1} x^n$$

$$= x(1 - 3x)^{-1} - xB(x)$$

It is always desirable to state initial conditions in terms of a_0, b_0, c_0. Solving our three recurrence relations for a_0, b_0, c_0 given $a_1 = b_1 = c_1 = 1$, we get $1 = b_1 = 3^0 - c_0$ or $c_0 = 0$. Similarly we find that $b_0 = 0$ and $a_0 = 1$. Then our functional equations are

$$A(x) - 1 = xA(x) + xB(x) + xC(x) \text{ or}$$

$$A(x) = \frac{x}{1 - x}(B(x) + C(x)) + \frac{1}{1 - x} \tag{7}$$

$$B(x) = x(1 - 3x)^{-1} - xC(x) \tag{8}$$

$$C(x) = x(1 - 3x)^{-1} - xB(x) \tag{9}$$

We can solve Equations (8) and (9) for $B(x)$ and $C(x)$ simultaneously. Multiplying Equation (9) by x and using this expression for $xC(x)$ in (8), we have

$$B(x) = x(1 - 3x)^{-1} - xC(x) = x(1 - 3x)^{-1} - (x^2(1 - 3x)^{-1} - x^2 B(x))$$

$$B(x)(1 - x^2) = (x - x^2)(1 - 3x)^{-1}$$

$$B(x) = \frac{(1 - x)x}{(1 - x^2)(1 - 3x)} = \frac{1 \cdot x}{(1 + x)(1 - 3x)} = \frac{\frac{1}{4}}{1 - 3x} - \frac{\frac{1}{4}}{1 + x}$$

The last step is again a partial fraction decomposition. The coefficient of x^n in $\frac{1}{4}(1 - 3x)^{-1}$ is $\frac{1}{4}3^n$ and in $-\frac{1}{4}(1 + x)^{-1}$ is $\frac{1}{4}(-1)^n$ (these coefficients are obtained by setting $y = 3x$ and $y = -x$, respectively, in $(1 - y)^{-1} = \sum_{n=0}^{\infty} y^n$). So b_n, the coefficient of x^n in $B(x)$, is $\frac{1}{4}(3^n - (-1)^n)$. The equations (8) and (9) are symmetric with respect to $B(x)$ and $C(x)$, and so $C(x) = B(x)$ and $c_n = b_n = \frac{1}{4}(3^n - (-1)^n)$. Now we can solve for $A(x)$ in (7):

$$A(x) = \frac{x}{1 - x}(B(x) + C(x)) + \frac{1}{1 - x} = \frac{2x}{1 - x}\left(\frac{\frac{1}{4}}{1 - 3x} - \frac{\frac{1}{4}}{1 + x}\right) + \frac{1}{1 - x}$$

$$= \frac{\frac{1}{2}x}{(1 - x)(1 - 3x)} - \frac{\frac{1}{2}x}{1 - x^2} + \frac{1}{1 - x}$$

$$= \left(\frac{\frac{1}{4}}{1 - 3x} - \frac{\frac{1}{4}}{1 - x}\right) - \frac{\frac{1}{2}x}{1 - x^2} + \frac{1}{1 - x}$$

The coefficient of x^n in $\frac{1}{4}(1 - 3x)^{-1}$ is $\frac{1}{4}3^n$, in $-\frac{1}{4}(1 - x)^{-1}$ is $-\frac{1}{4}$, in $-\frac{1}{2}x(1 - x^2)^{-1}$ is $-\frac{1}{2}$, n odd, or 0, n even, and in $(1 - x)^{-1}$ is 1. Collecting these terms, we get $a_n = \frac{1}{4}(3^n + 3)$, n even, and $=\frac{1}{4}(3^n + 1)$, n odd. ■

EXERCISES

1. Find generating functions whose coefficients satisfy the relations:
   (a) $a_n = a_{n-1} + 2$, $a_0 = 1$
   (b) $a_n = a_{n-1} + n(n - 1)$, $a_0 = 1$
   (c) $a_n = 3a_{n-1} - 2a_{n-2}$, $a_0 = a_1 = 1$
   (d) $a_n = 2a_{n-1} + 2^n$, $a_0 = 1$

2. Solve the recurrence relations in Exercise 1 using generating functions.

3. Find generating functions whose coefficients satisfy the relations:

   (a) $a_n = \displaystyle\sum_{i=0}^{n-1} a_i a_{n-1-i} (n \geqslant 1)$, $a_0 = 1$

   (b) $a_n = \displaystyle\sum_{i=2}^{n-2} a_i a_{n-i} (n \geqslant 3)$, $a_0 = a_1 = a_2 = 1$

   (c) $a_n = \displaystyle\sum_{i=1}^{n-1} 2^i a_{n-i} (n \geqslant 2)$, $a_0 = a_1 = 1$

4. Find a functional equation and solve it for the sequence of generating functions $F_n(x) = \displaystyle\sum_{k=0}^{n} a_{n,k} x^k$ whose coefficients satisfy (assume $F_0 = 1$ and $a_{n,0} = 1$):
   (a) $a_{n,k} = a_{n,k-1} - 2a_{n-1,k-1}$
   (b) $a_{n,k} = 2_{n-1,k} - 3a_{n,k-1}$
   (c) $a_{n,k} = a_{n,k-1} + 2^k a_{n-1,k}$

5. Verify the form of particular solutions to inhomogeneous recurrence relations in the table in Section 7.5.

6. Verify the statement in italics in Example 2.

7. Find a recurrence relation and solve it with generating functions for the number of ways to divide an n-gon into triangles with noncrossing diagonals. (*Hint:* Use reasoning similar to Example 4.)

8. Find a recurrence relation and associated generating function for the number of n-digit ternary sequences that end with "012."

9. Find a recurrence relation and associated generating function for the number of different binary trees with n leaves.

10. (a) Find a recurrence relation for a_n, the number of ways to partition n distinct objects among n indistinguishable boxes (some boxes may be empty).

(b) Let $g(x) = \sum_{n=0}^{\infty} a_n x^n / n!$, where $a_0 = 1$. Show that $g(x)$ satisfies the differential equation $g'(x) = g(x)e^x$. Solve this equation for $g(x)$.

11. Find a recurrence relation for $a_{n,k}$, the number of k-subsets of an n set with repetition. Find an equation for $F_n(x) = \sum_{k=0}^{\infty} a_{n,k} x^k$ and solve for $F_n(x)$ and $a_{n,k}$.

12. (a) Let $a_{n,k}$ be the probability that k successes occur in an experiment with n trials if each trial has probability p of success. Find a recurrence relation for $a_{n,k}$. Use this relation to find and solve an equation for $F_n(x) = \sum_{k=0}^{n} a_{n,k} x^k$.

(b) Repeat this problem with $a_{n,k}$, the probability that the kth success occurs on the nth trial.

13. (a) Find a recurrence relation for $a_{n,r}$, the number of r-permutations of n elements.

(b) Show that $F_n(x) = \sum_{r=0}^{n} a_{n,r} x^r$ satisfies the equation

$$F_n(x) = (1 + x)F_{n-1}(x) + x^2 F'_{n-1}(x).$$

(c) Find a functional equation for $G_n(x) = \sum a_{n,r} x^r / r!$.

14. (a) Define $s_{n,r}$ as numbers such that $\sum_{r=0}^{n} s_{n,r} x^n = x(x - 1)(x - 2) \cdots (x - r + 1)$. Find a recurrence relation for $s_{n,r}$.

(b) Find a differential equation for $F_n(x) = \sum_{r=0}^{n} s_{n,r} x^r / r!$.

15. Find and solve a recurrence relation for $a_{n,k}$, the number of n-digit quaternary sequences with an odd number of 1's and an odd number of 2's.

16. Find and solve simultaneous recurrence relations for determining the number of n-digit ternary sequences whose sum of digits is a multiple of 3.

17. Find and solve simultaneous recurrence relations for determining the number of n-coin sequences of pennies, nickels, and dimes if there are two types of pennies, three types of nickels, and two types of dimes, and if no two coins of the same value appear consecutively. (*Hint:* Use symmetry.)

7.7 SUMMARY AND REFERENCES

In this chapter we saw that recurrence relations are one of the simplest ways to solve counting problems. Without fully understanding the combinatorial process, as was required in Chapter 5, we now need only express a given problem for n objects in terms of the problem posed for fewer numbers of objects. Once a recurrence relation has been found, then starting with a_1 (the solution for one object), we can successively compute the solutions for 2, 3, . . . up to any (moderate) value of n. Or we can try one of the techniques in the later sections of this chapter to solve the recurrence relation explicitly.

The first recurrence relation in mathematical writings was the Fibonacci relation. In his work *Liber abaci*, published in 1220, Leonardo di Pisa, known also as Fibonacci, posed a counting problem about the growth of a rabbit population (Exercise 7 in Section 7.1). The number a_n of rabbits after n months was shown to satisfy the Fibonacci relation $a_n + a_{n-1} + a_{n-2}$. As mentioned in Section 6.6, DeMoivre gave the first solution of this relation 500 years later using the generation function derivation given in Example 2 of Section 7.6. The Fibonacci relation and numbers have proven to be amazingly ubiquitous. For example, it has been shown that the ratios of Fibonacci numbers provide an optimal way (in a certain sense) to divide up an interval when searching for the minimum of a function in this interval (see Kiefer [3]). The appearance of Fibonacci numbers in rings of leaves around flowers is discussed in Adler [1].

The methods for solving recurrence relations appeared originally in the development of the theory of difference equations, mathematical cousins of differential equations. For a survey of difference equations methods, see Levy and Lessman [4]. For a nice presentation of applications of difference equations, see Goldberg [2].

1. I. Adler, "The consequence of constant pressure in phyllotaxis," *J. Theoretical Biology* **65** (1977), 29–77.

2. S. Goldberg, *Introduction to Difference Equations*, John Wiley & Sons, New York, 1958.

3. J. Kiefer, "Sequential Minimax Search for a Maximum," *Proceedings of American Math. Society* **4** (1953), 502–506.

4. H. Levy and F. Lessman, *Finite Difference Equations*, Macmillan, New York, 1961.

Chapter Eight
Inclusion-Exclusion

8.1 COUNTING WITH VENN DIAGRAMS

In this chapter we develop a set-theoretic formula for solving a broad class of counting problems in which several interacting properties either all must hold or none must hold. An example is counting all 5-card hands with no void in any suit, or equivalently, 5-card hands with at least one card in each suit. In the process of solving such problems, we have to count the subsets of outcomes in which various combinations of the properties hold. We use Venn diagrams to depict these different combinations. Check Appendix A1 for a review of the essentials of sets and Venn diagrams.

Let us begin with a one-property Venn diagram, and then progress to two- and three-property problems. In Figure 8.1 we show a set A within a universe $\mathcal{U}$. The complementary set $\bar{A}$ consists of all elements of $\mathcal{U}$ not in A. Let $N(S)$ denote the number of elements in set S. We also define $N = N(\mathcal{U})$. Clearly $N(\bar{A}) = N - N(A)$, or $N(A) = N - N(\bar{A})$. Stated in probabilistic terms, we are saying that the probability of an event E is one minus the probability of $\bar{E}$.

For example, if there is a "universe" of 100 students in a math course, and if there are 30 students who are not math majors in the course, then the number of math majors in the course $N(A) = N - N(\bar{A}) = 100 - 30 = 70$.

Next let us consider a problem with two sets. Let the universe $\mathcal{U}$ be all students in a school, let F be the set of students taking French, and let L be the set of students taking Latin. See Figure 8.2. We want formulas for the number

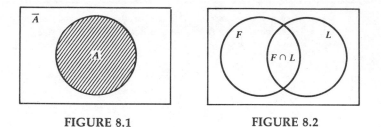

FIGURE 8.1   FIGURE 8.2

of students taking French or Latin $N(F \cup L)$ and the number taking neither language $N(\bar{F} \cap \bar{L})$ in terms of N, $N(F)$, $N(L)$, and $N(F \cap L)$. Note that $N(F \cup L)$ is not simply $N(F) + N(L)$, because $N(F) + N(L)$ counts each student taking both languages two times. Thus we must know how many students take both languages, the number $N(F \cap L)$. Subtracting $N(F \cap L)$ from $N(F) + N(L)$ corrects the double counting of students taking two languages. That is,

$$N(F \cup L) = N(F) + N(L) - N(F \cap L) \tag{1}$$

By de Morgan's Law (Equation BA3 of Appendix A.1), $\bar{F} \cap \bar{L} = \overline{F \cup L}$, and so

$$N(\bar{F} \cap \bar{L}) = N(\overline{F \cup L}) = N - N(F \cup L)$$

Combining this equation with (1), we have

$$N(\bar{F} \cap \bar{L}) = N - N(F \cup L) = N - N(F) - N(L) + N(F \cap L) \tag{2}$$

Example 1

If a school has 100 students with 50 students taking French, 40 students taking Latin, and 20 students taking both languages, how many students take no languages?

If this problem, $N = 100$, $N(F) = 50$, $N(L) = 40$, and $N(F \cap L) = 20$. We need to determine $N(\bar{F} \cap \bar{L})$. By (2), we have

$$N(\bar{F} \cap \bar{L}) = 100 - 50 - 40 + 20 = 30 \qquad \blacksquare$$

Example 2

How many arrangements of the digits 0, 1, 2, . . . , 9 are there in which the first digit is greater than 1 and the last digit is less than 8?

Let $\mathscr{U}$ be all arrangements of 0, 1, 2, . . . , 9. Let F be the set of all arrangements with a 0 or 1 in the first digit, and let L be the set of all arrangements with an 8 or 9 in the last digit. Then the number of arrangements with first digit greater than 1 and the last digit less than 8 is $N(\bar{F} \cap \bar{L})$.

We have $N = 10!$, $N(F) = 2 \cdot 9!$ (two choices for the first digit followed by any arrangement for the remaining 9 digits), and $N(L) = 2 \cdot 9!$. Similarly $N(F \cap L) = 2 \cdot 2 \cdot 8!$. Then by (2),

$$N(\bar{F} \cap \bar{L}) = 10! - 2 \cdot 9! - 2 \cdot 9! + 2 \cdot 2 \cdot 8!$$

∎

Now let us consider a problem with three sets. We extend Figure 8.2 with the additional set G of students taking German, as shown in Figure 8.3. We want formulas for $N(F \cup L \cup G)$ and $N(\bar{F} \cap \bar{L} \cap \bar{G})$. This time we will concentrate on the formula for $N(\bar{F} \cap \bar{L} \cap \bar{G})$. The first guess might be

$$N(\bar{F} \cap \bar{L} \cap \bar{G}) \stackrel{?}{=} N - N(F) - N(L) - N(G)$$

As in Figure 8.2, this formula double counts (that is, subtracts twice) the students in two of the sets, F, L, and G. We can correct this first formula by adding the number of students taking two languages. Thus

$$N(\bar{F} \cap \bar{L} \cap \bar{G}) \stackrel{?}{=} N - N(F) - N(L) - N(G)$$
$$+N(F \cap L) + N(L \cap G) + N(F \cap G) \qquad (3)$$

Figure 8.4 shows how many times a student will be added and subtracted in parts of this formula. A student taking no languages is counted once (by N)—such students are exactly the ones we want to count. The trick is to make sure that all other students are counted a net of 0 times. The students taking one language are counted once by N and subtracted once by the term $-(N(F) + N(L) + N(G))$, for a net count of 0. The students taking two languages are

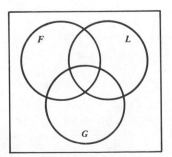

FIGURE 8.3

Number of Languages Taken by Student	N	$-(N(F) + N(L) + N(G))$	$+(N(F \cap L) + N(L \cap G) + N(F \cap G))$	$-N(F \cap L \cap G)$
0	+1	0	0	0
1	+1	−1	0	0
2	+1	−2	+1	0
3	+1	−3	+3	−1

FIGURE 8.4

counted once by N, subtracted twice by $-(N(F) + N(L) + N(G))$ (since they are in exactly two of the three sets), and then added once by the term $+(N(F \cap L) + N(L \cap G) + N(F \cap G))$ (since they are in exactly one of the three pairwise intersections), for a net count of 0. Finally, we consider the students taking all three languages. They are counted once by N, then subtracted three times by the sum of the three sets (since they are in all three sets), then added three times by the pairwise intersections (since they are in all three of these subsets). This yields a net count of $1 - 3 + 3 = 1$. Then we must correct formula (3) by subtracting $N(F \cap L \cap G)$ to make the net count of students with all three languages 0:

$$N(\bar{F} \cap \bar{L} \cap \bar{G}) = N - (N(F) + N(L) + N(G)) + (N(F \cap L)$$
$$+ N(L \cap G) + N(F \cap G)) - N(F \cap L \cap G) \qquad (4)$$

For general sets A_1, A_2, A_3, we rewrite (4) as

$$N(\bar{A}_1 \cap \bar{A}_2 \cap \bar{A}_3) = N - \sum_i N(A_i) + \sum_{i,j} N(A_i \cap A_j)$$
$$- N(A_1 \cap A_2 \cap A_3) \qquad (5)$$

where the sums are understood to run over all possible i and all i,j pairs, respectively.

Example 3

If a school has 100 students with 40 taking French, 40 taking Latin, and 40 taking German, if 20 students are taking any given pair of languages, and 10 students are taking all three languages, then how many students are taking no languages?

Here $N = 100$, $N(F) = N(L) = N(G) = 40$, $N(F \cap L) = N(L \cap G) = N(F \cap G) = 20$, and $N(F \cap L \cap G) = 10$. Then by (4), the number of students taking no languages is

$$N(\bar{F} \cap \bar{L} \cap \bar{G}) = 100 - (40 + 40 + 40) + (20 + 20 + 20) - 10 = 30 \qquad \blacksquare$$

Example 4

How many integers less than 70 are relatively prime to 70?

Let $\mathcal{U}$ be the set of integers between 1 and 70. The phrase "relatively prime" means have no common divisors. The prime divisors of 70 are 2, 5, and 7. Then we want to count the number of integers which do not have 2 or 5 or 7 as divisors. Let A_1 be the set of integers in $\mathcal{U}$ divisible by 2, A_2 integers divisible by 5, and A_3 integers divisible by 7. The problem asks for $N(\bar{A}_1 \cap \bar{A}_2 \cap \bar{A}_3)$.

Clearly $N = 70$.

$$N(A_1) = 70/2 = 35, \ N(A_2) = 70/5 = 14, \text{ and } N(A_3) = 70/7 = 10$$

The integers divisible by 2 and 5 are simply the integers divisible by 10. Thus $N(A_1 \cap A_2) = 70/10 = 7$. By similar reasoning, $N(A_2 \cap A_3) = 70/(5 \cdot 7) = 2$, $N(A_1 \cap A_3) = 70/(2 \cdot 7) = 5$, $N(A_1 \cap A_2 \cap A_3) = 70/(2 \cdot 5 \cdot 7) = 1$. So by (5),

$$N(\bar{A}_1 \cap \bar{A}_2 \cap \bar{A}_3) = 70 - (35 + 14 + 10) + (7 + 2 + 5) - 1 = 24 \quad \blacksquare$$

Example 5

How many n-digit ternary (0, 1, 2) sequences are there with at least one 0, at least one 1, and at least one 2?

Instead of numbering the sets A_1, A_2, A_3, let us use A_0, A_1, A_2, where A_i is the number of n-digit ternary sequences with no i's. Let $\mathcal{U}$ be the set of all n-digit ternary sequences. Then the number of sequences with at least one of each digit will be $N(\bar{A}_0 \cap \bar{A}_1 \cap \bar{A}_2)$.

The number of n-digit ternary sequences is $N = 3^n$. The number of n-digit ternary sequences with no 0's is simply the number of n-digit sequences of 1's and 2's. Thus $N(A_0) = 2^n$. Similarly $N(A_1) = N(A_2) = 2^n$. The only n-digit sequence with no 0's or 1's is the sequence of all 2's. Then $N(A_0 \cap A_1) = 1$; also $N(A_1 \cap A_2) = N(A_0 \cap A_2) = 1$. Finally, there is no ternary sequence with no 0's or 1's or 2's. Then by (5),

$$N(\bar{A}_0 \cap \bar{A}_1 \cap \bar{A}_2) = 3^n - (2^n + 2^n + 2^n) + (1 + 1 + 1) - 0$$
$$= 3^n - 3 \cdot 2^n + 3$$

$\blacksquare$

Observe that whereas polynomial algebra was used in Chapter 6 to model counting problems and recurrence relations were used in Chapter 7, now we are using a set-theoretic model. This approach does not eliminate combinatorial enumeration as the other models did. We still must solve the subproblems of finding $N(A_i)$, $N(A_i \cap A_j)$, and so forth, but these are much easier problems.

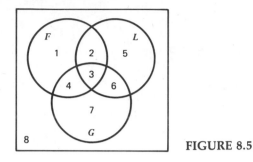

FIGURE 8.5

Example 6

Suppose there are 100 students in a school and there are 40 students taking each language, French, Latin, and German. Twenty students are taking only French, 20 taking only Latin, and 15 taking only German. In addition, 10 students are taking French and Latin. How many students are taking all three languages? No languages?

We draw the Venn diagram for this problem and number each region as shown in Figure 8.5. Let N_i denote the number of students in region i, for $i = 1, 2, \ldots, 8$. Students taking only French are the subset $F \cap \bar{L} \cap \bar{G}$, region 1; so $N_1 = 20$. Similarly $N_5 = 20$ and $N_7 = 15$. Students taking both French and Latin are the subset $F \cap L = (F \cap L \cap \bar{G}) \cup (F \cap L \cap G)$, regions 2 and 3. So $N_2 + N_3 = 10$. The set of students taking French is F, which consists of regions 1, 2, 3, and 4. So

$$40 = N(F) = N_1 + (N_2 + N_3) + N_4 = 20 + 10 + N_4, \text{ or } N_4 = 10$$

Similarly, set L consists of regions 2, 3, 5, and 6, and so $40 = N(L) = (N_2 + N_3) + N_5 + N_6 = 10 + 20 + N_6$, or $N_6 = 10$.

Since G consists of regions 3, 4, 6, and 7 and since we were given $N_7 = 15$ and have just found that $N_4 = N_6 = 10$, then

$$40 = N(G) = N_3 + N_4 + N_6 + N_7 = N_3 + 10 + 10 + 15, \text{ or } N_3 = 5$$

But region 3 is the subset $F \cap L \cap G$ of students taking all three languages. Thus there are five tri-lingual students.

The general line of attack in these problems is to break the Venn diagram into the eight regions shown in Figure 8.5. For each subset whose size is given, write that number as a sum of N_i's regions i in that subset. By combining these equations (and sometimes solving them simultaneously), one can eventually determine the number of elements in each region. Then the size of any subset is readily found as the sum of the sizes of the regions in that subset. For example, in the preceding paragraph we determined all N_i's except N_2 and N_8. But $N_2 + N_3 = 10$ and N_3 was found to be 5; thus $N_2 = 5$.

$N_8 = N(\bar{F} \cap \bar{L} \cap \bar{G})$ is the number of students taking no languages. Since all regions total to N, then

$$N_8 = N - \sum_{i=1}^{7} N_i$$

$$= 100 - (20 + 5 + 5 + 10 + 20 + 10 + 15)$$

$$= 100 - 85 = 15 \qquad \blacksquare$$

EXERCISES

SUMMARY OF EXERCISES   The first 24 exercises are similar to the examples in this section. The last five exercises involve more complicated Venn diagram arguments; see the last paragraph in Example 6 for the strategy for solving these problems.

1. How many 5-letter "words" using the 26-letter alphabet (letters can be repeated) either begin or end with a vowel?

2. How many 9-digit Social Security numbers are there with repeated digits?

3. How many n-digit ternary sequences are there in which at least one pair of consecutive digits are the same?

4. What is the probability that at least two heads (not necessarily consecutive) will appear when a coin is flipped 8 times?

5. What is the probability that a 5-card hand has at least one pair (possibly two pairs, three of a kind, full house, or four of a kind)?

6. If n people of different heights are lined up in a queue, what is the probability that at least one person is just behind a taller person?

7. Suppose 60% of all families own a dishwasher, 30% own a trash compacter, and 20% own both. What percent of all families own at least one of these two appliances?

8. Suppose a bookcase has 200 books, 70 in French and 100 about mathematics. How many non-French books not about mathematics are there if:
    (a)   There are 30 French mathematics books?
    (b)   There are 60 French nonmathematics books?

9. Among 600 families, 100 families have no children, 200 have only boys, and 200 have only girls. How many families have boy(s) and girl(s)?

10. How many arrangements of the 26 different letters are there that:
    (a)   Contain either the sequence "the" or the sequence "aid"?
    (b)   Contain neither the sequence "the" nor the sequence "math"?

11. How many arrangements of the digits 0, 1, . . . , 9 are there that do not end with a 8 and do not begin with a 3?

12. How many Secret Codes are there in Mastermind (see Appendix A.5) with exactly one red peg and no other color's pegs used exactly two times?

13. A school has 200 students with 80 students taking each of the three subjects: trigonometry, probability, and basket-weaving. There are 30 students taking any given pair of these subjects, and 15 students taking all three subjects.
    (a) How many students are taking none of these three subjects?
    (b) How many students are taking only probability?

14. Suppose 60% of all college professors like tennis, 65% like bridge, and 50% like Mastermind; 45% like any given pair of recreations.
    (a) Should you be suspicious if told 20% like all three recreations?
    (b) What is the largest percent who could like all three recreations?

15. How many ways are there to assign 20 different people to three different rooms with at least one person in each room?

16. How many n-digit numbers are there with at least one of the digits 1 or 2 or 3 absent? (*Hint:* This is a union problem; find $N(A_1 \cup A_2 \cup A_3)$.)

17. How many arrangements are there of *MURMUR* with no pair of consecutive letters the same?

18. How many numbers between 1 and 30 are relatively prime to 30?

19. How many numbers between 1 and 280 are relatively prime to 280?

20. If three couples are seated around a circular table, what is the probability that no wife and husband are beside one another?

21. How many arrangements are there of *TAMELY* with either *T* before *A*, or *A* before *M*, or *M* before *E*? By "before", we mean anywhere before, not just immediately before. (*Hint:* This is a union problem; find $N(A_1 \cup A_2 \cup A_3)$.)

22. How many arrangements are there of *MATHEMATICS* with both *T*'s before both *A*'s, or both *A*'s before both *M*'s, or both *M*'s before the *E*? Note "before" is used as in Example 21.

23. The Bernsteins, Hendersons, and Smiths each have 5 children. If the 15 children of these three families camp out in five different tents, where each tent holds 3 children, and the 15 children are randomly assigned to the five tents, what is the probability that every family has two (or more) of its members in the same tent?

24. Suppose 45% of all newspaper readers like wine, 60% like orange juice, and 55% like tea. Suppose 35% like any given pair of these beverages and 25% like all three beverages.
    (a) What percent of the readers like only wine?
    (b) What percent of the readers like exactly two of these three beverages?

25. Suppose that among 40 toy robots, 28 have a broken wheel or are rusted but not both, 6 are not defective, and the number with a broken wheel equals the number with rust. How many robots are rusted?

26. Suppose a school with n students offers two languages, PASCAL and BASIC. If 30 students take no language, 70 students do not take just PASCAL (i.e., either they do not take PASCAL or they take both languages), 80 students do not take just BASIC, and 20 students take both languages, determine n.

27. Suppose a school with 120 students offers Yoga and Karate. If the number of students taking Yoga alone is twice the number taking Karate (and possibly Yoga also), if 25 more students study neither skill than study both skills, and if 75 students take at least one skill, then how many students study Yoga?

28. In a class of 30 children, 20 take Latin, 14 take Greek, and 10 take Hebrew. If no child takes all three languages and 8 children take no language, how many children take Greek and Hebrew?

29. Suppose among 150 people on a picnic, 90 bring salads or sandwiches, 80 bring sandwiches or cheese, 100 bring salads or cheese, 50 bring cheese and at least one other of the above foods, 45 bring cheese but not both other foods, 60 bring at least two foods, and 20 bring all three foods. How many people bring just one of the three foods?

8.2 INCLUSION-EXCLUSION FORMULA

In this section we generalize the formula for counting $N(\bar{A}_1 \cap \bar{A}_2 \cap \bar{A}_3)$ to n sets $A_1, A_2, \ldots, A_n$. To simplify notation, we will omit the intersection symbol "$\cap$" in expressions and write intersected sets as a product. For example, $A_1 \cap A_2 \cap \bar{A}_3$ would be written $A_1 A_2 \bar{A}_3$. Using this new notation, the number of elements in none of the sets $A_1, A_2, \ldots, A_n$ would be written $N(\bar{A}_1 \bar{A}_2 \cdots \bar{A}_n)$. The following formula is known as the inclusion-exclusion formula because of the way it successively includes (adds) and excludes (subtracts) the various k-tuple intersections of sets.

Theorem 1—Inclusion-Exclusion Formula

Let $A_1, A_2, \ldots, A_n$ be n sets in a universe $\mathcal{U}$ of N elements. Let S_k denote the sum of the sizes of all k-tuple intersections of the A_i's. Then

$$N(\bar{A}_1 \bar{A}_2 \cdots \bar{A}_n)$$
$$= N - S_1 + S_2 - S_3 + \cdots + (-1)^k S_k + \cdots + (-1)^n S_n \quad (1)$$

Proof

To clarify the definition of the S_k's, $S_1 = \Sigma_{i=1}^n N(A_i)$, $S_2 = \Sigma_{i,j} N(A_iA_j)$, S_k is the sum of the $N(A_{i_1}A_{i_2} \cdots A_{i_k})$'s for all sets of k A_i's, and finally $S_n = N(A_1A_2 \cdots A_n)$. We prove this formula by the same method used for $N(\overline{FLG})$ in the previous section: we shall show that the net effect of (1) is to count any element in none of the sets A_i once and to count elements in one or more A_i's a net of 0 times.

If an element is in none of the A_i's, that is, is in $\overline{A}_1\overline{A}_2 \cdots \overline{A}_n$, then it is counted once in the right-hand side of (1) by the term N and is not counted in any of the S_k's. So the count is 1 for each element in $\overline{A}_1\overline{A}_2 \cdots \overline{A}_n$, as required. An element is exactly one A_i is counted once by N, is subtracted once by S_1 (since it is in one of the A_i's), and is counted in none of the other S_k's—for a net count of 0, as required. Now more generally let us show that an element x in exactly m of the A_i's has a net count of 0 in (1). Element x is counted once by N, is counted m times by S_1 (since x is in m A_i's), is counted $C(m,2)$ times by S_2 (since x is in the intersection A_iA_j for all $C(m,2)$ pairs of sets containing x), . . . , is counted $C(m,k)$ times by S_k, $k \leq m$. It is not counted in S_k when $k > m$. So the net count of x in (1) is

$$1 - \binom{m}{1} + \binom{m}{2} - \binom{m}{3} + \cdots + (-1)^k \binom{m}{k} + \cdots + (-1)^m \binom{m}{m}$$

But this alternating sum of binomial coefficients can be evaluated from the binomial expansion

$$(1 + x)^m = 1 + \binom{m}{1} x + \binom{m}{2} x^2 + \cdots + \binom{m}{k} x^k + \cdots + \binom{m}{m} x^m$$

by setting $x = -1$. So the net count of x equals $(1 + (-1))^m = 0^m = 0$. ∎

A convenient shorthand way to write formula (1) is

$$N(\overline{A}_1\overline{A}_2 \cdots \overline{A}_n) = \sum_{X' \subseteq X} (-1)^{|X'|} N(X')$$

where X' ranges over all possible subsets of the set X of all properties.

Corollary

Let $A_1, A_2, \ldots, A_n$ be sets in the universe $\mathcal{U}$. Then

$$N(A_1 \cup A_2 \cup \cdots \cup A_n)$$
$$= S_1 - S_2 + S_3 - \cdots + (-1)^{k-1}S_k \cdots (-1)^{n-1}S_n \qquad (2)$$

Proof

Formula (1) can be written

$$N(\bar{A}_1\bar{A}_2 \cdots \bar{A}_n) = N - (S_1 - S_2 + S_3 + \cdots + (-1)^{n-1}S_n) \qquad (3)$$

But the number of elements in no set is clearly the total number of elements minus the number of elements in one or more sets, that is, $N(\bar{A}_1\bar{A}_2 \cdots \bar{A}_n) = N - N(A_1 \cup A_2 \cup \cdots \cup A_n)$. Hence the expression in parentheses in (3) must be $N(A_1 \cup A_2 \cup \cdots \cup A_n)$. ∎

Before giving examples of the inclusion-exclusion formula, we want to emphasize an important logical point about applying this formula. To use this formula in a counting problem, one must select a universe $\mathcal{U}$ and a collection of sets A_i in that universe such that the objects to be counted are the subset of elements in $\mathcal{U}$ that are in *none* of the A_i's. That is, the A_i's represent properties *not* satisfied by the objects being counted.

Example 1

How many ways are there to select a 5-card hand from a regular 52-card deck such that the hand contains at least one card in each suit?

The universe $\mathcal{U}$ should be the set of 5-card hands. We need to define the sets A_i such that hands with at least one card in each suit are in none of the A_i's. With a moment's thought, we see that at least one card in a suit is equivalent to no void in the suit. Thus we let A_1 be the set of 5-card hands with a void in spades; A_2 hands with a void in hearts; A_3 a void in diamonds; and A_4 a void in clubs. Now the question asks for $N(\bar{A}_1\bar{A}_2\bar{A}_3\bar{A}_4)$, and we can use the inclusion-exclusion formula.

We must next calculate N, S_1, S_2, S_3, and S_4. As noted in Chapter 5, a 5-card hand is simply a subset of 5 cards, and so $N = C(52,5)$. The hands in A_1 are simply subsets of 5 cards chosen from the $52 - 13 = 39$ nonspade cards. So $N(A_1) = C(39,5)$. Likewise, $N(A_i) = C(39,5)$, $i = 2, 3, 4$, and $S_1 = 4 \cdot C(39,5)$. The hands in A_1A_2 are hands chosen from the 26 non-(spades or hearts) cards, and so $N(A_1A_2) = C(26,5)$. There are $C(4,2) = 6$ different intersections of two out of the four sets, and so $S_2 = 6 \cdot C(26,5)$. There are $C(4,3) = 4$ different triple intersections of the sets and each has $C(13,5)$ hands. So $S_3 = 4 \cdot C(13,5)$. Finally, a hand cannot be void in all four suits, and so $S_4 = 0$. Then by (1),

$$N(\bar{A}_1\bar{A}_2\bar{A}_3\bar{A}_4) = \binom{52}{5} - 4\binom{39}{5} + 6\binom{26}{5} - 4\binom{13}{5} + 0$$

∎

We note that in general, S_k is a sum of $C(n,k)$ different k-tuple intersections of the n A_i's.

Example 2

How many ways are there to distribute r distinct objects into five (distinct) boxes with at least one empty box?

Let $\mathcal{U}$ be all distributions of r distinct objects into five boxes. Let A_i be the set of distributions with a void in box i. Then the required number of distributions with at least one void is $N(A_1 \cup \cdots \cup A_5)$. We find $N = 5^r$, $N(A_i) = 4^r$ (distributions with each object going into one of the other 4 boxes), $N(A_i A_j) = 3^r$, and so forth. Thus by (2)

$$N(A_1 \cup A_2 \cdots \cup A_5) = S_1 - S_2 + S_3 - S_4 + S_5$$

$$= \binom{5}{1} 4^r - \binom{5}{2} 3^r + \binom{5}{3} 2^r - \binom{5}{4} 1^r + 0$$

■

Example 3

How many different integer solutions are there to the equation

$$x_1 + x_2 + x_3 + x_4 + x_5 + x_6 = 20, \quad 0 \le x_i \le 8$$

This type of problem was solved with generating functions in Section 6.2. Let $\mathcal{U}$ be all integer solutions with $x_i \ge 0$, and let A_i be the set of integer solutions with $x_i > 8$, or equivalently $x_i \ge 9$. Then the number of solutions with $0 \le x_i \le 8$ will be $N(\bar{A}_1 \cdots \bar{A}_6)$.

Recalling the formulas for integer solutions from Section 5.5 (Example 4), we have

$$N = C(20 + 6 - 1, 20) = C(25, 20),$$

$$N(A_i) = C((20 - 9) + 6 - 1, (20 - 9)) = C(16, 11),$$

$$N(A_i A_j) = C((20 - 9 - 9) + 6 - 1, (20 - 9 - 9)) = C(7, 2)$$

For a solution to be in three or more A_i's the sum of the respective x_i's would exceed 20—impossible. So $S_j = 0$ for $j \ge 3$, and

$$N(\bar{A}_1 \cdots \bar{A}_6) = N - S_1 + S_2 = \binom{25}{20} - \binom{6}{1}\binom{16}{11} + \binom{6}{2}\binom{7}{2}$$

■

Note that using generating functions to solve the preceding problem, we would seek the coefficient of x^{20} in

$$(1 + x + \cdots + x^8)^6 = [(1 - x^9)/(1 - x)]^6 = (1 - x)^{-6}(1 - x^9)^6$$

The coefficient of x^{20} in this product is $a_{20}b_0 + a_{11}b_9 + a_2b_{18}$, where a_k is the coefficient of x^k in $(1 - x)^{-6}$ and b_k is the coefficient of x^k in $(1 - x^9)^6$. This coefficient of x^{20} turns out to be exactly the above expression for $N(\bar{A}_1 \cdots \bar{A}_6)$. The $(1 - x^9)^6$ factor does the inclusion-exclusion task of subtracting cases where one x_i is at least 9 and adding back cases where two x_i's are at least 9. By using generating functions to solve this problem, we did not need to know anything about the inclusion-exclusion complexities of this problem. Generating functions automatically performed the required combinatorial logic!

Example 4

What is the probability that if n people randomly reach into a dark closet to retrieve their hats, no person will pick their own hat?

The probability will be the fraction of outcomes in which no person gets their own hat. Our universe $\mathcal{U}$ is all ways for the n people to successively select a different hat. So $N = n!$ Let A_i be the set of outcomes in which person i gets their own hat. Then $N(A_i) = (n - 1)!$, since given that person i gets their hat, the number of possible outcomes is all ways for the other $n - 1$ people to select hats. Similarly $N(A_i A_j) = (n - 2)!$, and generally $N(A_{i_1} A_{i_2} \cdots A_{i_k}) = (n - k)!$ for k-way intersections. Since S_k is a sum of $C(n,k)$ k-way terms, we obtain by (1)

$$N(\bar{A}_1 \bar{A}_2 \cdots \bar{A}_n) = N - S_1 + S_2 + \cdots + (-1)^k S_k + \cdots + (-1)^n S_n$$

$$= n! - \binom{n}{1}(n - 1)! + \binom{n}{2}(n - 2)! \cdots (-1)^k \binom{n}{k}(n - k)!$$

$$\cdots (-1)^n \binom{n}{n} 0! = \sum_{k=0}^{n} (-1)^k \binom{n}{k}(n - k)! \tag{4}$$

Recalling that $\binom{n}{k} = n!/(k!(n - k)!)$, we see that $\binom{n}{k}(n - k)! = n!/k!$ So

$$N(\bar{A}_1 \bar{A}_2 \cdots \bar{A}_n) = \sum_{k=0}^{n} \frac{(-1)^k n!}{k!} = n! \sum_{k=0}^{n} \frac{(-1)^k}{k!} \tag{5}$$

and now the probability no person gets their own hat is

$$\frac{N(\bar{A}_1\bar{A}_2\cdots\bar{A}_n)}{N} = \left(n!\sum_{k=0}^{n}\frac{(-1)^k}{k!}\right)/n!$$

$$= 1 - 1 + \frac{1}{2!} - \frac{1}{3!} + \cdots + \frac{(-1)^n}{n!} \quad (6)$$

This alternating series is the first $n + 1$ terms of $\sum_{k=0}^{\infty}(-1)^k/k!$, which is the power series for e^x when $x = -1$. The series converges very fast. The difference between e^{-1} and (6) is always less than $1/n!$. For example, $e^{-1} = 0.367879\ldots$ and for $n = 8$, the series in (6) equals 0.367888 (even for $n = 5$, it is 0.366). Thus for all but very small n, the desired probability is essentially e^{-1}. The answer is virtually independent of $n!$ ■

The problem treated in Example 4 is equivalent to asking for all permutations of the sequence 1, 2, . . . , n such that no number is left fixed, that is, no number i is still in the ith position. Such rearrangements of a sequence are called **derangements**. The symbol D_n is used to denote the number of derangements of n integers. From Example 4, we have

$$D_n = n!\sum_{k=0}^{n}\frac{(-1)^k}{k!} \cong n!e^{-1}$$

For exact values of D_n, it is usually easier to use the following simple recurrence relation:

$$D_n = nD_{n-1} + (-1)^n, \quad n \geqslant 2$$

See Exercise 33 for details on deriving this recurrence.

Example 5

How many ways are there to color the four vertices in the graph shown in Figure 8.6 with n colors (such that vertices with a common edge must be different colors).

We label the edges e_1, e_2, e_3, e_4, e_5, as shown in Figure 8.6. The universe should be the set of all ways to color the vertices. So $N = n^4$ (n choices for each vertex). The situation we must avoid is having adjacent vertices the same color. Then for each edge e_i, we define the set A_i to be all colorings of the graph in which the vertices at either end of e_i have the same color. Now $N(A_i) = n^3$ (since the two vertices connected by e_i get one color and the two other vertices each get any color), and $S_1 = 5 \cdot n^3$. Similarly, $N(A_iA_j) = n^2$ and $S_2 = C(5,2)n^2$. A little care must be used with three-way intersections. Specifically, edges e_1, e_2, e_3 interconnect three, not four, vertices, and so $N(A_1A_2A_3) = n^2$ (one color for vertices x_1, x_2, x_3 and one color for vertex x_4). Likewise, $N(A_3A_4A_5) = n^2$. Any other set of three edges interconnect all four vertices,

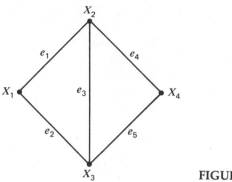

FIGURE 8.6

and so the associated $N(A_i A_j A_k)$ equals n. Then $S_3 = 2n^2 + (C(5,3) - 2)n$. Four or five edges interconnect all four vertices, and so $S_4 = C(5,4)n$ and $S_5 = n$. Then the answer to our problem is

$$N(\bar{A}_1 \cdots \bar{A}_5) = n^4 - 5n^3 + \binom{5}{2} n^2 - \left[2n^2 + \left(\binom{5}{3} - 2 \right) n \right] + \binom{5}{4} n - n$$

$$= n^4 - 5n^3 + 8n^2 - 4n \tag{7}$$

∎

Note that the expression (7) is the chromatic polynomial of this graph (see Example 9 in Section 5.2).

We conclude this section with two generalizations of the inclusion-exclusion formula. *Many readers may want to skip this material.*

Theorem 2

If $A_1, A_2, \ldots, A_n$ are n sets in a universe $\mathcal{U}$ of N elements, then the number N_m of elements in exactly m sets and the number N_m^* of elements in at least m sets are given by:

$$N_m = S_m - \binom{m + 1}{m} S_{m+1} + \binom{m + 2}{m} S_{m+2} + \cdots + (-1)^{k-m} \binom{k}{m} S_k$$

$$\cdots (-1)^{n-m} \binom{n}{m} S_n \tag{8}$$

$$N_m^* = S_m - \binom{m}{m - 1} S_{m+1} + \binom{m + 1}{m - 1} S_{m+2} + \cdots + (-1)^{k-m} \binom{k - 1}{m - 1} S_k$$

$$\cdots (-1)^{n-m} \binom{n - 1}{m - 1} S_n \tag{9}$$

Proof

The formula for N_m can be proved in a fashion similar to the proof of the inclusion-exclusion formula. We must show that elements in more than m sets are counted a net of 0 times in (8). In this formula the count is slightly trickier to sum. Readers who dislike such technicalities should skip this proof.

The count for an element in r sets, $m \leq r \leq n$, is $C(r,k)$ in S_k, and so the net count of this element in (8) is

$$\binom{r}{m} - \binom{m+1}{m}\binom{r}{m+1} \cdots (-1)^{k-m} \binom{k}{m}\binom{r}{k} \cdots (-1)^{r-m} \binom{r}{m}\binom{r}{r}$$

$$(10)$$

Remember that the element is not counted in S_k for $k > r$. Recall from Example 1 of Section 5.5 that the number of ways to pick k objects from r and then pick m special objects from those k is equal to the number of ways to pick the m special objects from the r first and then pick $k - m$ more from the remaining $r - m$. So

$$\binom{k}{m}\binom{r}{k} = \binom{r}{m}\binom{r-m}{k-m}$$

Using this substitution, (10) now becomes

$$\binom{r}{m} - \binom{r}{m}\binom{r-m}{1} \cdots (-1)^{k-m} \binom{r}{m}\binom{r-m}{k-m} \cdots (-1)^{r-m} \binom{r}{m}\binom{r-m}{r-m}$$

$$= \binom{r}{m}\left[1 - \binom{r-m}{1} \cdots (-1)^{k-m} \binom{r-m}{k-m} \cdots (-1)^{r-m} \binom{r-m}{r-m}\right]$$

As in Theorem 1, the expression in brackets here is just the expansion of $(1 + x)^{r-m}$ with $x = -1$. So (10) sums to 0, as required.

The formula for N_m^* can be verified with induction by showing that formula (9) for N_m^* satisfies $N_m^* = N_m + N_{m+1}^*$: "in at least m sets" means "in exactly m sets" or "in at least $m + 1$ sets." ∎

Example 6

Find the number of 4-digit ternary sequences with exactly two 1's and the number with at least two 1's.

Let $\mathcal{U}$ be all 4-digit ternary sequences. Let A_i be 4-digit ternary sequences with a 1 in position i. Then $N = 3^4$ and $S_1 = C(4,1)3^3$, $S_2 = C(4,2)3^2$, $S_3 = C(4,3)3^1$, and $S_4 = C(4,4)3^0 = 1$ Now by formulas (8) and (9), the desired numbers are

$$N_2 = S_2 - \binom{3}{2} S_3 + \binom{4}{2} S_4 = \binom{4}{2} 3^2 - \binom{3}{2}\binom{4}{3} 3^1 + \binom{4}{2} 1 = 24$$

and

$$N_2^* = S_2 - \binom{2}{1} S_3 + \binom{3}{1} S_4 = \binom{4}{2} 3^2 - \binom{2}{1}\binom{4}{3} 3^1 + \binom{3}{1} 1 = 33 \qquad \blacksquare$$

The observant reader should recall that N_2 and N_2^* can be computed directly by a simple combinatorial argument.

EXERCISES

SUMMARY OF EXERCISES   These exercises require a combination of inclusion-exclusion modeling and Chapter-5-type enumeration skills (to solve the subproblems, some of which are tricky). Exercises 34–44 use Theorem 2.

1. How many n-digit decimal sequences (using digits 0, 1, 2, . . . , 9) are there in which digits 1, 2, 3 all appear?

2. How many ways are there to roll 10 distinct dice so that all 6 faces appear?

3. What is the probability that a 10-card hand has at least one 4-of-a-kind?

4. How many positive integers ≤ 420 are relatively prime to 420?

5. What is the probability that a 13-card bridge hand has:
    (a)   At least one card in each suit?
    (b)   At least one void?
    (c)   At least one of each type of face card (face cards are Aces, Kings, Queens, and Jacks)?

6. Given five pairs of gloves, how many ways are there for five people to each choose two gloves with no one getting a matching pair?

7. How many arrangements are there of $a, a, a, b, b, b, c, c, c$ without three consecutive letters the same?

8. Given $2n$ letters, two of each of n types, how many arrangements are there with no pair of consecutive letters the same?

9. How many permutations of the 26 letters are there which contain none of the sequences "MATH," "RUNS," "FROM," or "JOE"?

10. How many integer solutions of $x_1 + x_2 + x_3 + x_4 = 30$ are there with:
    (a) $0 \leq x_i \leq 10$?
    (b) $-10 \leq x_i \leq 20$?
    (c) $0 \leq x_i, x_1 \leq 5, x_2 \leq 10, x_3 \leq 15, x_4 \leq 21$?

11. How many ways are there to distribute 25 identical balls into six distinct boxes with at most 6 balls in any of the first three boxes?

12. How many ways are there for a child to take 12 pieces of candy from a stranger with four types of candy if the child does not take exactly two pieces of:
    (a) Any type of candy?
    (b) Exactly two of the types of candy?

13. Santa Claus has five toy airplanes of each of n plane models. How many ways are there to put one airplane in each of r ($r \geq n$) identical stockings such that all models of planes are used?

14. Determine N_2 and N_2^* by a direct combinatorial argument.

15. A wizard has five friends. During a long wizard's conference, it met any given friend at dinner 10 times, any given pair of friends 5 times, any given threesome of friends 3 times, any given foursome 2 times, and all five friends together once. If in addition it ate alone 6 times, determine how many days the wizard's conference lasted.

16. How many secret codes can be made by assigning each letter of the alphabet a (unique) different letter?

17. How many ways are there to distribute 10 books to 10 children (one to a child) and then collect the books and re-distribute them with each child getting a new one?

18. How many arrangements of 1, 2, . . . , n are there in which only the odd integers must be deranged (even integers may be in their own positions)?

19. Consider the following game with a pile of n cards numbered 1 through n. Successively pick a different (random) number between 1 and n and remove all cards in the pile down to, and including, the card with this number until the pile is empty. If the chosen card number has already been removed, pick another number. What is the approximate probability that at some stage the number chosen is the card at the top of the pile? (*Hint:* Approach as arrangement with repetition.)

20. How many ways are there to assign each of five professors in a math department to two courses in the fall semester (i.e., 10 different math courses) and then reassign each professor two courses in the spring semester such that no professor teaches the same 2 courses both semesters?

21. (a) How many arrangements of the integers 1 through n are there in which i is never immediately followed by $i + 1$, for $i = 1, 2, . . . ,$ $n - 1$?
    (b) Show that your answer equals $D_n + D_{n-1}$.

22. Repeat Exercise 21a now considering n to be followed by 1.

23. How many ways are there to seat n couples around a circular table such that no couple sits next to each other?

24. If two identical dice are rolled n successive times, how many sequences of outcomes contain all doubles (a pair of 1's, of 2's, etc.)?

25. How many arrangements of $a, a, a, b, b, b, c, c, c$ have no adjacent letters the same? (*Hint:* This is *tricky*—not a normal inclusion-exclusion problem.)

26. (a) If there are n families with five members each, how many ways are there to seat all $5n$ people at a circular table so that each person sits beside another member of their family?

(b) Repeat with four members in each family.

27. Suppose that a person with seven friends invites a different subset of three friends to dinner every night for one week (seven days). How many ways can this be done so that all friends are included at least once?

28. The rooms in the circular house plan shown below are to be painted with eight colors such that rooms with a common doorway must be different colors. In how many ways can this be done?

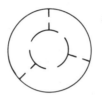

29. How many ways are there to color properly the vertices in the following graphs?

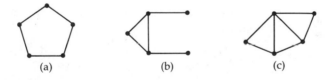

(a)   (b)   (c)

30. The four walls and ceiling of a room are to be painted with five colors available. How many ways can this be done if bordering sides of the room must have different colors?

31. How many ways are there to distribute r distinct objects into n indistinguishable boxes with no box empty? This number is called a *Stirling number of the second kind.*

32. Find the number of ways to give each of six different people seated in a circle one of m different types of entrees if adjacent people must get different entrees.

33. (a) Show that D_n satisfies the recurrence $D_n = (n - 1)(D_{n-1} + D_{n-2})$.

(b) Rewriting the recurrence in part (a) as $D_n - nD_{n-1} = -(D_{n-1} - (n - 1)D_{n-2})$, iterate backwards to obtain the recurrence $D_n = nD_{n-1} + (-1)^n$.

(c) Use part (b) to make a list of D_n values up to $n = 10$.

(d) Use part (a) to show that $D(x) = e^{-x}(1 - x)^{-1}$ is the exponential generating function for D_n.

34. How many ways are there to distribute r distinct objects into n distinct boxes with exactly three empty boxes? With at least three empty boxes?

35. How many ways are there to deal a 6-card hand with at most one void in a suit?

36. How many ways are there to arrange the letters in *INTELLIGENT* with at least two consecutive pairs of identical letters?

37. If n balls labeled $1, 2, \ldots, n$ are successively removed from an urn, a *rencontre* is said to occur if the ith ball removed is numbered i. If the n balls are removed in random order, what is the probability that exactly k rencontres occur? Show that this probability is about $(1/k!)e^{-1}$.

38. Show that the number of ways to place r different balls in n different cells with m cells having exactly k balls is

$$\frac{(-1)^m n! r!}{m!} \sum_{j=m}^{n} (-1)^j \frac{(n-j)^{r-jk}}{(j-m)!(n-j)!(r-jk)!(k!)^j}$$

39. (a) If $g(x)$ is the (ordinary) generating function for N_m (see Theorem 2), show that $g(x) = \sum_{k=0}^{n} S_k(x-1)^k$. This $g(x)$ is called the *hit polynomial*.
    (b) Show that $\frac{1}{2}(g(1) + g(-1))$ is the number of elements in an even number of A_i's.
    (c) Use part (b) to determine the number of ternary sequences with an even number of 0's. Simplify your answer with a binomial expansion summation.

40. Show that $S_m = \sum_{k=m}^{n} \binom{k}{m} N_m$.

41. Use a combinatorial argument (with inclusion-exclusion) to prove:

    (a) $\sum_{k=0}^{m} (-1)^k \binom{m}{k} = \binom{n-m}{n-r}$, $m + r \leq n$

    (b) $\sum_{k=0}^{m} (-1)^k \binom{n}{k}\binom{n-k}{m-k} = 0$

    (c) $\sum_{k=m}^{n} (-1)^{k-m} \binom{n}{k} = \binom{n-1}{m-1}$

    (d) $\sum_{k=0}^{n} (-1)^k \binom{n}{k}\binom{n-k+r-1}{r} = \binom{r-1}{n-1}$

42. Use Theorem 2 to show that

$$\binom{n}{m} = \sum_{k=m}^{n} (-1)^{k-m} \binom{k}{m}\binom{n}{k} 2^{n-k}$$

43. Prove formula (9) in Theorem 2.

44. In Example 4, let the random variable X = no. of people who get own hat. Find $E(X)$.

8.3 RESTRICTED POSITIONS AND ROOK POLYNOMIALS

In this section we consider the special problem of counting arrangements of n objects when various objects can only appear in certain positions. We will solve one such problem involving five objects in careful detail. At the same time we will indicate how to generalize our analysis to any restricted-positions problem.

Consider the problem of finding all arrangements of a, b, c, d, e with the restrictions indicated in Figure 8.7. That is, a may *not* be put in position 1 or 5; b may not be put in 2 or 3; c not in 3 or 4; and e not in 5. There is no restriction on d. A permissible arrangement can be represented by picking five un-marked squares in Figure 8.7, with one square in each row and each column. For example, the permissible arrangement *badec* corresponds to picking squares $(a,2)$, $(b,1)$, $(c,5)$, $(d,3)$, $(e,4)$.

When viewed in terms of the 5×5 array of squares, the arrangement problem can be thought of as a matching problem. In Section 4.4 we used 0's for darkened squares and 1's for white squares in matrix representations of matching problems. In the associated bipartite matching graph $G = (X,Y,E)$, the rows stood for X-vertices, the columns for Y-vertices, and each white square would have represented an edge. Before we worried about whether or not there was a complete matching. Now our concern is how many complete matchings are there.

We use the inclusion-exclusion formula, expression (1) of the previous section, to count the number of permissible arrangements for Figure 8.7. Let $\mathcal{U}$ be the set of all arrangements of the five letters without restrictions. So $N = 5!$ Let A_i be the set of arrangements with a forbidden letter in position i. The number of permissible arrangements will then be $N(\bar{A}_1\bar{A}_2\bar{A}_3\bar{A}_4\bar{A}_5)$ (note

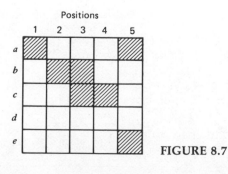

FIGURE 8.7

that we could equally well have defined the properties in terms of the *i*th letter being in a forbidden position). In terms of Figure 8.7, A_i is the set of all collections of five squares, each in a different row and column such that the square in column i is a darkened square. We obtain $N(A_i)$ by counting the ways to put a forbidden letter in position i times the 4! ways to arrange the remaining four letters in the other four positions (we do not worry about forbidden positions for these letters). Then $N(A_1) = 1 \cdot 4!$, $N(A_2) = 1 \cdot 4!$, $N(A_3) = 2 \cdot 4!$, $N(A_4) = 1 \cdot 4!$, and $N(A_5) = 2 \cdot 4!$, and collecting terms we obtain

$$S_1 = \sum_{i=1}^{5} N(A_i) = 1 \cdot 4! + 1 \cdot 4! + 2 \cdot 4! + 1 \cdot 4! + 2 \cdot 4!$$

$$= (1 + 1 + 2 + 1 + 2)4! = 7 \cdot 4!$$

Observe that $(1 + 1 + 2 + 1 + 2) = 7$ is just the number of the darkened squares in Figure 8.7. Since each choice of a darkened square (i.e., some letter in some forbidden position) leads to 4! possibilities, then

$$S_1 = \text{(number of darkened squares)} \cdot 4!$$

for any restricted-positions problem with a 5 × 5 family of darkened squares as in Figure 8.7.

Next $N(A_i A_j)$ will be the number of ways to put (different) forbidden letters in positions i and j times the 3! ways to arrange the remaining three letters. Or equivalently, the ways to pick two darkened squares, one in column i and one in column j (but in different rows), times 3! Then

$$N(A_1 A_2) = 1 \cdot 3!, \qquad N(A_1 A_3) = 2 \cdot 3!, \qquad N(A_1 A_4) = 1 \cdot 3!$$

$$N(A_1 A_5) = 1 \cdot 3!, \qquad N(A_2 A_3) = 1 \cdot 3!, \qquad N(A_2 A_4) = 1 \cdot 3!$$

$$N(A_2 A_5) = 2 \cdot 3!, \qquad N(A_3 A_4) = 1 \cdot 3!, \qquad N(A_3 A_5) = 4 \cdot 3!$$

$$N(A_4 A_5) = 2 \cdot 3!$$

Collecting terms, we obtain

$$S_2 = \sum_{i,j} N(A_i A_j) = (1 + 2 + 1 + 1 + 1 + 1 + 2 + 1 + 4 + 2)3! = 16 \cdot 3!$$

The number 16 represents the number of ways to select two darkened squares, each in a different row and column. Generalizing, we will have

$$S_k = \binom{\text{number of ways to pick } k \text{ darkened squares}}{\text{each in a different row and column}} \cdot (5 - k)! \qquad (1)$$

Since letter d's row in Figure 8.7 has no darkened squares, there is no way to pick five darkened squares, each in a different row (and column). Thus $S_5 = 0$. On the other hand, tedious case-by-case counting apparently awaits us for S_3 and S_4. Instead, let us try to develop a theory for determining the number of ways to pick k darkened squares, each in a different row and column.

This darkened squares selection problem can be restated in terms of a recreational mathematics question about a chess-like game. A chess piece called a **rook** can capture any opponent's piece in the same row or column of the rook (provided there are no intervening pieces). Instead of using a normal 8×8 chessboard, suppose we "play chess" on the "board" consisting solely of the darkened squares in Figure 8.7.

Counting the number of ways to place k mutually noncapturing rooks on this board of darkened squares is equivalent to our original problem of counting the number of ways to pick k darkened squares to Figure 8.7, each in a different row and column. The phrase "k mutually noncapturing rooks" is simpler to say and more pictorial.

A common technique in combinatorial analysis is to break a big messy problem into smaller manageable subproblems. We will now develop two breaking-up operations to help us count noncapturing rooks on a given board B. The first operation applies to a board B that can be decomposed into **disjoint** subboards B_1 and B_2, that is, subboards involving different sets of rows and columns. Often a board has to be properly rearranged before the disjoint nature of the two subboards can be seen.

When the rows and columns of Figure 8.7 are rearranged as shown in Figure 8.8, it is obvious that the three darkened squares in rows a and e and columns 1 and 5 are disjoint from the four darkened squares in rows b and c and columns 2, 3, and 4. Let B be the board of darkened squares in Figure 8.8, let B_1 be the three darkened squares in rows a and e, and let B_2 be the four darkened squares in rows b and c.

Let $r_k(B)$ denote the number of ways to place k noncapturing rooks on board B, and let $r_k(B_i)$ denote the ways on subboard B_i. Then it is easy to see that $r_1(B_1) = 3$ and $r_2(B_1) = 1$, while $r_1(B_2) = 4$ and $r_2(B_2) = 3$. It will be convenient to define $r_0 = 1$ for all boards. Clearly $r_k(B_i) = 0$ for $k > 2$, since each subboard has only two rows. Observe next that since B_1 and B_2 are disjoint,

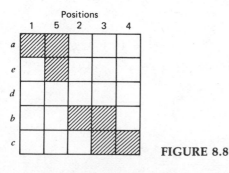

FIGURE 8.8

placing, say, two noncapturing rooks on the whole board B can be broken into three cases: placing two noncapturing rooks on B_1 (and none on B_2), placing one rook on each subboard, or placing two noncapturing rooks on B_2. Thus we see that

$$r_2(B) = r_2(B_1)r_0(B_2) + r_1(B_1)r_1(B_2) + r_0(B_1)r_2(B_2)$$
$$= 1 \cdot 1 + 3 \cdot 4 + 1 \cdot 3 = 16$$

This is the same answer we obtained earlier when summing all $N(A_iA_j)$'s. This reasoning applies to $r_k(B)$ for any k and for any board B that decomposes into disjoint subboards.

Lemma

If B is a board of darkened squares that decomposes into the two disjoint subboards B_1 and B_2, then

$$r_k(B) = r_k(B_1)r_0(B_2) + r_{k-1}(B_1)r_1(B_2) + \cdots + r_0(B_1)r_k(B_2) \tag{2}$$

The observant reader may notice that (2) is very similar to formula (6) in Section 6.2 for the coefficient of a product of two generating functions. That is, if $f(x) = \Sigma a_r x^r$ and $g(x) = \Sigma b_r x^r$, then the coefficient of x^k in $h(x) = f(x)g(x)$ is $a_k b_0 + a_{k-1} b_1 + \cdots + a_0 b_k$. We will now exploit this similarity. We define the **rook polynomial** $R(x, B)$ of the board B of darkened squares to be

$$R(x, B) = r_0(B) + r_1(B)x + r_2(B)x^2 + \cdots$$

Remember that $r_0(B) = 1$ for all B. Note that the rook polynomial depends only on the darkened squares, not on the size of the original assignment diagram. Then for B_1 and B_2 as defined above, we found that

$$R(x,B_1) = 1 + 3x + 1x^2 \quad \text{and} \quad R(x,B_2) = 1 + 4x + 3x^2$$

Moreover, by the correspondence between (2) and the formula for the product of two generating functions, we see that $r_k(B)$ is simply the coefficient of x^k in the product $R(x,B_1)R(x,B_2)$. That is,

$$R(x,B) = R(x,B_1)R(x,B_2) = (1 + 3x + 1x^2)(1 + 4x + 3x^2)$$
$$= 1 + (3 \cdot 1 + 1 \cdot 4)x + (1 \cdot 1 + 3 \cdot 4 + 1 \cdot 3)x^2$$
$$+ (1 \cdot 4 + 3 \cdot 3)x^3 + 1 \cdot 3x^4$$
$$= 1 + 7x + 16x^2 + 13x^3 + 3x^4$$

This product relation is true for any such B, B_1, and B_2.

Theorem 1

If B is a board of darkened squares that decomposes into the two disjoint subboards B_1 and B_2, then

$$R(x,B) = R(x,B_1)R(x,B_2)$$ ∎

Without meaning to belittle the role of generating functions in Theorem 1, we should observe that the generating functions were used here solely because polynomial multiplication corresponds to the subboard composition rule given in (2). That is, plugging the $r_k(B_i)$'s into two polynomials and multiplying them together is a more familiar way to organize the computation required by (2). Later in this section when we decompose a board B into nondisjoint subboards, rook polynomials play a truly essential role.

Now that $R(x,B)$ has been determined for the board of darkened squares in Figure 8.8 and hence in Figure 8.7, we can solve our original problem about the permissible arrangements of a, b, c, d, e. Expression (1) for S_k can now be rewritten

$$S_k = r_k(B)(5 - k)!$$

By the inclusion-exclusion formula, the number of permissible arrangements is

$$N(\bar{A}_1\bar{A}_2\bar{A}_3\bar{A}_4\bar{A}_5) = N - S_1 + S_2 - S_3 + S_4 - S_5$$
$$= 5! - r_1(B)4! + r_2(B)3! - r_3(B)2! + r_4(B)1! - r_5(B)0!$$
$$= 5! - 7 \cdot 4! + 16 \cdot 3! - 13 \cdot 2! + 3 \cdot 1! - 0 \cdot 0!$$
$$= 120 - 168 + 96 - 26 + 3 - 0 = 25$$

The values for the $r_k(B)$'s came from the rook polynomial $R(x,B)$ computed above. This rook-polynomial-based variation on the inclusion-exclusion formula is valid for any arrangement problem with restricted positions.

Theorem 2

The number of ways to arrange n distinct objects when there are restricted positions is equal to

$$n! - r_1(B)(n - 1)! + r_2(B)(n - 2)! + \cdots + (-1)^k r_k(B)(n - k)!$$
$$\cdots (-1)^n r_n(B)0! \qquad (3)$$

where the $r_k(B)$'s are the coefficients of the rook polynomial $R(x,B)$ for the board B of forbidden positions. ∎

Let us summarize the little theory we have developed.

1. Given a problem of counting arrangements or matchings with restricted positions, display the constraints in an array with darkened squares for forbidden positions, as in Figure 8.7.

2. Try to rearrange the array so that the board B of darkened squares can be decomposed into disjoint subboards B_1 and B_2.

3. By inspection, determine the $r_k(B_i)$'s, the number of ways to place k non-capturing rooks on subboard B_i.

4. Use the $r_k(B_i)$'s to form the rook polynomials $R(x,B_1)$ and $R(x,B_2)$, and multiply $R(x,B_1) \cdot R(x,B_2)$ to obtain $R(x,B)$.

5. Insert the coefficients $r_k(B)$ of $R(x,B)$ in formula (3).

Example 1

How many ways are there to send six different birthday cards, denoted C_1, C_2, C_3, C_4, C_5, C_6, to three aunts and three uncles, denoted $A_1, A_2, A_3, U_1, U_2, U_3$, if aunt A_1 would not like cards C_2 and C_4; if A_2 would not like C_1 or C_5; if A_3 likes all cards; if U_1 would not like C_1 or C_5; if U_2 would not like C_4; and if U_3 would not like C_6?

The forbidden positions information is displayed in Figure 8.9. We rearrange the board by putting together rows with a darkened square in the same column, and putting together columns with a darkened square in the same row. For example, rows C_1 and C_5 both have darkened squares in columns A_2 and U_1, and so we put C_1 and C_5 together; similarly for C_2 and C_4, and for A_2 and U_1 and for A_1 and U_2. We get the rearrangement shown in Figure 8.10. Thus the original board B of darkened squares decomposes into the two disjoint subboards B_1 in rows C_1 and C_5, and B_2 in rows C_2, C_4, and C_6. Actually B_2 itself decomposes into two disjoint subboards B_2' and B_2'', where B_2'' is the single square (C_6, U_3). By inspection we see that

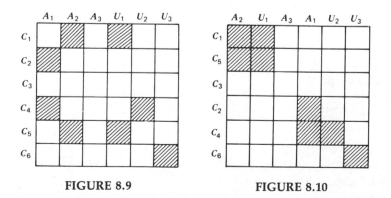

FIGURE 8.9   FIGURE 8.10

$$R(x,B_1) = 1 + 4x + 2x^2$$

$$R(x,B_2) = R(x,B_2')R(x,B_2'') = (1 + 3x + x^2)(1 + x)$$

So

$$R(x,B) = R(x,B_1)R(x,B_2)$$
$$= (1 + 4x + 2x^2)(1 + 3x + x^2)(1 + x)$$
$$= 1 + 8x + 22x^2 + 25x^3 + 12x^4 + 2x^5$$

Then the answer to the card mailing problem is

$$\sum_{k=0}^{6} (-1)^k r_k(B)(6 - k)!$$

$$= 6! - 8 \cdot 5! + 22 \cdot 4! - 25 \cdot 3! + 12 \cdot 2! - 2 \cdot 1! + 0 \cdot 0! = 160$$

∎

Now let us consider the problem of determining the coefficients of $R(x,B)$ when the board B does not decompose into two disjoint subboards. Consider the board B shown in Figure 8.11. Let us break the problem of determining $r_k(B)$ into two cases, depending on whether or not a certain square s is one of the squares chosen for the k noncapturing rooks. Let B_s be the board obtained from B by deleting square s, and let B_s^* be the board obtained from B by deleting all squares in the same row or column as s. If s is the square indicated in Figure 8.11, then B_s and B_s^* are as shown in Figure 8.12. If square s is not used, we must place k noncapturing rooks on B_s. If square s is used, then we must place $k - 1$ noncapturing rooks on B_s^*. Hence we conclude that

$$r_k(B) = r_k(B_s) + r_{k-1}(B_s^*) \qquad (4)$$

Using the generating function methods introduced in Section 7.6 for turning a recurrence relation into a generating function, we obtain

$$R(x,B) = R(x,B_s) + xR(x,B_s^*)$$

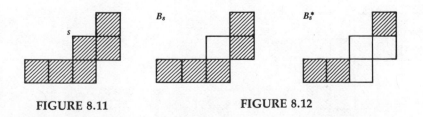

FIGURE 8.11   FIGURE 8.12

Verification is left as an exercise. Observe that B_s and B_s^* both break into disjoint subboards whose rook polynomials are easily determined by inspection:

$$R(x,B_s) = (1 + 3x)(1 + 2x) = 1 + 5x + 6x^2$$
$$R(x,B_s^*) = (1 + 2x)(1 + x) = 1 + 3x + 2x^2$$

Thus

$$R(x,B) = R(x,B_s) + xR(x,B_s^*) = (1 + 5x + 6x^2) + x(1 + 3x + 2x^2)$$
$$= 1 + 6x + 9x^2 + 2x^3$$

These results apply to any board B and any square s in B.

Theorem 3

Let B be any board of darkened squares. Let s be one of the squares of B, and let B_s and B_s^* be as defined above. Then

$$R(x,B) = R(x,B_s) + xR(x,B_s^*)$$ ∎

Theorem 3 provides a way to simplify any board's rook polynomial. If the boards B_s and B_s^* do not break up into disjoint subboards, we can reapply Theorem 3 to B_s and B_s^*. It is important that the square s be chosen to split up B as much as possible.

Example 2

In Example 1, suppose that the tastes of uncle U_1 change and now he would not like card C_2 but would like C_1. The new board of forbidden positions is shown in Figure 8.13. The square in the bottom left corner, call it t, is still disjoint from the other squares, call them board B_1. Board B_1 cannot be decomposed into disjoint subboards, and so we must use Theorem 3. The square s that breaks up B_1 most evenly is (C_2, U_1), indicated in Figure 8.13.

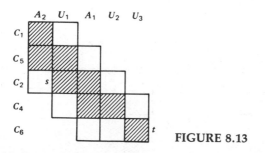

FIGURE 8.13

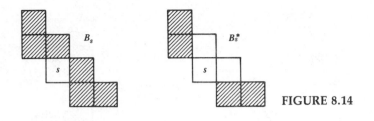

FIGURE 8.14

Boards B_s and B_s^*, shown in Figure 8.14, both decompose into simple disjoint subboards:

$$R(x,B_s) = (1 + 3x + x^2)(1 + 3x + x^2) = 1 + 6x + 11x^2 + 6x^3 + x^4$$
$$R(x,B_s^*) = (1 + 2x)(1 + 2x) - 1 + 4x + 4x^2$$

Then

$$R(x,B_1) = R(x,B_s) + xR(x,B_s^*)$$
$$= (1 + 6x + 11x^2 + 6x^3 + x^4) + x(1 + 4x + 4x^2)$$
$$= 1 + 7x + 15x^2 + 10x^3 + x^4$$

and

$$R(x,B) = R(x,B_1)R(x,t)$$
$$= (1 + 7x + 15x^2 + 10x^3 + x^4)(1 + x)$$
$$= 1 + 8x + 22x^2 + 25x^3 + 11x^4 + x^5$$

Now by Theorem 2, the number of ways to send birthday cards is

$$6! - 8 \cdot 5! + 22 \cdot 4! - 25 \cdot 3! + 11 \cdot 2! - 1 \cdot ! + 0 \cdot 0! = 159$$

Note that uncle U_1's change in tastes only changed the final rook polynomial slightly: $r_4(B)$ changed from 12 to 11 and $r_5(B)$ changed from 2 to 1. ■

EXERCISES

SUMMARY OF EXERCISES   The first six exercises are similar to the examples in this section; the next eight exercises develop theory about rook polynomials and combinatorial theory based on rook polynomials.

1.   Describe the associated chessboard of darkened squares for finding all derangements of 1, 2, 3, 4, 5.

2.   Find the rook polynomial for the following boards:

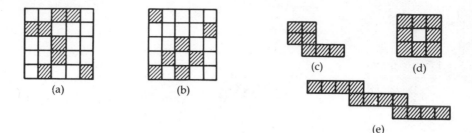

(a)   (b)   (c)   (d)

(e)

3. Seven dwarfs $D_1, D_2, D_3, D_4, D_5, D_6, D_7$ each must be assigned to one of seven jobs in a mine, $J_1, J_2, J_3, J_4, J_5, J_6, J_7$. D_1 cannot do jobs J_2 or J_3; D_2 cannot do J_1 or J_5; D_4 cannot do J_3 or J_6; D_5 cannot do J_2 or J_7; D_7 cannot do J_4. D_3 and D_6 can do all jobs. How many ways are there to assign the dwarfs to different jobs?

4. A pair of two distinct dice are rolled six times. Suppose none of the ordered pairs of values (1,5), (2,6), (3,4), (5,5), (5,3), (6,1), (6,2) occur. What is the probability that all six values on the first die and all six values on the second die occur once in the six rolls of the two dice?

5. Suppose five officials O_1, O_2, O_3, O_4, O_5 are to be assigned five different city cars, a VW, a Pinto, a Vega, a Datsun, and a Rolls Royce. O_1 will not drive a VW or Vega; O_2 will not drive a Datsun; O_3 will not drive a Pinto or Rolls; O_4 will not drive a Pinto; and O_5 will not drive a VW or Vega. If a feasible assignment of cars is chosen randomly, what is the probability that:
   (a) O_1 gets the Rolls?
   (b) O_2 or O_5 get the Rolls (model with an altered board)?

6. A computer dating service wants to match four women each with one of five men. If woman 1 is incompatible with men 3 and 5; woman 2 is incompatible with men 1 and 2; woman 3 is incompatible with man 4; and woman 4 is incompatible with men 2 and 4, how many matches of the four women are there?

7. Find the rook polynomial for a full $n \times n$ board.

8. An *ascent* in a permutation is a consecutive pair of the form $i, i + 1$. The ascents in the following permutation are underlined: 12534.
   (a) Design a chessboard to represent a permutation of 1, 2, 3, 4, 5 so that a check in entry (i,j) means i is followed immediately by j in the permutation. Darken entries so as to exclude all ascents.
   (b) Find the rook polynomial for the darkened squares in part a.
   (c) Find the number of permutations of 1, 2, 3, 4, 5 with k ascents. (*Hint:* See Theorem 2 in Section 8.2.)

9. Use the recurrence relation $r_k(B) = r_k(B_s) + r_{k-1}(B_s^*)$ to prove that $R(x,B) = R(x,B_s) + xR(x,B_s^*)$.

10. Let $R_{n,m}(x)$ be the rook polynomial for an $n \times m$ chessboard (n rows, m columns, all squares may have rooks).
   (a) Show that $R_{n,m}(x) = R_{n-1,m}(x) + mxR_{n-1,m-1}(x)$.
   (b) Show that $\dfrac{d}{dx} R_{n,m}(x) = nmR_{n-1,m-1}(x)$. (*Hint:* Use identity (4) in Section 5.5.)

11. Find two different chessboards (not row or column rearrangements of one another) that have the same rook polynomial.

12. Consider all permutations of $1, 2, \ldots, n$ in which i appears in neither position i nor $i + 1$ (n not in n or 1). Such a permutation is called a *menage*. Let $M_n(x)$ be the rook polynomial for the forbidden squares in a menage. Let $M_n^*(x)$ be the rook polynomial when n may appear in position 1, and let $M_n^0(x)$ be the rook polynomial when both 1 and n may appear in position 1.
   (a) Show that $M_n^*(x) = xM_{n-1}^*(x) + M_n^0(x)$, $M_n^0(x) = xM_{n-1}^0(x) + M_{n-1}^*(x)$, and $M_n(x) = M_n^*(x) + xM_{n-1}^*(x)$.
   (b) Using the initial conditions $M_1^*(x) = 1 + x$, $M_1^0(x) = 1$, show by induction that

   $$M_n^*(x) = \sum_{k=0}^{n} \binom{2n - k}{k} x^k \quad \text{and} \quad M_n^0(x) = \sum_{k=0}^{n-1} \binom{2n - k - 1}{k} x^k$$

   (c) Find $M_n(x)$.

13. Given an $n \times m$ chessboard $C_{n,m}$ (see Exercise 10) and a board C of darkened squares in $C_{n,m}$, the *complement* C' of C in $C_{n,m}$ is the board of nondarkened squares.
   (a) Show that

   $$r_k(C') = \sum_{j=0}^{k} (-1)^j \binom{n - j}{k - j}\binom{m - j}{k - j} (k - j)! r_k(C)$$

   (b) Show that $R(x,C') = x^n R(1/x,C)$.

14. Use Theorem 2 in Section 8.2 to derive a formula for counting arrangements when exactly k elements appear in forbidden positions.

8.4 SUMMARY AND REFERENCES

Frequently the combinatorial complexities in the counting problems in Chapter 2 arose from simultaneous constraints such as "with at least one card in each suit." In this chapter, we formed a simple set-theoretic model for such problems and solved once and for all the combinatorial logic of this model.

The resulting formula, the inclusion-exclusion formula, was then applied to various counting problems. This formula requires the proper set-theoretic restatement of a problem and the solution of some fairly straightforward subproblems, but in return the formula eliminates all the worry about logical decomposition (and worry about counting some outcomes twice). After having no help in their problem-solving in Chapter 5, readers should find it easy to appreciate fully the power of a formula that does the reasoning for them. This is what mathematics is all about!

The last section on rook polynomials provides a nice mini-theory about organizing the inclusion-exclusion computations in arrangements with restricted positions.

The inclusion-exclusion formula was obtained by J. Sylvester about 100 years ago, although in the early eighteenth century the number of derangements had been calculated by Montmort and a (noncounting) set union and intersection version of the formula was published by De Moivre. Rook polynomials were not invented until the mid-twentieth century (see Riordan [1]).

1. J. Riordan, *An Introduction to Combinatorial Analysis,* John Wiley & Sons, New York, 1958.

PART THREE

ADDITIONAL TOPICS

Chapter Nine

Polya's Enumeration Formula

9.1   EQUIVALENCE AND SYMMETRY GROUPS

In this chapter we examine a special class of counting problems. Consider the ways of coloring the corners of a square black, green, or white: there are three color choices at each of the four corners, giving $3 \cdot 3 \cdot 3 \cdot 3 = 81$ different colorings. Suppose however that any physical motion of our square is possible (like a square molecule in a liquid). Now how many different corner colorings are there? The unoriented figure being colored could be a n-gon or a cube, and the edges or faces could be colored instead of the corners. The floating square problem is equivalent to finding the number of different necklaces of four beads colored black, green, or white. Polya's motivation in developing his formula for counting distinct colorings when motions are allowed came from a problem in chemistry, the enumeration of isomers.

The difficulty in these problems comes from the geometric symmetries of the figure being colored. We develop a special formula, based on this set of symmetries, to count all distinct colorings of a figure. With a little more work, we also obtain a generating function that gives a **pattern inventory** of the distinct colorings. For example, the pattern inventory of black-white colorings of the corners of a cube with all physical motions allowed is

$$b^8 + b^7w + 3b^6w^2 + 3b^5w^3 + 7b^4w^4 + 3b^3w^5 + 3b^2w^6 + bw^7 + w^8$$

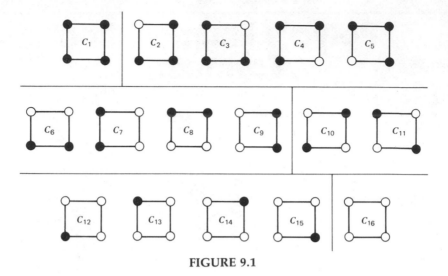

FIGURE 9.1

where the coefficient of $b^i w^j$ is the number of nonequivalent colorings with i black corners and j white corners.

We develop our formula and associated theory around the sample problem of coloring the corners of an unoriented square (floating in three dimensions) with black and white. It is impossible to draw an unoriented object; any picture shows it in a fixed position. Thus we start our analysis with the $2^4 = 16$ black-white colorings of the fixed square (see Figure 9.1). We can partition these 16 colorings into subsets of colorings that are equivalent when the square is floating. There are six such subsets (shown in Figure 9.1), and so there are six different 2-colorings of the floating square. (The set of fixed colorings would be too large if a harder sample problem were used, such as the 3-colorings of the square with $3^4 = 81$ fixed colorings.)

We seek a theory and formula to explain why there are six such distinct 2-colorings of the square. Note that the subsets of equivalent colorings vary in size.

To define the partition of a set into subsets of equivalent elements, we must first define the equivalence of two elements a and b. We write this equivalence as $a \sim b$. The fundamental properties of an **equivalence relation** are:

(i)   Transitivity: $a \sim b, b \sim c \Rightarrow a \sim c$.
(ii) Reflexivity: $a \sim a$.
(iii) Symmetry: $a \sim b \Rightarrow b \sim a$.

All other properties of equivalence can be derived from these three. Any binary relation with these three properties is called an equivalence relation.

Such a relation defines a partition into subsets of mutually equivalent elements called **equivalence classes.**

Example 1

(a) For a set of people, being the same weight is an equivalence relation; all people of a given weight form an equivalence class.

(b) For a set of numbers, differing by an even number is an equivalence relation; the even numbers form one equivalence class and the odd numbers the other class.

(c) For a set of figures, having the same number of corners is an equivalence relation; for each n, an equivalence class consists of all n-corner figures.

∎

Next we turn our attention to the motions that map the square onto itself (see Figure 9.2). These motions are what make one of the above colorings equivalent to another one. Before we develop a theory about symmetries and their relation to coloring equivalence, let us take a closer geometric look at the symmetries of a square and some other figures.

Example 2

Figure 9.2 displays the set of motions that map the square onto itself, the symmetries of the square. How was this set obtained? More generally, what is the set of symmetries of an n-gon, for n even?

The symmetries of the square divide into two classes: rotations—circular

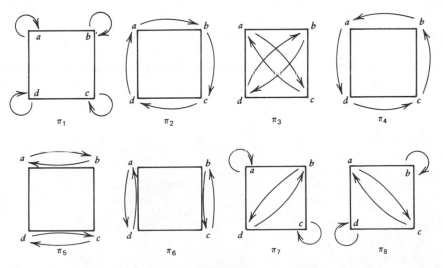

FIGURE 9.2

motions in the plane; and reflections (flips)—motions using the third dimension. The rotations, about the center of the square, are easy to find. Each rotation is an integral multiple of the smallest (nonzero) rotation. They are $\pi_2 = 90°$ rotation, $\pi_3 = 180°$ rotation, $\pi_4 = 270°$ rotation, and $\pi_1 = 360°$ (or $0°$) rotation.

The reflections are a little harder to "see" for the simple reason that three-dimensional motion is more difficult to picture mentally than rotational motion. The reflections are $\pi_5 =$ reflection about the vertical axis, $\pi_6 =$ reflection about the horizontal axis, $\pi_7 =$ reflection about opposite corners a and c, and $\pi_8 =$ reflection about opposite corners b and d.

In a regular n-gon, the smallest rotation is $(360/n)°$. There are n rotations in all. There are two types of reflections for a regular even-n-gon: flipping about two opposite sides and flipping about two opposite corners. Since there are $n/2$ pairs of opposite sides and $n/2$ pairs of opposite corners, a regular even-n-gon will have $n/2 + n/2 = n$ reflections. Summing rotations and reflections, we find that a regular even-n-gon has $2n$ symmetries. ■

Example 3

Describe the symmetries of a pentagon, and more generally, of an n-gon for odd n.

As noted in Example 2, a pentagon will have five rotational symmetries of $0°$, $72°$, $144°$, $216°$, and $288°$. However, the symmetric reflections about opposite sides or opposite corners do not exist in the pentagon. Instead, we flip about a corner and the middle of an opposite side (see Figure 9.3). There are five such flips, for a total of 10 symmetries.

In a regular odd n-gon, there are n such flips. Summing rotations and reflections, we find that a regular odd n-gon also has $2n$ symmetries. We leave it as an exercise for the reader to show that these $2n$ symmetries are all distinct and that there are no other symmetries of an n-gon besides rotations and flips. ■

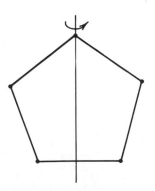

FIGURE 9.3 Symmetric reflection of a pentagon

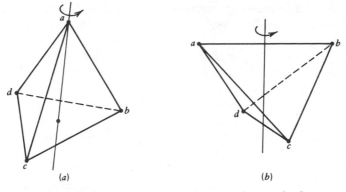

FIGURE 9.4 Symmetric revolutions of a tetrahedron

Example 4

Describe the symmetries of a tetrahedron.

A tetrahedron consists of four equilateral triangles that meet at six edges and four corners (see Figure 9.4). Besides the motion leaving all corners fixed, call it the 0° motion, we can revolve 120° or 240° about a corner and the center of the opposite face (see Figure 9.4a), or we can revolve 180° about the middle of opposite edges (see Figure 9.4b). Since there are four pairs of a corner and opposite face and three pairs of opposite edges, we have a total of 1 (0° motion) + 4 · 2 + 3 = 12 symmetries. The reader should check that these 12 symmetries are distinct and that no other symmetries exist.

The symmetries of a square are naturally characterized by the way they permute the corners of the square. Thus the motion π_3 (see Figure 9.2) can be described as the corner permutation: $a \to c$, $b \to d$, $c \to a$, $d \to b$; in tabular form, we write $\begin{pmatrix} abcd \\ cdab \end{pmatrix}$. Any permutation can be considered as a product of disjoint cyclic permutations or **cycles** (proof of this claim is an exercise). For example, $\pi_3 = (ac)(bd)$ (or π_3 could be written as $(bd)(ca)$), $\pi_4 = (adcb)$, and $\pi_7 = (a)(bd)(c)$. The reader should find cycle decompositions for the other π_i's. The depiction of a motion as in Figure 9.2, with arrows indicating the mapping at each corner, will make it easier to trace out cycles in later calculations.

In permuting the corners, the symmetries create permutations of the colorings of the corners. For example, if C_i is the ith square in Figure 9.1, then π_3 is the following permutation of colorings:

$$\pi_3 = \begin{pmatrix} C_1 & C_2 & C_3 & C_4 & C_5 & C_6 & C_7 & C_8 & C_9 & C_{10} & C_{11} & C_{12} & C_{13} & C_{14} & C_{15} & C_{16} \\ C_1 & C_4 & C_5 & C_2 & C_3 & C_8 & C_9 & C_6 & C_7 & C_{10} & C_{11} & C_{14} & C_{15} & C_{12} & C_{13} & C_{16} \end{pmatrix}$$

The point is that while a symmetry π_i is easily visualized by how it moves the corners of the square, what we are really interested in is the way π_i takes one coloring into another (making them equivalent). Thus we formally define our coloring equivalence as follows:

> colorings C and C' are *equivalent*, or $C \sim C'$,
> if there exists a π_i such that $\pi_i(C) = C'$

The properties of the set G of symmetries that interest us are the ones that make the relation, $C \sim C'$ (some π_i takes C to C'), an equivalence relation. These properties of G are (here $\pi_i \cdot \pi_j$ means applying motion π_i followed by motion π_j):

(a) *Closure:* If $\pi_i, \pi_j \in G$, then $\pi_i \cdot \pi_j \in G$; for example, in Figure 9.2, $\pi_3 \cdot \pi_5 = \pi_6$.

(b) *Identity:* G contains a motion π_I such that $\pi_I \cdot \pi_i = \pi_i$ and $\pi_i \cdot \pi_I = \pi_i$; in Figure 9.2, π_I is π_1.

(c) *Inverses:* For each $\pi_i \in G$, there exists an inverse in G, denoted π_i^{-1}, such that $\pi_i \cdot \pi_i^{-1} = \pi_I$ and $\pi_i^{-1} \cdot \pi_i = \pi_I$; for example, in Figure 9.2, $\pi_2^{-1} = \pi_4$.

Observe that closure makes our coloring relation $\sim$ satisfy transitivity (property i of an equivalence relation). For suppose $C \sim C'$ and $C' \sim C''$. Since $C \sim C'$, there must exist $\pi_i \in G$ such that $\pi_i(C) = C'$. Similarly there is a $\pi_j \in G$ with $\pi_j(C') = C''$. Then by closure, there exists $\pi_k = \pi_i \cdot \pi_j \in G$ with $\pi_k(C) = (\pi_i \cdot \pi_j)(C) = C''$. Thus $C \sim C''$. Similarly, properties (b) and (c) of the symmetries imply our coloring relation satisfies properties (ii) and (iii) of an equivalence relation (see Exercise 16).

Any collection G of mathematical objects that satisfies properties (a), (b), and (c) is called a **group**.* Thus we have:

Theorem

Let G be a group of permutations of the set S (corners of a square) and T be any collection of colorings of S (2-colorings of the corners). Then G induces a partition of T into equivalence classes with the relation on T, $C \sim C' \Leftrightarrow$ some $\pi \in G$ takes C to C'.   ∎

Note that T could be any possible collection of colorings. The following simple lemma about groups lies at the heart of the counting formula developed in the next section.

---

* Actually we left out one further property, the group's binary operation must be associative. This property always holds for symmetries.

Lemma

For any two permutations π_i, π_j in a group G, there exists a unique permutation π_k in G such that $\pi_i \cdot \pi_k = \pi_j$; namely, $\pi_k = \pi_i^{-1} \cdot \pi_j$.

Proof

Clearly if $\pi_k = \pi_i^{-1} \cdot \pi_j$, then $\pi_i \cdot \pi_k = \pi_i \cdot (\pi_i^{-1} \cdot \pi_j) = \pi_j$, as claimed. Suppose there also exists a permutation π_k' such that $\pi_i \cdot \pi_k' = \pi_j$. Then $\pi_i \cdot \pi_k = \pi_i \cdot \pi_k'$, and hence $\pi_i^{-1} \cdot (\pi_i \cdot \pi_k) = \pi_i^{-1} \cdot (\pi_i \cdot \pi_k')$ or $\pi_k = \pi_k'$. ∎

EXERCISES

SUMMARY OF EXERCISES   The first 13 exercises continue the examples of equivalence and symmetries given in this section. The remaining exercises develop basic aspects of group theory associated with symmetries. Prior experience with modern algebra is needed for most of these problems.

1. Which of the following relations are equivalence relations? State your reasons.
    (a) $\leq$ (less than or equal to), for a set of numbers.
    (b) = (equal to), for a set of numbers.
    (c) "Difference is odd," for a set of numbers.
    (d) "Being blood relations," for a group of people.
    (e) "Having a common friend," for a group of people.

2. Which of the following collections with given operations are groups? For those collections that are groups, which elements are the identities?
    (a) The nonnegative integers 0, 1, 2, . . . with addition.
    (b) The integers 0, 1, 2, . . . , $n - 1$ with addition modulo n.
    (c) All polynomials (with integer coefficients) with polynomial addition.
    (d) All nonzero fractions with regular multiplication.
    (e) All invertible 2 × 2 real-valued matrices with matrix multiplication.

3. Find all symmetries of the following figures (indicate with arrows where the corners move in each symmetry, as in Figure 9.2):

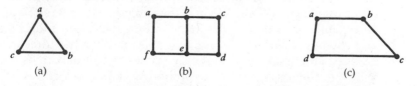

(a)   (b)   (c)

4. Write the following symmetries or permutations as a product of cyclic permutations:

(a) π_2 (c) π_6 (e) $\pi_4 \cdot \pi_7$ (g) $\begin{pmatrix} 1234567 \\ 3546712 \end{pmatrix}$

(b) π_3 (d) π_1 (f) $\pi_7 \cdot \pi_4$

5. For the symmetries of the square listed in Figure 9.2, give the associated permutation of 2-colorings for:
   (a) π_1 (b) π_2 (c) π_5 (d) π_7

6. (a) List all 2-colorings of the three corners of a triangle.
   (b) For the following symmetries of a triangle, give the associated permutation of 2-colorings for: (i) $\pi = 120°$ rotation and (ii) $\pi =$ flip about vertical axis.

7. (a) Show that the 8 symmetries of a square listed in Figure 9.2 are all distinct.
   (b) Show that there are no other symmetries of a square besides those listed in Figure 9.2.

8. (a) Show that the $2n$ symmetries of an n-gon (even or odd) mentioned in Examples 2 and 3 are all distinct.
   (b) Show that there are exactly $2n$ symmetries of an n-gon.

9. (a) Show that the 12 symmetries of a tetrahedron listed in Example 4 are all distinct.
   (b) Show that there are exactly 12 symmetries of a tetrahedron.

10. Find the symmetry of the square equal to the following products:
    (a) $\pi_2 \cdot \pi_4$ (b) $\pi_5 \cdot \pi_2$ (c) $\pi_7 \cdot (\pi_2 \cdot \pi_8)$
    (d) $(\pi_7 \cdot \pi_6) \cdot \pi_3$

11. (a) Write out the 6 × 6 multiplication table for the product of all pairs of symmetries of a triangle.
    (b) Repeat part a for integers 1, 2, 3, 4 with multiplication modulo 5.
    (c) Repeat part a for the following group of permutations of 1, 2, 3, 4:

$$\begin{pmatrix} 1234 \\ 1234 \end{pmatrix} \quad \begin{pmatrix} 1234 \\ 2143 \end{pmatrix} \quad \begin{pmatrix} 1234 \\ 3412 \end{pmatrix} \quad \begin{pmatrix} 1234 \\ 4321 \end{pmatrix}$$

12. Find two symmetries of the square π_i, π_j such that $\pi_i \cdot \pi_j \neq \pi_j \cdot \pi_i$ (this means the group of symmetries of a square is noncommutative).

13. Many organic compounds consist of a basic structure formed by carbon atoms (carbon atoms are the corners of a floating figure), plus submolecular groups called radicals that are attached to each carbon atom (the carbons are like corners and the radicals like colors). Suppose such an organic molecule has six carbon atoms, with four radicals of type A and two radicals of type B. Suppose this molecule has three *isomers*, that is, three different ways that the two types of radicals can be distributed. Which of the following two hypothetical carbon structures would have three isomers with four A's and two B's?

(a)   (b)

14. Prove that an equivalence relation partitions a set into disjoint subsets of mutually equivalent elements.

15. Give a procedure for decomposing any permutation into a product of cycles.

16. Show that properties (a) and (b) of a group G of permutations imply properties (ii) and (iii), respectively, of the associated equivalence relation defined in the theorem.

17. Let S be a set and G a group of permutations of S. For any two subsets S_1, S_2 of S, define $S_1 \sim S_2$ to mean that for some $\pi \in G$, $S_1 = \pi(S_2)$ $(=\{\pi(s) | s \in S_2\})$. Show that $\sim$ is an equivalence relation.

18. Prove that a collection G of permutations is a group if and only if for every pair $\pi_i, \pi_j \in G$, there exist unique elements $\pi_p, \pi_r \in G$ such that $\pi_p \cdot \pi_i = \pi_j$ and $\pi_i \cdot \pi_r = \pi_j$.

19. Prove that the set of permutations of the 2-colorings of a square forms a group.

20. Prove that for any prime p, the integers $1, 2, \ldots, p - 1$ with multiplication modulo p form a group.

21. A subset of elements in a group G is said to *generate* the group if all elements in G can be obtained as (repeated) products of elements in the subset.
    (a) Which of the following subsets generate the group of symmetries of a square? (i) π_1, π_2, π_3 (ii) π_2, π_5 (iii) π_3, π_6 (iv) π_6, π_7.
    (b) Show that the group of symmetries of a regular n-gon can be generated by a subset of two elements.

22. If a subset G' of elements in a group G is itself a group, then G' is called a *subgroup* of G.
    (a) Show that $G' = \{\pi_1, \pi_2, \pi_3, \pi_4\}$ and $G'' = \{\pi_1, \pi_7\}$ are subgroups of the group G of symmetries of a square.
    (b) Find another 4-element subgroup of G containing π_3.
    (c) Find all subgroups of the group G of symmetries of a square.

23. (a) How many different binary relations on n elements are possible?
    (b) How many symmetric binary relations are possible?

24. Show that $\pi_i \sim \pi_j$ if there exists $\pi \in G$ such that $\pi_i = \pi^{-1}\pi_j\pi$ is an equivalence relation.

25. A *transposition* is a permutation that interchanges the positions of just

two elements (and leaves all other elements fixed). Show by induction that any permutation of n elements can be written as a composition of (several) transpositions.

26. For a given group G of n elements, define the function f_π on G as follows: for each $\pi' \in G$, $f_\pi(\pi') = \pi \cdot \pi'$.
    (a) Show that f_π is a one-to-one mapping for any $\pi \in G$.
    (b) If we define $f^*_{\pi_1} f_{\pi_2} = f_{\pi_1 \cdot \pi_2}$, show that the set $\{f_\pi\}$ forms a group.

9.2 BURNSIDE'S LEMMA

We now derive a method for counting the number of different nonequivalent 2-colorings of the square. More generally, in a set T of c colorings we seek the number N of equivalence classes in the partition induced by a group G of s symmetries.

Let E_x be the equivalence class containing the coloring x. Then E_x consists of x and all colorings y equivalent to x, that is, all y's such that for some $\pi \in G$, $\pi(x) = y$. If each of the s π's takes x to a different coloring $\pi(x)$, then E_x has s colorings (note that for a given x, the $\pi(x)$'s include x since $\pi_I(x) = x$, where π_I is the identity symmetry). And if every equivalence class is like this with s colorings, then $sN = c$, or $N = c/s$.

Consider, for example, the $c = n!$ fixed seatings of n people around a table. There are $s = n$ cyclic rotations of the seatings, and each equivalence class consists of n seatings. Thus the number of equivalence classes (cyclicly different seatings) is $N = n!/n = (n - 1)!$.

On the other hand, suppose we have a small round table with three fixed positions (each 120° apart) for chairs, and a pile of red and black chairs is available. There are $2^3 = 8$ ways to place a red or black chair in each position. There are three cyclic rotations of the table possible, 0°, 120°, and 240°. We have $c = 8$ "colorings" and $s = 3$ symmetries, but the number of equivalence classes cannot be $N = \frac{8}{3}$, a fraction!

It is true that the three arrangements of one black and two red chairs (or vice versa) form an equivalence class, since 0°, 120°, and 240° rotations move the one black chair to different positions. However, an arrangement of three black chairs (or three red chairs) forms an equivalence class by itself. Any rotation maps this arrangement into itself, that is, leaves it fixed.

We need to correct the numerator in the formula $N = c/s$ by adding the multiplicities of an arrangement, when several cyclic motions map the arrangement to itself instead of to other arrangements. This way every equivalence class still will have s members when multiplicities are counted. Since two symmetries, besides the 0° symmetry, leave the all-black-chair arrangement fixed and similarly for the all red, we have $N = (8 + 2 + 2)/3 = 4$.

The "multiplicity" correction is even more complicated for 2-colorings of a square. Here the size of an equivalence class E_x may be 1 or 2 or 4, but never 8

(= the number of symmetries of the square). The first problem is that several π's, besides the identity symmetry π_I, may leave x fixed—that is, $\pi(x) = x$. The other problem is that if y is another coloring in E_x, there may be several π's all taking x to y. For example, the coloring C_{10} (see Figure 9.1) is fixed by symmetries $\pi_1, \pi_3, \pi_7, \pi_8$, and is mapped to C_{11} by symmetries $\pi_2, \pi_4, \pi_5, \pi_6$.

Let $\pi_1', \pi_2', \ldots, \pi_k'$ be the set of π's that take x to y. By the lemma in Section 9.1, these π_i''s can be written in the form $\pi_i' = \pi_1' \cdot \pi_i''$, where π_i'' is a symmetry that must leave y fixed (since π_1', which takes x to y, followed by π_i'' equals π_i', which takes x to y). Conversely, given any π° that leaves y fixed, $\pi_1'\pi^\circ$ takes x to y and so must be equal to one of the π_j'''s. Thus there is a 1-1 correspondence between the π's that take x to y and the π's that leave y fixed. Therefore, *to count the colorings in E_x with appropriate multiplicities* (i.e., coloring y in E_x has multiplicity 3 if three different π's take x to y), *it suffices for each coloring in E_x to sum the number of π's that leave the coloring fixed.*

As noted above, the coloring C_{10} (see Figure 9.1) is fixed by symmetries $\pi_1, \pi_3, \pi_7, \pi_8$, and is mapped to C_{11} by symmetries $\pi_2, \pi_4, \pi_5, \pi_6$. For example,

$$\pi_2 \cdot \pi_8 = \begin{pmatrix} abcd \\ bcde \end{pmatrix} \cdot \begin{pmatrix} abcd \\ cbad \end{pmatrix} = \begin{pmatrix} abcd \\ badc \end{pmatrix} = \pi_5$$

In terms of π_2, we can write the latter symmetries as $\pi_2 = \pi_2\pi_1$, $\pi_4 = \pi_2\pi_3$, $\pi_5 = \pi_2\pi_8$, $\pi_6 = \pi_2\pi_7$, where $\pi_1, \pi_3, \pi_7, \pi_8$ are all the symmetries leaving C_{11} fixed. So C_{10} and C_{11} each have multiplicity 4, and with these multiplicities, the size of their equivalence class including multiplicities is $4 + 4 = 8$ ($=s$), as required.

In general, when multiplicities are counted, each equivalence class will have s elements. If $\phi(y)$ denotes the number of π's that leave the coloring of x fixed, then $\Sigma_{y \in E_x} \phi(y) = s$. Summing over all equivalence classes, we obtain the following theorem, first proved by Burnside in 1897.

Theorem (Burnside)

Let G be a group of permutations of the set S (corners of a square). Let T be any collection of colorings of S (2-colorings of the corners) that is closed under G. Then the number N of equivalence classes is

$$N = \frac{1}{|G|} \sum_{x \in T} \phi(x) \quad \text{or}$$

$$N = \frac{1}{|G|} \sum_{\pi \in G} \Psi(\pi) \tag{*}$$

where $|G|$ is the number of permutations and $\Psi(\pi)$ is the number of colorings in T left fixed by π. ∎

By "closed under G," we mean that for all $\pi \in G$, $x \in T \Rightarrow \pi(x) \in T$. This closure property is automatic when T is the set of all corner 2-colorings of the square (in Section 9.4 we need to apply formula (*) to special subsets of S that are closed under G).

The two sums in the theorem both count all instances of some coloring being left fixed by some π, the first sums over the different colorings, the second sums over different π. Formula (*) is more useful in later computations.

We informally summarize the spirit behind formula (*) as follows. The total number c of all colorings is equal to $\Psi(\pi_I)$ since the identity leaves all colorings fixed. If there were s symmetries, then as a first approximation, we would expect to have s members of each equivalence class and hence a total $c/s = \Psi(\pi_I)/|G|$ equivalence classes. However, the s members of an element x's equivalence class may not be distinct (e.g., $\pi_i(x) = x$ or $\pi_j(x) = \pi_k(x)$). The terms $\Psi(\pi_i)$ in (*) add the "multiplicities" of repeated members so that each equivalence class implicitly has s members in the sum in (*).

Example 1

A baton is painted with n equal-sized cylindrical bands. Each band can be painted one of three colors. If the baton is unoriented as when spun in the air, how many different colorings of the baton are possible?

The only nontrivial symmetry is a 180° spin (revolution). We apply formula (*) for the group of the 0° and 180° spin. There are 3^n colorings of the fixed baton and so $\Psi(0°) = 3^n$. The number of colorings left fixed by a 180° spin depends on whether n is even or odd. If n is even, each of the $n/2$ bands on one half of the baton can be any color, $3^{n/2}$ choices, and then for the coloring to be fixed by a 180° spin, each of the symmetrically opposite bands must be the corresponding color. So $\Psi(180°) = 3^{n/2}$ and we have $N = \frac{1}{2}(3^n + 3^{n/2})$. To enumerate batons left fixed by a 180° spin when n is odd, we can use any color for the "odd" band in the middle of the baton, three choices. Each of the $(n-1)/2$ bands on one side of the middle band can be any color, $3^{(n-1)/2}$ choices, and again the other $(n-1)/2$ bands are colored symmetrically. So $\Psi(180°) = 3 \cdot 3^{(n-1)/2}$ and $N = \frac{1}{2}(3^n + 3^{(n+1)/2})$. ∎

Example 2

Suppose a necklace can be made from beads of three colors, black, white, and red. How many different necklaces with n beads are there?

When the n beads are positioned symmetrically about the circle, the beads occupy the positions of the corners of a regular n-gon. Thus our question asks for the number of corner colorings of an n-gon using three colors. The answer depends on what is meant by "different." If the beads are not allowed to move about the necklace, that is, the n-gon is fixed, the answer is $3 \cdot 3 \ldots \ldots \cdot 3 = 3^n$ (three color choices at each of n corners). A more realistic interpretation of our problem would allow the beads to move freely about the circle, that is, the n-gon rotates freely (but flips are not allowed). We employ formula (*) to count the number N of equivalence classes of these 3-colorings induced by the rotational symmetries of an n-gon. We shall do the calculations for $n = 2$ and $= 3$. A more general technique for larger n is developed in the next section.

(a) $n = 2$: There are $3 \cdot 3 = 9$ 3-colorings and two rotations, of $0°$ and $180°$. The $0°$ rotation leaves all colorings fixed, and so $\Psi(0°) = 9$. The $180°$ rotation cannot fix colorings in which the two corners have different colors. It fixes just the three monochromatic colorings; $\Psi(180°) = 3$. By formula (*), we have $N = \frac{1}{2}(9 + 3) = 6$.

(b) $n = 3$: There are $3^3 = 27$ 3-colorings and three rotations of $0°$, $120°$, $240°$. The $0°$ rotation leaves all colorings fixed, and so $\Psi(0°) = 27$. The $120°$ rotation cannot fix colorings in which some color occurs at only one corner. It follows that the $120°$ rotation fixes just the monochromatic colorings. Thus $\Psi(120°) = 3$. The $240°$ rotation is a reverse $120°$ rotation, and so $\Psi(240°) = 3$. By formula (*), we have $N = \frac{1}{3}(27 + 3 + 3) = 11$. ■

EXERCISES

SUMMARY OF EXERCISES The first 12 exercises continue the application of Burnside's Theorem to count colorings introduced in Examples 1 and 2. The remaining problems continue the algebraic theory started in the later exercises of Section 9.1 (prior experience with modern algebra is needed).

1. How many different n-bead necklaces are there using beads of red, white, blue, and green (assume necklaces can rotate but not flip over).
   (a) $n = 2$ (b) $n = 3$ (c) $n = 4$

2. How many different ways are there to place a diamond, sapphire, or ruby at each of the four vertices of this pin?

3. Fifteen balls are put in a triangular array as shown. How many different arrays can be made using balls of three colors if the array is free to rotate?

4. How many different ways are there to 2-color the 64 squares of an 8×8 chessboard that rotates freely?

5. A merry-go-round can be built with three different styles of horses. How many five-horse merry-go-rounds are there?

6. A domino is a thin rectangular piece of wood with two adjacent squares on one side (the other side is black). Each square is either blank or has 1, 2, 3, 4, 5, or 6 dots.
    (a) How many different dominoes are there?
    (b) Check your answer by modeling a domino as an (unordered) subset of 2 numbers chosen with repetition for 0, 1, 2, 3, 4, 5, 6.

7. Two n-digit decimal sequences consisting of digits 0, 1, 6, 8, 9 are vertically equivalent if reading one upside down produces the other, that is, 0068 and 8900. How many different (vertically inequivalent) n-digit (0,1,6,8,9) sequences are there?

8. How many different ways are there to color the five faces of an unoriented pyramid (with a square base) using red, white, blue, and yellow?

9. (a) Find the group of permutations of three objects.
    (b) Find the number of ways to distribute 12 identical balls in three indistinguishable boxes. (*Hint:* First let boxes be distinct, and then use part (a).)
    (c) Repeat part (b) with 4 red balls, 4 white balls, and 4 blue balls.

10. How many ways are there to 3-color the n bands of a baton if adjacent bands must have different colors?

11. How many ways are there to 3-color the corners of a square with rotations and reflections allowed if adjacent corners must have different colors?

12. (a) How many ways are there to distribute 12 red jelly beans to four children, a pair of identical female twins and a pair of identical male twins?
    (b) Repeat part a with the requirement that each child must have at least one jelly bean.

13. (a) Show that for any given i the subset G_i of symmetries of the square that leave coloring C_i (of Figure 9.1) fixed is a group (a subgroup of all symmetries).
    (b) Find G_4.
    (c) Find G_7.

14. For any subgroup of G' of a group G, define a *left coset* $\pi G'$, for some $\pi \in G$, to be the set $\pi G' = \{\pi \cdot \pi' | \pi' \in G')\}$. A *right coset* $G'\pi$ is similarly defined.
    (a) Show that for all $\pi \in G$, $|\pi G'| = |G'| = |G'\pi|$.
    (b) Show that for $\pi_i \neq \pi_j$, either $\pi_i G' = \pi_j G'$ or $\pi_i G' \cap \pi_j G' = \varnothing$.
    (c) Show that $\pi_i \sim \pi_j$ defined by $\pi_i G' = \pi_j G'$ is an equivalence relation.
    (d) Show that the size of any subgroup G' divides the size of G (this famous result in algebra is known as *Lagrange's Theorem*).

15. (a) Show that if G_i is the subgroup of G defined in Exercise 13 and if π' maps C_i to C_j, then the right coset $G_i\pi'$ (see Exercise 14) equals the

left coset of $\pi'G_j$ and both cosets are the set of all symmetries mapping C_i to C_j.

(b) Show that for any equivalence class E induced by G of 2-colorings of some figure, $\Sigma_{C_i \in E} |G_i| = |G|$, and hence the number N of equivalence classes equals $N = (1/|G|) \Sigma_{\text{all } C_i} |G_i|$ (*Hint:* Use the result in Exercise 14(a).)

16. If G' is a subgroup of G and $\pi G' = G'\pi$ for all $\pi \in G$, show that:
    (a) $(\pi_i G') \cdot (\pi_j G') = \{\pi \cdot \pi' | \pi \in \pi_i G', \pi' \in \pi_j G'\}$ equals $(\pi_i \cdot \pi_j)G'$.
    (b) The cosets $\pi G'$ form a group under $\cdot$.

9.3 THE CYCLE INDEX

Without further theory, we would find it very difficult to apply formula (*) to most coloring problems. If the set T were all 3-colorings of the corners of a 10-gon or a cube, then it would be close to impossible to determine $\Psi(\pi)$, the number of colorings left fixed by a symmetry of the figure. We show that $\Psi(\pi)$ can be determined simply from the structure of π. This approach leads to greatly shortened computations. As before, we develop the theory for this simplified calculation of $\Psi(\pi)$ with 2-colorings of a square.

Let us apply Burnside's formula (formula (*) in Section 9.2) to the 2-colorings of the square. Initially we determine the number of 2-colorings left fixed by π_i by inspection (using Figures 9.1 and 9.2). As we count $\Psi(\pi_i)$ for each π_i, we look for a pattern that would enable us to predict mathematically which colorings (and thus, how many colorings) must be left fixed by each π_i. It is helpful to make a table of the π_i's and the colorings that they leave fixed; see columns i and ii in Figure 9.5 (columns iii and iv are used later).

The $0°$ rotation π_1 leaves each corner fixed and hence it leaves each coloring fixed. So $\Psi(\pi_1) = 16$. The $90°$ motion π_2 cyclicly permutes corners a, b, c, d. Being left fixed by the $90°$ motion means that each corner in a coloring has the same color after the $90°$ motion as it did before. Since π_2 takes a to b, then a coloring left fixed by π_2 must have the same color at a as at b. Similarly such a coloring must have the same color at b as at c, the same color at c as at d, and the same at d as at a. Taken together, these conditions imply that only the colorings of all white or all black corners, C_1 and C_{16}, are left fixed. Thus $\Psi(\pi_2) = 2$. In general, *a coloring C will be left fixed by π if and only if for each corner v, the color at v is the same as the color at $\pi(v)$* (*thus keeping the color at $\pi(v)$ unchanged*).

Next we consider the $180°$ motion π_3. Looking at the depiction of π_3 in Figure 9.2, we see that π_3 causes corners a and c to interchange and corners b and d to interchange. It follows that a coloring left fixed by π_3 must have the same color at corners a and c and the same color at b and d (no further conditions are needed). With two color choices for a, c and with two color choices for b, d, we can construct $2 \cdot 2 = 4$ colorings that will be left fixed—namely, C_1, C_{10}, C_{11}, C_{16}. Hence $\Psi(\pi_3) = 4$.

The motion π_4 is similar to π_2, so $\Psi(\pi_4) = \Psi(\pi_2) = 2$. The horizontal flip π_5

(i) Motion π_i	(ii) Colorings Left Fixed by π_i^*	(iii) Cycle Structure Representation	(iv) Inventory of Colorings Left Fixed by π_i^*
π_1	16—all colorings	x_1^4	$(b+w)^4 = b^4 + 4b^3w + 6b^2w^2 + 4bw^3 + w^4$
π_2	2—C_1, C_{16}	x_4	$(b^4+w^4)^1 = b^4 + w^4$
π_3	4—$C_1, C_{10}, C_{11}, C_{16}$	x_2^2	$(b^2+w^2)^2 = b^4 + 2b^2w^2 + w^4$
π_4	2—C_1, C_{16}	x_4	$(b^4+w^4)^1 = b^4 + w^4$
π_5	4—C_1, C_6, C_8, C_{16}	x_2^2	$(b^2+w^2)^2 = b^4 + 2b^2w^2 + w^4$
π_6	4—C_1, C_7, C_9, C_{16}	x_2^2	$(b^2+w^2)^2 = b^4 + 2b^2w^2 + w^4$
π_7	8—$C_1, C_2, C_4, C_{10}, C_{11}, C_{13}, C_{15}, C_{16}$	$x_1^2 x_2$	$(b+w)^2(b^2+w^2) = b^4 + 2b^3w + 2b^2w^2 + 2bw^3 + w^4$
π_8	8—$C_1, C_3, C_5, C_{10}, C_{11}, C_{12}, C_{14}, C_{16}$	$x_1^2 x_2$	$(b+w)^2(b^2+w^2) = b^4 + 2b^3w + 2b^2w^2 + 2bw^3 + w^4$
Total		$P_G = \frac{1}{8}(x_1^4 + 2x_4 + 3x_2^2 + 2x_1^2 x_2)$	$= 8b^4 + 8b^3w + 16b^2w^2 + 8bw^3 + 8w^4$

FIGURE 9.5

interchanges a, b and c, d. Again we must have two pairs of like-colored corners when constructing the colorings that will be left fixed by π_5. Then like π_3, we have $\Psi(\pi_5) = 2 \cdot 2 = 4$ colorings fixed—namely, C_1, C_6, C_8, C_{16}.

A pattern is becoming clear. The reader should now be able quickly to predict that π_6 will also leave $2 \cdot 2 = 4$ colorings fixed, for again the motion interchanges two pairs of corners. Formally, an interchange is a cyclic permutation on two elements. All our enumeration of fixed colorings has been based on the fact that *if π_i cyclicly permutes a subset of corners* (*that is, the corners form a cycle of π_i*), *then those corners must be the same color in any coloring left fixed by π_i*.

As mentioned in Section 9.1, any π_i can be represented as a product of disjoint cycles. Thus all we need to do is get such a representation and count the number of cycles. For future use, let us also classify the cycles by their length. It will prove convenient to encode a motion's cycle information in the form of a product containing one x_1 for each cycle of 1 corner in π_i, one x_2 for each cycle of size 2, and so forth. This expression is called the **cycle structure representation** of a motion.

Now we add column iii to Figure 9.1 in which we write the cycle structure representation of each motion. So for π_2 and π_4 we enter x_4 and for π_3, π_5, π_6 we enter $x_2 x_2$ or x_2^2. But what about π_1? Previously, it sufficed to say that π_1 leaves all colorings fixed. Now is time to point out that a corner left fixed by a permutation is classified as a 1-cycle. Thus π_1 is really four 1-cycles. Its cycle structure representation is then x_1^4. We predict a posteriori that π_1 leaves $2^4 = 16$ colorings fixed.

For any π_i, the number of colorings left fixed will be given by setting each x_j equal to 2 (or, in general, the number of colors available) in the cycle structure representation of π_i—That is,

$$\Psi(\pi) = 2^{number\ of\ cycles\ in\ \pi}$$

Let us apply this theory to π_7 and π_8. For each of these motions, the cycle structure representation is seen to be $x_1^2 x_2$ (follow the arrows in Figure 9.2) and thus for each we can find $2^2 \cdot 2 = 8$ colorings that are left fixed.

To obtain the number of 2-colorings of the floating square with Burnside's formula, we sum the numbers in column ii of Figure 9.5 and divide by 8. There is a simpler way. First algebraically sum the cycle structure representations of each motion, collecting like terms, and then divide by 8. From column iii of Figure 9.5, we obtain $\frac{1}{8}(x_1^4 + 2x_4 + 3x_2^2 + 2x_1^2 x_2)$. This expression is called the **cycle index** $P_G(x_1, x_2, \ldots, x_k)$ for a group G of motions. By setting each $x_i = 2$, that is, $P_G(2, 2, \ldots, 2)$, we get the same answer as before (before, the steps were reversed: we first set $x_i = 2$ in each cycle structure representation and then added).

Another advantage to the cycle index approach is that for any m, $P_G(m, m, \ldots, m)$ will the number of m-colorings of an unoriented square. The argument used to derive this coloring counting formula with the cycle index is valid for colorings of any set with associated symmetries.

Theorem

Let S be a set of elements and G be a group of symmetries of S that acts to induce an equivalence relation on the set of m-colorings of S. Then the number of nonequivalent m-colorings of S is given by $P_G(m, m, \ldots, m)$. ∎

Example 1

Use this theorem to re-solve the necklace counting problems of Example 2 in Section 9.2.

Recall that beads can rotate freely around the circle but not flip, that three colors of beads are available, and that the number of different 3-colored strings of n beads is equal to the number of 3-colorings of a cyclicly unoriented n-gon. For $n = 2$, the rotations are of 0° and 180° with cycle structure representations of x_1^2 and x_2, respectively. Thus $P_G = (\frac{1}{2}(x_1^2 + x_2)$. The number of 3-colored strings of two beads is then $P_G(3,3) = \frac{1}{2}(3^2 + 3) = 6$.

For $n = 3$, the rotations are of 0°, 120°, and 240° with cycle structure representations of x_1^3, x_3, and x_3, respectively. Thus $P_G = \frac{1}{3}(x_1^3 + 2x_3)$. The number of 3-colored strings of three beads is $P_G(3,3,3) = \frac{1}{3}(3^3 + 2 \cdot 3) = 11$. The numbered of m-colored strings of three beads is $P_G(m,m,m) = \frac{1}{3}(m^3 + 2m)$.

Let us try a more complicated case: $n = 8$ (see Figure 9.6). The rotations are of 0°, 45°, 90°, 135°, 180°, 225°, 270°, and 315°. The 0° rotation consists of eight 1-cycles. The 45° rotation is the cyclic permutation $(abcdefgh)$. The 90° has the cycle decomposition $(aceg)(bdfh)$. The 135° rotation is the cyclic permutation $(adgbehcf)$. The 180° rotation has the cyclic decomposition (ae) (bf) (cg) (dh). The cycle structure representations are thus: 0° rotation, x_1^8; 45° rotation, x_8; 90° rotation, x_4^2; 135° rotation, x_8; and 180° rotation, x_2^4. The 225°, 270° and 315° rotations are reverse rotations of 135°, 90°, 45°, respectively, and thus have the corresponding cycle structure representations. Collecting terms, we obtain

$$P_G = \tfrac{1}{8}(x_1^8 + 4x_8 + 2x_4^2 + x_2^4)$$

The number of different m-colored necklaces of eight beads is

$$\tfrac{1}{8}(m^8 + 4m + 2m^2 + m^4)$$

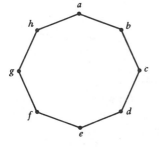

FIGURE 9.6

For $m = 3$, we have

$$\tfrac{1}{8}(3^8 + 4 \cdot 3 + 2 \cdot 3^2 + 3^4) = \tfrac{1}{8}(6561 + 12 + 18 + 81) = 834$$

■

Example 2

In Example 4 in Section 9.1 we listed the 12 symmetries of the tetrahedron (see Figure 9.4): the 0° motion, the eight revolutions of 120° and 240° about a corner and the middle of the opposite face, and the three revolutions of 180° about the middle of opposite edges. We now use our theorem to determine the number of corner 3-colorings of the floating tetrahedron.

The 0° motion has the cycle structure representation x_1^4. The 120° revolution about corner a and the middle of face bcd has the cyclic decomposition $(a)(bcd)$ and its cycle structure representation is $x_1 x_3$. By symmetry the other 120° and 240° revolutions have this same cycle structure representation. The 180° revolution about the middle of edges ab and cd has the cyclic decomposition $(ab)(cd)$ and its cycle structure representation is x_2^2. By symmetry, the other 180° revolutions have the same cycle structure representation. Thus we have

$$P_G = \tfrac{1}{12}(x_1^4 + 8x_1 x_3 + 3x_2^2)$$

The number of different corner 3-colorings is

$$P_G(3,3,3,3) = \tfrac{1}{12}(3^4 + 8 \cdot 3 \cdot 3 + 3 \cdot 3^2)$$
$$= \tfrac{1}{12}(81 + 72 + 27) = 15$$

■

EXERCISES

1. How many ways are there to color the corners of a floating square using four different colors?

2. How many ways are there to 4-color the corners of a pentagon that is:
   (a) Distinct with respect to rotations only?
   (b) Distinct with respect to rotations and flips?

3. How many ways are there to 3-color the corners of a hexagon that is:
   (a) Distinct with respect to rotations only?
   (b) Distinct with respect to rotations and flips?
   (c) Find two 3-colorings of the hexagon that are different in part (a) but equivalent in part (b).

4. How many different n-bead necklaces (cyclicly distinct) can be made from three colors of beads when:
   (a) $n = 7$ (b) $n = 9$ (c) $n = 10$ (d) $n = 11$

5. Find the number of different m-colorings of the vertices of the following flat figures when all physical motions are allowed.

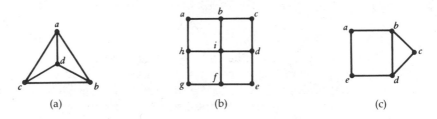

        (a)   (b)   (c)

6. (a) Find the cycle index for the group of symmetries of a square in terms of permutations of the edges, not corners.
   (b) How many ways are there to 3-color the edges of a floating square?
   (c) How many ways are there to 3-color both edges and corners of a floating square?
   (d) Why does not: (number of floating 3-colorings of corners) · (number of floating 3-colorings of edges) = (number of floating 3-colorings of edges and corners)? Explain.

7. How many ways are there to 3-color the *edges* of the floating figures in Exercise 5. (*Hint:* The cycle index now is for symmetries of the edges.)

8. (a) Find the number of different n-bead 3-colored necklaces (cyclicly distinct) in which each color appears at least once when (i) $n = 3$, (ii) $n = 4$, (iii) $n = 7$.
   (b) Repeat part a when necklaces may flip as well as rotate.

9. Find the number of different 2-sided dominoes (two squares of dots or a blank on each side of the domino)?

10. (a) Let G be the group of all 4! permutations of 1, 2, 3, 4. Find P_G.
    (b) Use part a to find the number of ways to paint four identical marbles each one of three colors (check your answer by modeling this problem as a selection-with-repetition problem).
    (c) Use part a to find the number of ways to put 12 balls chosen from three colors into four indistinguishable boxes with 3 balls in each box.

11. (a) Find the number of 2 × 4 chessboards distinct under rotation whose squares are colored red or black.
    (b) Suppose that two chessboards are also considered equivalent (besides for rotational symmetry) if one can be obtained from the other by complementing red and black colors. How many different 2 × 4 chessboards are there? (*Hint:* tricky—use part a.)

12. Show that if $x_1^{k_1}x_2^{k_2} \cdots x_m^{k_m}$ is the cycle index for a symmetry π of an n-gon (expressed in terms of a permutation of the corners), then $1k_1 + 2k_2 + \cdots + mk_m = n$.

13. In solving for the number of corner 2-colorings of some unoriented figure, suppose we are given the cycle index P_{G^*} of the group G^* of induced permutations of the 2-colorings, instead of the usual cycle index P_G of the group of symmetries of the figure. What integer values should be substituted now for each x_j in P_{G^*} to get the number of 2-colorings, or will no substitution work? Explain.

14. Let S be some set of n objects and G a group of permutations of S. For subsets S_1, S_2 of S, define the equivalence relation $S_1 \sim S_2$ if for some $\pi \in G$, $S_1 = \pi(S_2)$ $(=\{\pi(s)|s \in S_2\})$. Show that the number of equivalence classes equals $P_G(2,2, \ldots)$. (There are 2^n subsets of S in all.)

15. Find the number of m-colorings of the corners of a p-gon, where p is a prime larger than 2, if:
    (a) Only rotational symmetries are allowed.
    (b) Rotational and flip symmetries are allowed.

16. The *Euler phi function* ϕ is defined $\phi(n)$ = number of positive integers less than n and relatively prime to n. Using ϕ and summing over all divisors of n, find an expression for the number of m-colored n-bead necklaces (cyclicly unoriented).

9.4 POLYA'S FUNCTION

We are now ready to return to our original goal of a formula for the pattern inventory. The pattern inventory can be considered as giving the results of several formula (*)-type counting subproblems. In the case of 2-colorings of the floating square, we divide the colorings in Figure 9.1 into sets based on the numbers of black and white corners:

$$S_0 = \{C_1\}$$
$$S_1 = \{C_2, C_3, C_4, C_5\}$$
$$S_2 = \{C_6, C_7, C_8, C_9, C_{10}, C_{11}\}$$
$$S_3 = \{C_{12}, C_{13}, C_{14}, C_{15}\}$$
$$S_4 = \{C_{16}\}$$

In the pattern inventory, the coefficient of b^3w is the number of nonequivalent colorings with three blacks and one white. This number can be obtained from (*) if we let G, the group of symmetries of the square, act on just the set S_1 (we can apply (*) to the subset S_1, or to any S_i, because the S_i's are closed under G; see Burnside's theorem in Section 9.2). The coefficient of b^4 is the result when S_0 is the set. In general, *the coefficient of $b^{4-i}w^i$ is the result of (*) when S_i is the set on which G acts.*

Let us try to solve these five subproblems simultaneously. That is, we will list in a row the numbers of 2-colorings left fixed in each S_i by π_1, then below this row we will list the numbers left fixed in each subset by π_2, then by π_3, and so forth (see column iv in Figure 9.5). Then we total up the first column (the first number in each row) and divide by 8, total up the second column and divide by 8, and so forth.

Since the action of π_1 leaves all C's fixed, the first row is 1, 4, 6, 4, 1. Let us put this data in the same form as the pattern inventory. We write: $b^4 + 4b^3w + 6b^2w^2 + 4bw^3 + w^4$; this is an inventory of fixed colorings. For π_1, the inventory of fixed colorings is an inventory of all colorings. Observe that this inventory is simply $(b + w)^4 = (b + w) \cdot (b + w) \cdot (b + w)$, one $(b + w)$ for each corner. For π_2, the inventory is $b^4 + w^4$. For π_3, we find by observation that the inventory is $b^4 + 2b^2w^2 + w^4$. This expression factors into $(b^2 + w^2)^2$.

Just as we did before when counting the total number of colorings fixed by the action of some π, let us look for a "pattern" in the inventories of fixed colorings. Again the key to the "pattern" is the fact that in a coloring left fixed by π, all corners in a cycle of π must have the same color. Since π_2 has one cycle involving all four corners, the possibilities are thus all corners black or all corners white; hence the inventory is $b^4 + w^4$. The motion π_3 has two 2-cycles (ac) and (bd). Each 2-cycle used two blacks or two whites in a fixed coloring; hence the inventory of a cycle of size two is $b^2 + w^2$. The possibilities with two such cycles have the inventory $(b^2 + w^2)(b^2 + w^2)$.

The inventory of fixed colorings for π_i will be a product of factors $(b^j + w^j)$, one factor for each j-cycle of the π_i. So we need to know the number of cycles in π_i of each size. But this is exactly the information encoded in the cycle structure representation. Indeed, setting $x_j = (b^j + w^j)$ in the representation yields precisely the inventory of fixed colorings for π_i. By this method we compute the rest of the inventories of fixed colorings. See Figure 9.5. For π_7 especially, the inventory should be checked against the list of colorings in column ii. The pattern inventory is obtained by adding together the inventories of fixed colorings, collecting like-power terms, and dividing by 8.

As before in Section 9.3, we get a more compact formula and save some computation by first adding together the cycle structure representations and dividing by 8, and then setting each $x_i = (b^i + w^i)$ and doing the polynomial algebra all at once. Again the first step in this approach yields the cycle index $P_G(x_1, x_2, \ldots, x_k)$. Thus by setting $x_j = (b^j + w^j)$ in P_G, we obtain the pattern inventory.

If three colors, black, white, and green, were permitted, each cycle of size j would have an inventory of $(b^j + w^j + g^j)$ in a fixed coloring. So we would set $x_j = (b^j + w^j + g^j)$ in P_G. The preceding argument applies for any number of colors and any figure. In greater generality we have the following theorem.

Theorem (Polya's Enumeration Formula)

Let S be a set of elements and G be a group of permutations of S that acts to induce an equivalence relation on the colorings of S. The inventory of non-equivalent colorings of S using two colors is given by the generating function $P_G((b + w), (b^2 + w^2), (b^3 + w^3), \ldots, (b^k + w^k))$
  The inventory using colors $c_1, c_2, \ldots, c_m$ is

$$P_G\left(\sum_{j=1}^{m} c_j, \sum_{j=1}^{m} c_j^2, \ldots, \sum_{j=1}^{m} c_j^r\right)$$

∎

For a moment, let us return to the problem of counting the *total* number of 2-colorings left fixed. This number is simply the sum of the coefficients in the pattern inventory. To sum coefficients, we set the indeterminants, b and w (and hence their powers), equal to 1, or, equivalently, set $x_i = 2$ in P_G. If m colors were allowed, we would set $x_i = m$ in P_G, obtaining the same formula as in the theorem in Section 9.3.

As in many other generating function problems, actual expansion of the generating function for a pattern inventory can be quite tedious. We are expanding expressions of the form $(c_1^i + c_2^i + \cdots + c_m^i)^k$. When m and k get large, it is time to find a computer.

Example 1

Determine the pattern inventory for 3-bead necklaces distinct under rotations using black and white beads and using black, white, and red beads.
  From Example 1 of Section 9.3, we know $P_G = \frac{1}{3}(x_1^3 + 2x_3)$. Substituting $x_j = (b^j + w^j)$, we get

$$\frac{1}{3}((b + w)^3 + 2(b^3 + w^3)) = \frac{1}{3}((b^3 + 3b^2w + 3bw^2 + w^3) + (2b^3 + 2w^3))$$
$$= \frac{1}{3}(3b^3 + 3b^2w + 3bw^2 + 3w^3)$$
$$= b^3 + b^2w + bw^2 + w^3$$

This result could be obtained empirically. There is only one way to color all beads black or all white. If one bead is white and the others black, then by rotation the white bead can occur anywhere; thus there is only one necklace with one white and two blacks. The same is true by symmetry for a necklace with one black and two whites.
  Now consider 3-bead necklaces using three colors. Perhaps again each inventory term has coefficient one. There is a general test for whether all inventory coefficients are one: Compare N_1, the number of terms in the

pattern inventory, with N, the total number of patterns (i.e., $P_G(m, m, \ldots, m)$). Since N equals the sum of the coefficients in the inventory, then $N_1 = N$ if and only if each term has coefficient one. The number N_1 of terms in an inventory when n elements (corner, beads, etc.) are colored with m colors is $C(n + m - 1, n)$ (see Exercise 20). For the case at hand, $m = 3$, $n = 3$, and so $N_1 = \binom{3 + 3 - 1}{3} = 10$. From Example 1 of Section 9.3, we know that $N = 11$. Since $N_1 \neq N$, we know that all terms do not have coefficient 1.

On the other hand, the only way for $N_1 = N + 1$ is that nine terms in the inventory must have coefficient 1 and one term coefficient 2. But as argued above, there is only one necklace with all beads the same color and only one necklace with one bead of color A and the other two beads of color B. The only other possibility for a 3-bead necklace using three colors is to have one bead of each color. Thus there must be two necklaces with one bead of each color (the necklaces are any cyclic order of the three colors and the reverse cyclic order), and the pattern inventory for black, white and red 3-bead necklaces is

$$b^3 + w^3 + r^3 + b^2w + b^2r + w^2b + w^2r + r^2b + r^2w + 2bwr$$

■

Example 2

Find the pattern inventory for 7-bead necklaces distinct under rotations using three black and four white beads.

We need to determine the coefficient of b^3w^4 in the pattern inventory. Let us again use empirical methods to simplify this pattern inventory problem. Each rotation, except the 0° rotation, is a cyclic permutation when the number of beads is a prime (see Exercise 13(a)), so $P_G = \frac{1}{7}(x_1^7 + 6x_7)$. The pattern inventory is $\frac{1}{7}((b + w)^7 + 6(b^7 + w^7))$.

Since the factor $6(b^7 + w^7)$ in the pattern inventory contributes nothing to the b^3w^4 term, we can neglect it. Thus the number of 3-black, 4-white necklaces is simply

$$\frac{1}{7}(\text{coefficient of } b^3w^4 \text{ in } (b + w)^7) = \frac{1}{7}\binom{7}{3}$$

■

Example 3

Find the pattern inventory of black-white edge colorings of a tetrahedron.

Although we calculated the cycle index for corner symmetries of the tetrahedron in Example 2 of Section 9.3, we need a different cycle index for edge symmetries. Since the set of objects to be colored is the six edges, we need to consider the symmetries of the tetrahedron as permutations of the edges.

The 0° revolution clearly leaves all edges fixed and thus has cycle structure representation x_1^6. The 120° (or 240°) revolution about a corner and the middle of the opposite face cyclicly permutes the edges incident to that corner and cyclicly permutes the edges bounding the opposite face (see Figure 9.4). Thus the 120° revolution has cycle structure representation x_3^2.

The 180° revolution about opposite edges leaves those two edges fixed while the other two edges must pair off into two 2-cycles (see Figure 9.4; since two applications of a 180° revolution return the tetrahedron to its original position, edges not fixed must be in 2-cycles). Thus the 180° revolution has cycle structure representation $x_1^2 x_2^2$. Then $P_G = \frac{1}{12}(x_1^6 + 8x_3^2 + 3x_1^2 x_2^2)$.

Substituting $x_j = (b^j + w^j)$, we get

$$\frac{1}{12}((b + w)^6 + 8(b^3 + w^3)^2 + 3((b + w)^2(b^2 + w^2)^2)$$

$$= \frac{1}{12}((b^6 + 6b^5 w + 15b^4 w^2 + 20b^3 w^3 + 15b^2 w^4 + 6bw^5 + w^6)$$

$$+ (8b^6 + 16b^3 w^3 + 8w^6)$$

$$+ (3b^6 + 6b^5 w + 9b^4 w^2 + 12b^3 w^3 + 9b^2 w^4 + 6bw^5 + 3w^6))$$

$$= \frac{1}{12}(12b^6 + 12b^5 w + 24b^4 w^2 + 48b^3 w^3 + 24b^2 w^4 + 12bw^5 + 12w^6)$$

$$= (b^6 + b^5 w + 2b^4 w^2 + 4b^3 w^3 + 2b^2 w^4 + bw^5 + w^6)$$

∎

Example 4

Find the pattern inventory for corner 2-colorings of a floating cube.

The symmetries of the cube involve revolutions about opposite faces, about opposite edges, and about opposite corners. First, of course, there is the (fixed) 0° motion, with cycle structure representation x_1^8.

(a) *Opposite faces:* As a model, we revolve about the middle of the pair of opposite faces *abdc* and *efgh* (see Figure 9.7a). A 90° revolution yields the permutation (*abcd*) (*efgh*) with cycle structure representation x_4^2. A 270° revolution has the same structure. A 180° revolution yields the permutation (*ac*) (*bd*) (*eg*) (*fh*) with cycle structure representation x_2^4. There are three pairs of opposite faces and so the total contribution to the cycle index of opposite-face revolution is $6x_4^2 + 3x_2^4$.

(b) *Opposite edges:* As a model, we revolve about the middle of opposite edges *ab* and *gh* (see Figure 9.7b). A 180° revolution yields the permutation (*ab*)(*ce*)(*df*)(*gh*) with cycle structure representation x_2^4. There are six pairs of opposite edges, and so the total contribution of opposite-edge revolutions is $6x_2^4$.

(c) *Opposite corners:* As a model, we revolve about the opposite corners *a* and *g* (see Figure 9.7c). A 120° revolution yields the permutation (*a*) (*bde*) (*chf*)

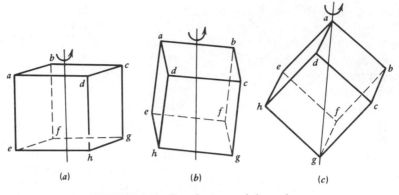

FIGURE 9.7   Revolutions of the cube
(a)   Revolution about opposite faces
(b)   Revolution about opposite edges
(c)   Revolution about opposite corners

(g) with cycle structure representation $x_1^2 x_3^2$. One way to see what this permutation does is by noting that the three corners b, d, e adjacent to a must be cyclically permutated in any motion that leaves a fixed (and similarly for the corners adjacent to g). A 240° revolution has the same structure. There are four pairs of opposite corners and so the contribution of opposite-corner revolutions is $8x_1^2 x_3^2$.

We leave it to the reader (see Exercise 17) to verify that we have enumerated all symmetries of the cube and that these symmetries are all distinct.

Collecting terms, we find $P_G = \frac{1}{24}(x_1^8 + 6x_4^2 + 9x_2^4 + 8x_1^2 x_3^2)$. The generating function of corner colorings of the cube using black and white is thus

$$\tfrac{1}{24}((b + w)^8 + 6(b^4 + w^4)^2 + 9(b^2 + w^2)^4 + 8(b + w)^2(b^3 + w^3)^2)$$

As before, the coefficients of the terms b^8, b^7w, bw^7, and w^8 will all be 1. The b^6w^2, b^5w^3, and b^4w^4 terms in the four factors of the generating function are $(\cdots + 28b^6w^2 + 56b^5w^3 + 70b^4w^4 + \cdots)$, $6(\cdots + 2b^4w^4 + \cdots)$, $9(\cdots + 4b^6w^2 + 6b^4w^4 + \cdots)$, and $8(\cdots + b^6w^2 + 2b^5w^3 + 4b^4w^4 + \cdots)$, respectively. Summing and dividing by 24 we have $\frac{1}{24}(\cdots + 72b^6w^2 + 72b^5w^3 + 168b^4w^4 + \cdots) = 3b^6w^2 + 3b^5w^3 + 7b^4w^4$. (It is easy to detect errors in these calculations, since most errors will result in a noninteger coefficient.) By symmetry, we fill out the pattern inventory to obtain

$$b^8 + b^7w + 3b^6w^2 + 3b^5w^3 + 7b^4w^4 + 3b^3w^5 + 3b^2w^6 + bw^7 + w^8$$

EXERCISES

SUMMARY OF EXERCISES   The first sixteen exercises use Polya's enumeration formula, the remaining problems involved associated theory. Note that "floating" means all physical rotations and flips are allowed.

1. Find the pattern inventory for black and white corner colorings of a floating pentagon.

2. Find an expression for the pattern inventory for black-white, n-bead necklaces (cyclicly distinct) and find the number of necklaces with 3 white beads and the rest black:
    (a) $n = 6$ (b) $n = 9$ (c) $n = 10$ (d) $n = 11$

3. Find the pattern inventory for black, white, and red corner colorings of a floating square.

4. Find an expression for the pattern inventory for the 2-colorings (cyclicly distinct) of the 16 squares in a 4 × 4 chessboard.

5. Find an expression for the pattern inventory for black-white corner colorings of the following floating figures:

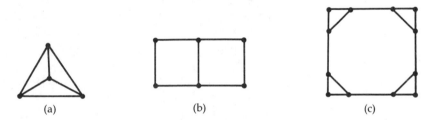

(a)   (b)   (c)

6. (a) Find the pattern inventory for corner 2-colorings of a floating pyramid (with a square base).
    (b) Repeat part a for edge colorings.
    (c) Repeat part a for face colorings.

7. Find an expression for the pattern inventory for edge 2-colorings of the floating figures in parts a and b in Exercise 5.

8. Find an expression for the pattern inventory for edge 2-colorings of a floating cube and find the number of edge 2-colorings with 3 white and 9 black edges.

9. Find an expression for the pattern inventory for face 2-colorings of:
    (a) A floating tetrahedron.
    (b) A floating cube.

10. (a) Find the pattern inventory for the corner 3-colorings of a floating pentagon with adjacent corners different colors.

    (b)   Repeat part 1 for a floating tetrahedron.

    (c)   Repeat part a for a floating cube.

11.   Find an expression for the pattern inventory for black-white colorings of four indistinguishable balls (see Exercise 10 in Section 9.3).

12.   Give an empirical argument (without use of the cycle index) to show that there are $\left\lfloor \dfrac{n}{2} \right\rfloor$ different n-bead necklaces with two white beads and n-2 black beads ($\lfloor r \rfloor$ is the largest integer $\leqslant r$).

13.   Given that p is a prime and p-bead necklaces are made of black and white beads.

    (a)   Show that each rotation except $0°$ is a cyclic permutation of the corners.

    (b)   What is the number of such necklaces with exactly k white beads?

14.   (a)   Suppose that instead of coloring the corners of a floating square, we attach 0 or 1 or 2 identical jelly beans at a corner. Find a generating function, and expand it, for the number of different squares with a total of k beans at its corners. (*Hint:* In the inventory of patterns left fixed, the exponents within a factor will vary.)

    (b)   Repeat part a with each corner getting a red jelly bean or a white jelly bean or a red and a white jelly bean. Now find a generating function for the number of squares with j whites and k reds.

15.   Suppose n rods with small holes drilled through their midpoints are strung along a piece of wire and each end of each baton is painted one of three possible colors (red, white, blue). How many indistinguishable configurations are there for:

    (a) $n = 2$ (b) $n = 3$ (c) $n = 4$.

16.   How many distinct (non-isomorphic) unlabeled graphs are there with four vertices. (*Hint:* Each possible edge has two possible colors: "edge present" and "edge absent.")

17.   Show that the 24 symmetries of a cube listed in Example 4 are all distinct and include all symmetries of a cube.

18.   Characterize the geometric figures which when floating have only one coloring with one black corner and the others white.

19.   Show that the set S_k of all black-white colorings of the corners of a fixed figure in which exactly k corners are white is closed under the group of symmetries of the figure.

20.   Show that there are $C(n + m - 1, n)$ different terms in the pattern inventory for m-colorings of corners of an n-gon.

21.   Let $a_{k,n}$ denote the minimum number of coefficients in the pattern inventory of all k-colorings of the corners of an n-gon needed so that using symmetry all coefficients in the pattern inventory are known. Find a generating function $g_n(x)$ for $a_{k,n}$.

9.5 SUMMARY AND REFERENCES

Polya's enumeration formula is important, practically because it solves important problems, mathematically because it is an elegant marriage of group theory and generating functions, and pedagogically because it is the only truly powerful combinatorial formula we will see in this book (where a difficult problem can be solved by "plugging" the right numbers into one special formula). Of equal importance is the manner in which Polya's formula has been developed: Sections 9.3 and 9.4 could be read as a case study in the experimental derivation of a mathematical theory. Students interested in a more rigorous development of this theory and its extensions are referred to [1, 2, 4]. The curious student with facility in the German language is encouraged to read Polya's original paper [5], which contains many chemical applications.

In a more rigorous development, one defines a coloring as a function f from a set S (of corners) to a set R (of colors). The function f_6, corresponding to C_6 (in Figure 9.1), would be written in tabular form as $\left(\dfrac{abcd}{wwbb} \right)$. We define the composition $\pi \cdot f$ of a motion and a color function to be a new color function f'; for example, $\pi_2 \cdot f_6$ maps $a \to b \to w$, $b \to c \to b$, $c \to d \to b$, $d \to a \to w$, or $\pi_2 \cdot f_6 = \left(\dfrac{abcd}{wbbw} \right) = f_7$. The colorings permutation induced by a symmetry π maps each f to $\pi \cdot f$. Although tedious, a formal calculation of the coloring permutation is thus possible (before we did it by inspection). We define coloring equivalence by $f \sim f'$ if and only if there exists π such that $\pi \cdot f = f'$. Our theory can readily be restated in terms of this new definition.

Polya's enumeration formula has an important application to another field of combinatorial mathematics. It is used to enumerate families of graphs (see Exercise 16 in Section 9.4). This application was pioneered by F. Harary. See Harary and Palmer [3].

1. N. G. DeBruijn, "Polya's Theory of Counting" in *Applied Combinatorial Mathematics* (E. F. Beckenbach, ed.), John Wiley, New York, 1964.

2. N. G. DeBruijn, "Enumerative Combinatorial Problems Concerning Structures," *Nieuw Archief Wiskunde* (3) **11** (1963), 142–161.

3. F. Harary and E. Palmer, *Graphical Enumeration*, Academic Press, New York, 1973.

4. C. L. Liu, *Introduction to Combinatorial Mathematics*, Chap. 5, McGraw-Hill, New York, 1968.

5. G. Polya, "Kombinatorische Anzahlbestimmungen fur Gruppen, Graphen und chemische Verbindungen," *Acta Math.* **68** (1937), 145–254.

Chapter Ten

Combinatorial Modeling in Theoretical Computer Science

10.1 FORMAL LANGUAGES AND GRAMMARS

This chapter is meant as a bridge from combinatorial problem-solving to theoretical computer science. Three topics are presented: formal languages, finite-state machines, and NP-completeness. Rather than study formal languages and finite-state machines to formalize programming language structure, we will use these topics to model combinatorial problems that were previously analyzed by standard counting methods. The essence of the theory of NP-completeness is building combinatorial models, and so here our interests and those of computer scientists coincide. There is a fourth natural topic—logical circuit design—that we have omitted; the interested reader should refer to any discrete structures text, such as Gill [3], for an introduction to logic design.

A **grammar** $G = \langle N, T, P, s \rangle$ consists of:

(i)   a finite set N of **nonterminal** elements;

(ii)   a finite set T of **terminal** elements;

(iii)   a finite set P of **production rules** of the form $\alpha \rightarrow \beta$, where α and β are strings of elements from $N \cup T$; and

(iv)   a special start element $s \in N$.

362

We say that sequence $\gamma\alpha\delta$ **directly derives** $\gamma\beta\delta$, written $\gamma\alpha\delta \rightarrow \gamma\beta\delta$ if $\alpha \rightarrow \beta$ is a production of G and γ, δ are strings of elements from $N \cup T$. We say ω_1 **derives** ω_n if we have a succession of direct derivations $\omega_1 \Rightarrow \omega_2 \Rightarrow \cdots \Rightarrow \omega_n$. A **sentence** of G is any sequence ω of terminals derived from s, the start symbol. The **language L(G) derived from** G is the set of all sentences of G.

This grammatical structure was designed with programming and spoken languages in mind. In spoken language, a sentence of the grammar is a normal sentence, terminal elements are words of the language, and nonterminal elements are parts of speech and sentence components—for example, adjective and subject. The Backus-Naur form of a programming language is a grammar, whose sentences are grammatically correct programs. However, for our combinatorial purposes, sentences will be possible outcomes in some combinatorial collection, such as 5-card poker hands with two pairs.

Example 1

Construct a grammar G whose language $L(G)$ consists of all binary sequences with no pair of consecutive 0's.

Clearly the set of terminal elements is $\{0,1\}$. The start element s must be in the set of nonterminals. We will introduce other nonterminals as needed. If all binary sequences were to be generated (including the empty sequence $\emptyset$), the production rules should be $s \rightarrow 1s$, $s \rightarrow 0s$, $s \rightarrow \emptyset$. To avoid two 0's in a row, we need to replace the second production rule with rules that force any 0 to be followed by a 1 (or for the 0 to be the end of the sequence). One set of production rules that works is

$$G_1: \quad s \rightarrow 1s, \quad s \rightarrow 01s, \quad s \rightarrow 0, \quad s \rightarrow \emptyset$$

Another possible set (with start symbol s') is

$$G_2: \quad s' \rightarrow 1s', \quad s' \rightarrow 0t', \quad t' \rightarrow 1s', \quad s' \rightarrow 1, \quad s' \rightarrow 0, \quad t' \rightarrow 1$$

For example, the sentence 01101 would be derived in G_1,

$$s \Rightarrow 01s \Rightarrow 011s \Rightarrow 01101s \Rightarrow 01101$$

and derived in G_2,

$$s' \Rightarrow 0t' \Rightarrow 01s' \Rightarrow 011s' \Rightarrow 0110t' \Rightarrow 01101$$

Observe that the first two production rules of G_1 naturally suggest a recurrence relation for a_n, the number of n-digit binary sequences with no consequence 0's. The relation is

$$a_n = a_{n-1} + a_{n-2} \quad \blacksquare$$

The recursive form of the grammars in Example 1 is in the spirit of modern programming language grammars, which contain recursion with ⟨statement⟩, the start symbol, such as

$$\langle \text{statement} \rangle \rightarrow \langle \text{IF statement} \rangle$$

$$\langle \text{IF statement} \rangle \rightarrow \text{IF} \langle \text{condition} \rangle \text{ THEN } \langle \text{statement} \rangle$$

Example 2

Construct a grammar G whose language $L(G)$ consists of all ternary sequences with the first 0 before the first 1; possibly no 1 appears but at least one 0 must appear. Also give G_n whose language $L(G_n)$ consists of such sequences of length n.

The desired grammar for all such sequences must create sentences that start with a subsequence (possibly empty) of 2's followed by a 0. After 0 appears, there is no longer any restriction on the following digits.

$$G_3: \quad s \rightarrow 2s, \quad s \rightarrow 0t, \quad t \rightarrow 0t, \quad t \rightarrow 1t, \quad t \rightarrow 2t, \quad t \rightarrow \emptyset$$

For example, the sentence 22021 would be derived

$$s \Rightarrow 2s \Rightarrow 22s \Rightarrow 220t \Rightarrow 2202t \Rightarrow 22021t \Rightarrow 22021$$

From this grammar, it is clear that the number of n-digit ternary sequences with a 0 before the first 1 is

$$\sum_{k=0}^{n-1} 3^{n-k-1}$$

where k represents the number of 2's before the first 0 (after which the remaining $n - k - 1$ digits can be picked arbitrarily).

Now consider how we can alter the preceding grammar to force all its sentences to have a given length n. The grammar G_n must be made to "count to n." To do this, we require an indexed family of nonterminals s_i, t_i, $i = 2$, . . . , n.

$$G_n: \quad s \rightarrow 2s_2, \quad s \rightarrow 0t_2, \quad s_i \rightarrow 2s_{i+1}, \quad s_i \rightarrow 0t_{i+1}, \quad i = 2, \ldots, n-1$$

$$t_i \rightarrow 0t_{i+1}, \quad t_i \rightarrow 1t_{i+1}, \quad t_i \rightarrow 2t_{i+1}, \quad i = 2, \ldots, n-1$$

$$s_n \rightarrow 0, \quad t_n \rightarrow 0, \quad t_n \rightarrow 1, \quad t_n \rightarrow 2$$

■

Grammars are similar to generating functions in the sense that they naturally generate outcomes to a whole family of enumeration problems—that is, outcomes of all sizes rather than outcomes of a given length. Note that for finite enumeration problems one could always make up a huge grammar with each of the possible outcomes as the right-hand side of a production (with s on the left).

Example 3

Construct a grammar G whose language $L(G)$ consists of all 5-card hands with a full house (three of one kind and a pair of a second kind).

This grammar is fairly simple to build, but it requires a substantial number of terminal and nonterminal elements. The hand naturally breaks into two parts, a 3-of-a-kind and a pair. Let us agree to generate the hand with the 3-of-a-kind first. The first type of production rules is

$$s \rightarrow t_i p_j, \quad \text{for all possible } i \neq j, \quad i, j = 1, 2, \ldots, 13$$

where the subscripts range over the 13 kinds of cards. Next we must derive from each t_i all 3-of-a-kinds using the ith kind and derive from each p_j all pairs of the jth kind. The simplest approach is just to have one production rule for each possible 3-of-a-kind

$$t_i \rightarrow c_{is}c_{ih}c_{id}, \quad t_i \rightarrow c_{is}c_{ih}c_{ic}, \quad t_i \rightarrow c_{is}c_{id}c_{ic}, \quad t_i \rightarrow c_{ih}c_{id}c_{ic}$$

where c_{ix} is the card of the ith kind in suit x: *spades, hearts, diamonds, clubs*. And similarly for the pairs.

The formula for the number of 5-card hands with a full house is readily seen from these production rules to be $13 \cdot 12 \cdot$ (number of 3-of-a-kind) $\cdot$ (number of pairs of one kind) $= 13 \cdot 12 \cdot C(4,3) \cdot C(4,2)$. ∎

As a general principle, formal grammars are not efficient for generating (unordered) subsets of collections with many types of terminal elements. Also as noted above, it is cumbersome to make grammars generate sequences of a given length. On the other hand, many combinatorial collections are easily generated by grammars. And when a grammar for a combinatorial collection is found, it frequently points the way to a recurrence relation or combinatorial formula for the number of such outcomes of a given length.

There are many special types of grammars, whose production rules have a specified form. See Hopcroft and Ullman [4] for a good survey of types of grammars. We shall discuss just one special type of grammar, called a **regular grammar**. The language generated by a regular grammar is called a **regular language**. In a regular grammar, all production rules are of the form

$$t \rightarrow at' \quad \text{or} \quad t \rightarrow a, \quad t, t' \in N, \quad a \in T \text{ (possibly } t = t')$$

Regular grammars are part of a more general class of grammars called **context-**

free grammars in which the left-hand side of production rules always consists of a single nonterminal element.

The grammars G_2 in Example 1 and G_n in Example 2 are both regular. Note that grammar G_1 in Example 1 is equivalent to G_2 although G_1 is not regular. **Equivalent** grammars produce the same language. Look carefully at how G_2 eliminates the non-regular productions $s \rightarrow 01s$ and $s \rightarrow \emptyset$ in G_1.

Can G_3 be converted into an equivalent regular grammar G_3'? The answer is yes. The only nonregular production in G_3 is $t \rightarrow \emptyset$. This rule would be the last production used in the derivation of any sequence of G_3. To replace $t \rightarrow \emptyset$, we must "back up" one step in the derivation. See the example of the derivation of 22021 in Example 2. The next-to-last production used in the derivation of 22021 is $t \rightarrow 1t$. The new rule $t \rightarrow 1$ has the same effect as $t \Rightarrow 1t \Rightarrow 1$. We must make similar new rules for the other instances where t appears on the right-hand side of a production: $t \rightarrow 0t, t \rightarrow 1t, t \rightarrow 2t, s \rightarrow 0t$. The required new grammar is

$$G_3': \quad s \rightarrow 2s, \quad s \rightarrow 0t, \quad s \rightarrow 0,$$

$$t \rightarrow 0t, \quad t \rightarrow 1t, \quad t \rightarrow 2t,$$

$$t \rightarrow 0, \quad t \rightarrow 1, \quad t \rightarrow 2$$

Regular grammars and languages have a very special property that will be developed in the next section, and so it is important to know whether a language generated by a given grammar has an equivalent regular grammar. The following is a standard example of a simple language with no regular grammar.

Example 4

Construct a grammar to generate the language L whose sentences are all binary sequences consisting of n 0's followed by n 1's, for all possible n. Show that no regular grammar exists for this language.

The idea behind building the appropriate grammar for this language is to derive sequences from the middle outward:

$$G_4: \quad s \rightarrow 0s1, \quad s \rightarrow 01$$

The problem with any attempt to generate such sequences from left to right, as required by a regular grammar, is that the grammar must somehow remember, through the use of different nonterminals, how many 0's it made when it starts making 1's. At first thought, the following grammar, similar to G_n, appears to work:

$$G_5: \quad s \rightarrow 0s_1, \quad s_i \rightarrow 0s_{i+1}, \quad s_i \rightarrow 0t_{i+1}, \quad t_i \rightarrow 1t_{i-1}, \quad s_1 \rightarrow 1, \quad t_1 \rightarrow 1$$

The index i has an infinite range—there is no limit on the possible size of n—but there can only be a finite number of productions. ∎

We conclude this section by observing that there is a natural graphical representation for context-free grammars G. Recall that in context-free grammars the left-hand side of a production consists of a single nonterminal. A sentence of G may be represented by a rooted tree as follows. The start symbol s is the root of the tree, nonterminals are the internal vertices and terminals are the leaves. An internal vertex labeled with the nonterminal t will have children $c_1, c_2, \ldots, c_n$ (indexed in left-to-right order) if $t \to c_1 c_2 \cdots c_n$ is a production rule of G. For example, the derivation of the sentence 000111 in $L(G_4)$ (from Example 4) has the following tree:

The trees associated with sentences of a regular grammar have a very simple form. They are binary trees in which the internal vertices form a simple path branching to the right with single leaves branching off to the left, as shown below.

EXERCISES

SUMMARY OF EXERCISES   The first three exercises require only derivations of sentences in given grammars. Exercises 4–18 involve constructing grammars, with regular grammars required in Exercises 10–18. The last two exercises concern the impossibility of building grammars for certain languages.

1. Give a derivation of the following sentences and draw the associated tree.
   (a) 101101 in G_1 (of Example 1).
   (b) 220102 in G_3 (of Example 2).
   (c) 20202 in G_n, $n = 5$ (of Example 2).
   (d) $c_{3s} c_{3h} c_{3c} c_{8h} c_{8d}$ in Example 3.
   (e) 00001111 in G_5 (of Example 4).

2. Give a derivation of each of the following sentences in the grammar: $s \rightarrow tt$, $t \rightarrow 01t$, $t \rightarrow 10t$, $t \rightarrow 0$, $t \rightarrow 1$. Also draw the associated tree.
   (a) 0010   (b) 1001   (c) 001100

3. Give three derivations of each of the following sentences in the grammar in Exercise 2.
   (a) 010101   (b) 010100

4. Construct a grammar to generate all binary sequences starting with a 1 and ending with a 0.

5. Construct a grammar to generate all binary sequences consisting of a sequence of 0's followed by a sequence of 1's (the number of 0's need not equal the number of 1's).

6. Construct a grammar to generate all 5-digit binary sequences.

7. (a) Construct a grammar to generate all possible Secret Codes in the game of Mastermind (see Appendix A.5).
   (b) Repeat part a for Secret Codes with exactly one Red.
   (c) Repeat part a for Secret Codes with at least one Red.
   (d) Repeat part a for Secret Codes with no repeated colors.

8. Construct a grammar to generate all 5-card hands with:
   (a) a four-of-a-kind.   (b) a straight.

9. (a) Construct a grammar to generate all subsets of six objects chosen with repetition from piles of 3 A's, 3 B's and 3 C's.
   (b) Repeat part a with at least one of each letter.

10. Convert the grammar G_1 in Example 1 into a regular grammar.

11. Convert the grammar in Exercise 2 into a regular grammar.

12. Construct a regular grammar to generate all ternary sequences that repeat the subsequence 012 one or more times—for example, 012012012.

13. Construct a regular grammar to generate all outcomes of a sequence of 5 flips of a coin in which the third Head occurs on the last flip.

14. Construct a regular grammar to generate all 6-digit binary sequences with exactly two 1's.

15. (a) Construct a regular grammar to generate all sequences built on the letters A, H, M, T that contain the consecutive 4-letter subsequence $MATH$ (this subsequence might occur more than once).
    (b) Construct a regular grammar to generate all sequences built on the letters A, H, M, T that do *not* contain the 4-letter subsequence $MATH$.

16. Construct a regular grammar to generate all arrangements of a, e, i, x, x, x, x, x with no consecutive pair of vowels.

17. Construct a regular grammar to generate all 10-digit binary sequences with at least twice as many 1's as 0's.

18. Construct a grammar to generate all binary sequences which contain the same number of 0's and 1's. Show that no equivalent regular grammar exists for this language.

19. Construct a grammar to generate all binary sequences of the form $\omega\omega^{-1}$, where ω is any binary sequence and ω^{-1} is the inversion of ω (sequence reversed). Show that no equivalent regular grammar exists for this language.

20. Show that no grammar exists to generate all binary sequences of the form $\omega\omega$, where ω is any binary sequence.

10.2 FINITE-STATE MACHINES

In this section we present the simplest type of idealized computer, called a finite-state machine. A **finite-state machine** $M = \langle I, O, S, f, \theta \rangle$ consists of:

(i)   a finite set I of input elements;

(ii) a finite set O of output elements;

(iii) a finite set S of internal states;

(iv) a next-state function $f: I \times S \to S$; and

(v)   an output function $\theta: I \times S \to O$.

If a machine starts in some initial state s_1 and receives an input sequence $\alpha = a_1 a_2 \cdots a_n$ of elements in I, then after reading the first input element a_1, the machine prints the output element $\theta(a_1, s_1)$ and shifts to internal state $f(a_1, s_1)$. Successive input elements have similar effects. If a_i is read with the machine in state s_i, then element $\theta(a_i, s_i)$ is printed and the machine shifts to state $f(a_i, s_i)$. The machine stops after processing the last input element.

Example 1

Consider the finite-state machine M with $I = \{0,1\}$, $O = \{0,1,2,3\}$, $S = \{t_0, s_1, t_1, s_2, t_2, s_3\}$, and f and θ given by Table 10.1 (the left element in each entry in the table is the internal state and the right element the output). We claim that M counts (and prints out) the number of runs of 1's thus far in the input sequence, up to a maximum of three runs. Recall that a run of 1's is a consecutive set of 1's (preceded and followed by a 0). For example, the input sequence 0110010 would produce the following sequence of internal states and outputs, assuming M starts in state t_0,

$$(t_0, -) \xrightarrow{0} (t_0, 0) \xrightarrow{1} (s_1, 1) \xrightarrow{1} (s_1, 1) \xrightarrow{0} (t_1, 1) \xrightarrow{0} (t_1, 1) \xrightarrow{1} (s_2, 2) \xrightarrow{0} (t_2, 2)$$

∎

	0	1
t_0	$t_0/0$	$s_1/1$
s_1	$t_1/1$	$s_1/1$
t_1	$t_1/1$	$s_2/2$
s_2	$t_2/2$	$s_2/2$
t_2	$t_2/2$	$s_3/3$
s_3	$s_3/3$	$s_3/3$

TABLE 10.1

Example 2

Consider the finite-state machine M with $I = \{0,1\}$, $O = \{Y,N\}$, $S = \{s_0, s_1, s_2\}$, and f and θ given in Table 10.2. The machine starts in state s_0. This machine prints out Y until, if ever, it finds two consecutive 0's. If two 0's in a row occur then it prints N forever after. Thus if the input sequence contains no pair of consecutive 0's, the last output element printed will be a Y (for Yes), and otherwise the last output will be N (for No). ∎

The machine in Example 2 is said to recognize the collection of binary sequences with no consecutive 0's. In terms of the previous section, the machine recognizes the language generated by grammar G_1 or G_2 in Example 1 of Section 10.1. (Recognition is defined formally later in this section.)

Example 3

Design a finite-state machine that when presented with an input of a 3-subset of $\{1,2,3,4,5,6\}$ will print out the next 3-subset in lexicographic order.

Recall that if we write a subset of digits as a number with the digits written in increasing order, then the lexicographically next subset is simply the subset whose digits form the next larger number; for example, in our collection of 3-subsets, 235 is lexicographically followed by 236 which is followed by 245. An algorithm for finding the lexicographically next r-subset of an n-set was given in Section 5.6: the successor $b_1 b_2 \cdots b_r$ to the current subset $a_1 a_2 \cdots a_r$ is found by searching from *right to left* until the first $a_i \neq n - r + i$ is found; for this i, set $b_i = a_i + 1$ and for $i < j < r$, $b_j = b_{j-1} + 1$.

The lexicographically next 3-subset can usually be obtained by simply increasing the last digit by 1. However, if the last digit is a 6, then we cannot

	0	1
s_0	s_1/Y	s_0/Y
s_1	s_2/N	s_0/Y
s_2	s_2/N	s_2/N

TABLE 10.2

	–	1	2	3	4	5	6
t_0	$t_0/-$			$t_1/4$	$t_1/5$	$t_1/6$	$d_1/-$
t_1	$t_1/-$	$t_1/1$	$t_1/2$	$t_1/3$	$t_1/4$		
d_1	$d_1/-$		$s_3/4$	$s_4/5$	$s_5/6$	$d_2/-$	
d_2	$d_2/-$	$s_3'/4$	$s_4'/5$	$s_5'/6$			
s_2	$t_1/2$						
s_3	$t_1/3$						
s_4	$t_1/4$						
s_5	$t_1/5$						
s_3'	$s_2/3$						
s_4'	$s_3/4$						
s_5'	$s_4/5$						

TABLE 10.3

specify the new last digit until we look at the preceding digit(s). If the middle digit is a 5, then we must turn to the first digit. Because there may be a delay in printing the last, and possibly middle, digits, let us insert two blanks, written –, after each digit in the input. Also let us read the subset in reverse order (as required by the above algorithm) with the third (largest) digit first.

Table 10.3 presents the structure of the required machine, where t_0 is the starting state. If the last digit a_3 is not 6, then the last digit of the successor $b_3 = a_3 + 1$ and the machine goes to state t_1, where the other two digits of the current subset will be repeated unchanged. If a_3 is 6, then we go to state d_1 (delay one round) and print a blank. If the second digit a_2 is not a 5, then $b_3 = a_2 + 2$ and the new second digit $b_2 = a_2 + 1$. If $a_2 = 5$, then we go to state d_2 (delay two rounds) and now the new subset will be $a_1 + 1$, $a_1 + 2$, $a_1 + 3$.

For example, the subset 246 is processed as follows (the input is 6––4–– 2––):

$$(t_0,)\ \overset{6}{\rightarrow}\ (d_1,-)\ \overset{}{\rightarrow}\ (d_1,-)\ \overset{}{\rightarrow}\ (d_1,-)\ \overset{4}{\rightarrow}\ (s_5,6)\ \overset{}{\rightarrow}\ (l_1,5)\ \rightarrow$$
$$(t_1,-)\ \overset{2}{\rightarrow}\ (t_1,2)\ \overset{}{\rightarrow}\ (t_1,-)\ \overset{}{\rightarrow}\ (t_1,-)$$

∎

As n and r get large, a machine of this design to find the next r-subset of an n-set gets very complex. Not being able to store variables in a memory is often a giant handicap for finite-state machines. However finite-state machines are excellent for tasks of the type posed in Example 2—recognizing sentences in a regular language.

To be more precise, a finite-state machine M **recognizes** a language L if the following situation holds: when M processes an input string ω (starting from a specified starting state), then the last output element printed is a specified "recognition" element if and only if ω is a sentence of L.

Theorem

The following are equivalent statements about the collection L of sequences:

(a) $L = L(G)$ for some regular grammar G.

(b) L is recognized by some finite-state machine.

We shall only prove (a) → (b); (b) → (a) is left as an exercise. Our proof of part a breaks into two parts. First we prove the result for simple regular grammars. A regular grammar $G = (N,T,P,s)$ is **simple** if for each $t \; \varepsilon \; N$ and $a \; \varepsilon \; T$, there is at most one production in P of the form $t \to at'$ (there cannot be $t \to at'$ and $t \to at''$).

Lemma

If $L(G)$ is the language of a simple regular grammar G, there exists a finite-state machine M that recognizes $L(G)$.

Proof

We let the nonterminals of G be the states of M and the terminals be the inputs. The start symbol s of G will be the starting state of M. For a production rule $t_i \to a_k t_j$, we define $f(a_k, t_i) = t_j$. If, in addition, there is no production $t_i \to a_k$, then a sentence cannot be completed yet, and so we define $\theta(a_k, t_i) = 0$, a non-"recognition" output element. If there also is a production $t_i \to a_k$, then a sentence could (but need not) end at this point and we define $\theta(a_k, t_i) = *$, the "recognition" element. If there is a production $t_i \to a_k$ but no production of the form $t_i \to a_k t_j$ for some t_j, then still $\theta(a_k, t_i) = *$, but now $f(a_k, t_i) = r$, a special state M goes to when the sentence must have come to an end.

If input continues after M is in state r, then M prints a 0 and forever after stays in state r (printing 0). We also go to state r if in state $t_i M$ reads a_k, but G has no production with t_i on the left and a_k on the right. Note that f is well-defined since G is simple; that is, there is at most production with a given t_i on the left and a_k on the right. ∎

The machine in Example 2 is related to grammar G_2 of Section 10.1 in this fashion (the reader should verify this).

Optional

Let us now formally prove that the machine M constructed in the above manner will recognize $L(G)$. Our proof is by induction on n, the length of an input sequence.

For the initial step of $n = 1$, we must show that a_k is recognized as a sentence if and only if $s \to a_k$ is a production of G, but by the construction of

M, $\theta(s,a_k) = $ * if and only if $s \to a_k$ is a production. Furthermore, observe that after reading a_k, M is in state t_i if and only if $s \to a_k t_i$ is a production of G.

Now assume that for any input sequence ω of length $n - 1$, M recognizes ω if and only if ω is a sentence of G, and that if G has the derivation (starting from s), ωt_i, then after reading in ω, M is in state t_i (otherwise if there is no derivation of the form $\omega t'$, then after reading ω, M is in state r). Consider a sentence ω of G of length n, $\omega' = a_1 a_2 \cdots a_{n-1} a_n$ with $\omega = a_1 a_2 \cdots a_{n-1}$. Before the last production, call it $t_i \to a_n$, in deriving ω', G derived the sequence ωt_i. By the induction assumption, after reading ω, M is in state t_i. So when M reads in the last input element a_n, M prints out $\theta(t_i, a_n) = $ * and goes to state r (since $t_i \to a_n$ is a production of G), and thus M recognizes the sentence ω.

If ω' were not a sentence of G, then we have two cases to consider. First if there is no derivation of the form $\omega t'$, then by the induction assumption after reading ω, M is in state r and will stay there always printing out 0. Second, if G has a derivation of the form ωt_i but there is no production $t_i \to a_n$, then by the construction of M, $\theta(t_i, a_n) = 0$.

Finally, to complete the expanded induction hypothesis, we must verify that if G has a derivation of the form $\omega' t_j$, for some t_j, then after reading ω', M is in state t_j; otherwise M is in state r. Assume that the last production in getting to $\omega' t_j$ was $t_i \to a_n t_j$. Then again by induction after reading ω, M was in state t_i, and so next reading a_n moves M to $f(t_i, a_n) = t_j$. If G has no derivation of the form $\omega' t'$, then the argument is the same as given in the previous paragraph.

This completes the induction proof that M recognizes $L(G)$. ∎

To complete the proof of (a) → (b) of the theorem, we must extend the result of the lemma to general regular grammars. We shall do this by showing how to convert any regular grammar G into an equivalent simple regular grammar G'. First let us give an example of the conversion process.

G_1: $s \to 0s$, $s \to 0t$, G_1': $s \to 0\{s,t\}$,

       $t \to 0t$, $t \to 1t$, $t \to 1v$, $\{s,t\} \to 0\{s,t\}$, $\{s,t\} \to 1\{t,v\}$,

       $v \to 2$ $\{t,v\} \to 0\{s,t\}$, $\{t,v\} \to 1\{t,v\}$,

                                       $\{t,v\} \to 2$

Normally one would define new nonterminals in G_1', say, $s' = \{s,t\}$ and $t' = \{t,v\}$. However, to emphasize what is happening, we used subsets as nonterminals. In G_1, from the starting symbol s we can derive $0s$ or $0t$. To convert G_1 into a simple regular grammar, there can be only one production with s on the left and 0 on the right. Hence, we introduce the subset nonterminal $\{s,t\}$. The nonterminal $\{s,t\}$ has to take the place of s *and* t in all productions of G_1. (We also retain productions with the start symbol on the left to insure a unique start symbol.)

There is another pair of productions that cause trouble, $t \rightarrow 1t$ and $t \rightarrow 1v$, and so again we introduce a subset nonterminal $\{t,v\}$ and write $\{s,t\} \rightarrow 1\{t,v\}$. Next $\{t,v\}$ has to replace t and v in all productions of G_1. Observe that t is replaced by both $\{s,t\}$ and $\{t,v\}$. The resulting grammar for G_1'—see above—is simple as required. If G_1' were not simple because of a pair of productions such as $\{s,t\} \rightarrow 1\{s,t\}$ and $\{s,t\} \rightarrow 1\{t,v\}$, then we would iterate the previous process, making a new subset nonterminal $\{s,t,v\}$ and substituting it for $\{s,t\}$ and $\{t,v\}$ in all productions.

The general procedure for converting an arbitrary regular grammar $G = (N,T,P,s)$ into a simple regular grammar $G' = (N',T,P',s)$ can be summarized.

1. Initially N' consists of the singleton subset $\{s\}$ that has not been scanned, and P' is empty.

2. Let B be a subset in N' that has not been scanned yet. For each $a \in T$, determine the subset $B_a = \{t' \in N : \text{for some } t \in B, (t \rightarrow at') \in P\}$, place B_a in N' (unless B_a is already in N'), and place the production $B \rightarrow aB_a$ in P'.

3. Repeat step 2 until there are no unscanned subsets in N'.

4. For all instances of $B, B' \in N'$ where B' is a subset of B, eliminate B' from N' and eliminate all productions involving B' from P' (the one exception to this rule is the singleton subset $\{s\}$).

Using an induction proof similar to the one in the lemma above, we can rigorously show that this new procedure does produce a simple regular grammar G' equivalent to G (the proof is an exercise). Note that this procedure must terminate because there are only a finite number of subsets of the finite set N of nonterminals.

With this step we have completed the proof of one half of our theorem, that if G is a regular grammar there exists a finite-state machine that recognizes $L(G)$.

Although finite-state machines may seem somewhat artificial, they are a very helpful tool in the design of programming systems for problems ranging from the processing of arithmetic expressions on hand-held calculators to the construction of giant compilers. In compiler design today, finite-state machines are commonly used to recognize valid constructs, parse them, and even generate much of the assembly language code. Once compiler design was an extremely complex task taking a team of experts a year or more. By automating the construction of finite-state machines to recognize regular grammars, as described above, computer scientists can now build "compiler-compiler" programs that, when given as input the grammar of a programming language and a description of a target computer, can produce as output a compiler for the language on the specified computer.

We close this section by mentioning a famous extension of a finite-state machine called a **Turing machine**. The input of a Turing machine is placed on

a "tape," a two-way infinite string of positions that is initially blank except for
the input sequence. The Turing machine has a read/write "head" that reads in
the element at the position where the head is located. Like a finite-state
machine, the Turing machine has a next-state function $f(s,a)$ and an output
function $\theta(s,a)$ both based on the current state s and the element read a, but
now the output is written on the tape by the head, in the process erasing the
element just read. Implicitly the output set is the same as the input set for
Turing machines. There also is a head motion function $m(s,a)$ that tells the
Turing machine's head to move one position to the left or to the right.

The tape gives the Turing machine space for unlimited temporary storage—this was a major shortcoming of finite-state machines. Turing machines
can be constructed to mimic most of the operations of modern computers.
Like a modern computer, one can build a Turing machine whose input is a
program followed by input data. In particular, one can build a *universal Turing
machine T** that takes as input a program with the functions f, θ, m that defines
another Turing machine T followed by some input data. Then T^* will process
the data exactly as T would. See Hopcroft and Ullman [4] for more about
Turing machines and other automata with memory.

EXERCISES

SUMMARY OF EXERCISES   The first three exercises involve tracing the
action of machines with given inputs. Exercise 4 presents a simple but important graph model of a finite-state machine. Exercises 5–14 involve building
machines to perform various functions, most commonly recognizing a regular
language. Exercises 15 and 16 involve the conversion of a regular grammar to
a simple regular grammar. Exercises 17 and 18 involve the other half of the
theorem.

1.   Trace the successive internal states and outputs when the machine in
     Example 1 processes the following input:
     (a)   1100110   (b)   1010101

2.   Trace the successive internal states and outputs when the machine in
     Example 2 processes the following input:
     (a)   10101   (b)   1010011

3.   Trace the successive internal states and outputs when the machine in
     Example 3 computes the lexicographically next 3-subset in $\{1,2,3,4,5,6\}$
     following:
     (a)   135   (b)   136   (c)   356   (d)   456

4.   A finite-state machine can be represented with a directed multi-graph in
     which each internal state is represented by a vertex. If $f(s,a) = s'$ and
     $\theta(s,a) = o$, then there is a directed edge from s to s' labeled a,o. Draw the
     graph associated with the following finite-state machines:

    (a)   Machine in Example 1
    (b)   Machine in Example 2
    (c)   Machine in Example 3

5.   Build a finite-state machine that reads in a sequence of +1's and −1's and prints out the sum of the sequence thus far (assume that the sum is always between −3 and +3 inclusive).

6.   Build a finite-state machine that counts the number of occurrences of the subsequence 012 (up to a maximum of three occurrences) in an input ternary sequence.

7.   Build a finite-state machine that reads in a sequence of digits between 0 and 5 (inconclusive) and prints out the sum mod 6 of the numbers read thus far.

8.   Build a finite-state machine that reads in an integer in binary representation and prints out the next larger integer in binary representation. Assume the binary representation is read in reverse order (lowest order digit first).

9.   Build a finite-state machine that reads in an integer n in binary representation and prints out $n + 3$ in binary representation. Assume that the binary representation is read in reverse order.

10.   Build a finite-state machine to compute the lexicographically next outcome (in the manner of Example 3) in the collection.
    (a)   2-subsets of $\{1,2,3,4,5\}$.
    (b)   Permutations of 1, 2, 3.
    (c)   3-subcollections of $\{1,2,3\}$ with repetition allowed.
    (d)   5-digit binary sequences with exactly two 1's (lexicographically next here means next larger binary number).

11.   Build a finite-state machine to recognize the language of the regular grammar G_1 (appearing just after the proof of the lemma).

12.   Build a finite-state machine to recognize the languages of following regular grammars in the preceding section.
    (a) G_2 (b) G_n (c) G_3'

13.   Build a finite-state to recognize the language of the regular grammars requested in the following exercises from the preceding section.
    (a)   Exercise 12   (b)   Exercise 13(a)   (c)   Exercise 14
    (d)   Exercise 15   (e)   Exercise 16   (f)   Exercise 17

14.   (a)   Show that for the purposes of recognizing a language, it is not necessary for a finite-state machine to print output. Instead recognition can be indicated by having the machine's internal states divided into two classes, recognizing and non-recognizing, and looking at the state of the machine after the input sequence is read.
    (b)   Make the conversion indicated in part a on the machine in Example 2.

    (c)   Make the conversion indicated in part a on the machine built in (i) Exercise 11, (ii) Exercise 12(a), and (iii) Exercise 12(b).

15. Convert the following regular grammar into a simple regular grammar: $s \rightarrow 0s$, $s \rightarrow 0t$, $t \rightarrow 2v$, $t \rightarrow 2s$, $v \rightarrow 1s$, $v \rightarrow 2$.

16. Give a rigorous proof that the conversion given in this section from an arbitrary regular grammar to a simple regular grammar is correct.

17. Construct a regular grammar whose language is the collection of input sequences recognized by the following finite-state machines:
    (a)   Machine in Example 1 (with 2 as the recognition symbol).
    (b)   Machine to be built in Exercise 7 (with 3 as the recognition symbol).

18. Give a proof that the collection L of input sequences recognized by a given finite-state machine M (with a specified recognition symbol) is the language of some regular grammar G. You need to construct the desired regular grammar by reversing the construction in the proof of the Lemma. That is, internal states of M become the nonterminals of G, and if $f(t,a) = t'$ in M then $t \rightarrow at'$ is a production of G. A problem still to be solved is when to construct terminating productions such as $t \rightarrow a$.

10.3 LOGICAL PROPOSITIONS AND COMPUTATIONAL COMPLEXITY

One of the most important uses of counting methods in computer science and operations research occurs in the analysis of algorithms. To measure the efficiency of an algorithm one must determine how many steps (e.g., arithmetic operations, binary comparisons, or procedure calls) are required for an algorithm to solve a problem of a given size. Except for sorting, this text has stayed away from analysis of algorithms because good, fast algorithms usually require special data structures. However, in this section, we shall build some combinatorial and logical models that give important information about the computational effort required to solve many of the hard problems in graph theory.

    We were able to analyze sorting (in Section 3.5) because binary trees provided us with a simple model showing that any algorithm for sorting a random list of n items required at least $O(n \log n)$ binary comparisons. It was also a straightforward matter to show that certain sorting procedures (such as merge sort and heap sort) needed exactly $O(n \log n)$ comparisons. In terms of the general order of magnitude, we thus showed that those sorting algorithms were the fastest possible.

It is very rare that one can prove that no faster algorithm to solve a problem is possible; sorting is one of those rare exceptions. Usually, the best one can do is show that one algorithm is faster than any other *known* algorithm for solving a given problem. In this section we will study problems for which no reasonably good algorithms are known; the algorithms for these problems are so poor that it is hardly worth comparing them. This family of hard problems includes existence of a Hamilton circuit, minimal graph coloring, the Traveling Salesperson Problem, and existence of an edge cover of a given size. No approach is known for solving these problems that is significantly faster in the worst case that an exhaustive case-by-case examination of all potential solutions. In the n-city Traveling Salesperson Problem this means checking all $n!$ Hamilton circuits for the least costly one. In determining whether an n-vertex graph can be 3-colored, 3^n possible assignments of color to vertices must be tried. The computational effort of such searches grows exponentially, faster than any polynomial in n, the problem size.

Although little progress has been made on proving that no algorithm requiring polynomial effort can exist for solving any of the above problems, the theory of NP-completeness unifies the level of computational difficulty of these problems. Informally speaking, all NP-complete problems are equally hard to solve. That is, finding a polynomial-time algorithm for one NP-complete problem would yield polynomial-time algorithms for all NP-complete problems. This is because there exists a way to model any one NP-complete problem in terms of any other NP-complete problem. A formal definition of NP-completeness is given at the end of this section.

Because so many famous difficult problems have been shown to be NP-complete, it is widely accepted that no polynomial algorithm exists for these problems. Instead, one should look for heuristic methods to find approximate solutions or algorithms for special subcases. Thus showing that a problem is NP-complete has great practical significance to analysts who deal frequently with the problem in various applied settings.

On the other hand, it is the modeling and problem reformulation that is of paramount interest to us here. Throughout this text, we have been modeling counting problems in various ways, as coefficients of generating functions, with recurrence relations, or in terms of other counting problems. But these models were specially suited for the counting problems. It is quite another matter to model the Traveling Salesperson Problem in terms of a minimal coloring problem.

As a preliminary step, we recast all our problems as Yes or No decision problems. Is there a 3-coloring of this graph, is there a 4-coloring of this graph, and so on, rather than what is the minimal number of colors needed to color this graph.

To assist us in problem reformulation, we use a standard intermediate model. This model is determining whether a logical proposition is true or false. We want to reformulate a decision problem D as a question about the truth of a logical proposition. Later we shall reverse the process and model the truth of a logical proposition in terms of a problem like D.

The form of logical proposition we want is a conjunction of simple disjunctive terms such as $(x_1 \vee \bar{x}_2) \wedge (x_1 \vee \bar{x}_3 \vee x_4) \wedge (x_2 \vee \bar{x}_4)$. Any proposition using the operations $\wedge$, $\vee$, $-$ can be recast in this form using the laws of propositional calculus. See Appendix A1 for a review of basic facts about logical propositions. The x_i's each have two possible values, true or false. The question we want answered is, is there some choice of values for the x_i's that **satisfies** a given proposition—that is, makes it true. The only known procedure for testing satisfiability is building a truth table enumerating all 2^n possible collections of values for the n logical variables and seeing if any choice makes the proposition true.

Recall that for a conjunction of many terms to be true, every individual term must be true. So our intermediate problem can be restated as seeking to satisfy simultaneously a collection of simple disjunctive propositions.

Example 1

Reformulate the problem of deciding whether an n-vertex graph G can be 3-colored as a satisfiability problem (for a collection of simple disjunctive propositions).

The choices we have in graph coloring are which color the ith vertex should be, is it color 1 or color 2 or color 3. We define three logical variables for each of these possibilities: $x_{i,1}$, $x_{i,2}$, $x_{i,3}$ with $x_{i,j} = $ True if the ith vertex has color j, and $=$ False otherwise.

First we develop disjunctive propositions to represent the constraint that each vertex has exactly one color; that is, exactly one of $x_{i,1}$, $x_{i,2}$, $x_{i,3}$ is true. The following compound propositions express this constraint.

$$(x_{i,1} \wedge \bar{x}_{i,2} \wedge \bar{x}_{i,3}) \vee (\bar{x}_{i,1} \wedge x_{i,2} \wedge \bar{x}_{i,3}) \vee (\bar{x}_{i,1} \wedge \bar{x}_{i,2} \wedge x_{i,3}),$$
$$i = 1, 2, \ldots, n$$

By repeatedly applying the distributive law for disjunction and eliminating redundant terms, these compound propositions reduce to the following simple disjunctive propositions:

$$x_{i,1} \vee x_{i,2} \vee x_{i,3}, \quad \bar{x}_{i,1} \vee \bar{x}_{i,2}, \quad \bar{x}_{i,1} \vee \bar{x}_{i,3}, \quad \bar{x}_{i,2} \vee \bar{x}_{i,3},$$
$$i = 1, 2, \ldots, n \tag{1}$$

Next we need to express the coloring constraint that for each edge (v_i, v_j) of G, v_i, and v_j must not have the same color; that is, at least one of v_i and v_j does not have color k:

$$\bar{x}_{i,k} \vee \bar{x}_{j,k}, \quad k = 1, 2, 3, \quad i \neq j \quad i, j = 1, 2, \ldots, n \tag{2}$$

Satisfying the collection of simple disjunctive propositions in (1) and (2) is the desired logical reformulation of the decision problem, can G be 3-colored.

■

Example 2

Reformulate the problem of deciding whether an n-vertex graph has a Hamilton path, starting at a given vertex v_1, as a satisfiability problem.

The choices we shall make are which vertex comes next, that is, which vertex is second on the Hamilton path, which vertex is third, and so on. Let $x_{i,k}$ be true if vertex v_i is the kth vertex on the Hamilton path, $i, k = 2, 3, \ldots,$ n (recall that v_1 is given to be first). First we express the constraint that each vertex must be visited:

$$x_{i,1} \vee x_{i,2} \vee \cdots \vee x_{i,n} \qquad i = 2, 3, \ldots, n \qquad (3)$$

Next two vertices cannot both be the kth vertex.

$$\bar{x}_{i,k} \vee \bar{x}_{j,k} \text{ for all pairs } i, j \text{ and all } k > 1 \qquad (4)$$

Finally, we build the Hamilton path following sequence of edges. We must start at v_1; that is, $x_{1,1}$ is true. In general, if v_i is the kth vertex on the path, then a vertex in $N(v_i)$ must be the $(k + 1)$st vertex [$N(v)$ is the set of neighbors of v]. Recall that the proposition "if p then q" is equivalent to the simple proposition $\bar{p} \vee q$. Thus we want propositions of the form

$$\bar{x}_{i,k} \vee (x_{j_1, k+1} \vee x_{j_2, k+1} \vee \cdots \vee x_{j_m, k+1}), \qquad j_h \in N(v_i),$$
$$i, k = 1, 2, \ldots, n - 1 \qquad (5)$$

Satisfying the collection of simple disjunctive propositions (3), (4), (5), plus $x_{1,1}$ is equivalent to the existence of a Hamilton path starting at v_1. ∎

Building the propositional models of these graphs problems clearly requires some ingenuity. The following example shows that sometimes straightforward brute-force models will work.

Example 3

Reformulate the problem of finding a complete subgraph K_q of size q in an n-vertex graph G as a satisfiability problem.

Let x_i be true if v_i is in the K_q. We will give a large proposition that is not the conjunction of simple disjunctions, but can be converted into that form using the laws of propositional calculus. First, if (v_i, v_j) is *not* an edge of G, we require that not both vertices are in the K_q.

$$\bar{x}_i \vee \bar{x}_j \qquad \text{all } i, j, \text{ where } (v_i, v_j) \text{ is not an edge of } G \qquad (6)$$

Next we want q x_i's corresponding to some q-subset of vertices in G to be true, without violating (6) (not violating (6) means that the q vertices form a K_q). To avoid any confusion, we want the other $n - q$ x_i's to be false. We must try all $C(n,q)$ possible q-subsets of G. So we write a huge disjunction ranging over all possible q-subsets of G. For example, if $n = 4$ and $q = 3$, we would want

$$(x_1 \wedge x_2 \wedge x_3 \wedge \bar{x}_4) \vee (x_1 \wedge x_2 \wedge \bar{x}_3 \wedge x_4) \vee (x_1 \wedge \bar{x}_2 \wedge x_3 \wedge x_4) \vee$$
$$(\bar{x}_1 \wedge x_2 \wedge x_3 \wedge x_4) \tag{7}$$

Then for $n = 4$, $q = 3$, (6) and (7) would be the propositions to be satisfied. For large n and q, (7) becomes enormous. ∎

Note that all of the preceding propositional models can be constructed using a number of variables and a number of steps that are polynomial functions of n (assuming that q is fixed in Example 3).

Thus far we have only considered the modeling problem of reformulating a graph question in terms of satisfiability of a logical proposition. The reverse modeling problem of reformulating the satisfiability of an arbitrary proposition in terms of the existence of a k-coloring of some graph or other graph questions is much harder. However, this reverse modeling is the critical step in NP-completeness. If satisfiability questions can be reformulated (with polynomial effort) as the existence of, say, a clique of some size in some graph, then k-colorability and existence of a Hamilton path can also be reformulated as a clique problem using the results in Examples 1 and 2, and similarly for any other graph question that can be modeled as a satisfiability problem.

There is a convenient fact of the propositional calculus that we shall use. For any collection C of simple disjunctions, there is a logically equivalent collection C' of simple disjunctions each of which contains just three logical variables.

Example 4

Show that any satisfiability problem (of a collection of simple disjunctions) can be modeled by the existence of a 3-coloring of some graph.

We assume, as just mentioned, that the disjunctions to be satisfied each contain three variables. Let $x_1, \ldots , x_n$ be the variables in the disjunctions. For each x_i we create two vertices v_i and $\bar{v}_i$ joined together by an edge $(v_i, \bar{v}_i)$. We next create special vertices s, t: s is adjacent to just t, while t is adjacent to all the v_i and $\bar{v}_i$. Thus far we have build the graph H shown in Figure 10.1. Observe that in a 3-coloring of H in which t is color 2, exactly one of the pair v_i, $\bar{v}_i$ can have the color 1. Let s have color 1. If v_i has color 1, this will model x_i being true, and $\bar{v}_i$ being color 1 will model x_i false (i.e., $\bar{x}_i$ true); if either vertex is not color 1, then it must be color 3.

For each disjunction D_j, we create a 10-vertex subgraph G_j connecting the three vertices in H corresponding to the three variables in D_j: v_i for x_i and $\bar{v}_i$ for $\bar{x}_i$. D_j is shown in Figure 10.2, where we suppose $D_j = x_1 \vee \bar{x}_2 \vee x_4$. We call the v and $\bar{v}$ vertices in G_j the *input* vertices of G_j.

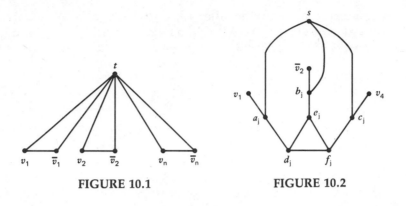

FIGURE 10.1   FIGURE 10.2

We claim that the desired graph G is the union of the vertices and edges in H and the G_j's. We noted above that, assuming t gets color 2 and s color 1, then the v_i's and $\bar{v}_i$'s must be colors 1 or 3. If all the input vertices in a G_j have color 3, then since s is color 1, a_j, b_j, c_j must all be color 2. This leaves only colors 1 and 3 available for coloring the three vertices in the triangle (d_j, e_j, f_j) — an impossibility. Hence, G is not 3-colorable if all input variables of some G_j are color 3. On the other hand, if one or more of the input vertices of G_j is color 1 (this corresponds to disjunction D_j beings satisfied), then it is easy to check that G is 3-colorable.

Thus a choice of colors for the input variables v_i and $\bar{v}_i$ makes G 3-colorable if and only if the corresponding truth values for the logical variables x_i satisfy all the disjunctions D_j. ∎

Example 5

Show that any satisfiability problem can be modeled by the question of whether there exists an edge cover of some specified size in a graph.

A set V' of vertices in an undirected graph G is an edge cover if every edge in G is incident to at least one vertex of V'. Edge covers were introduced in Section 4.4. We assume the disjunctions each contain exactly three logical variables. As in the previous example, for each logical variable x_i, we create two vertices v_i and $\bar{v}_i$ joined by an edge, and for each 3-variable disjunction D_j, we build a subgraph G_j, shown in Figure 10.3, with input vertices corresponding to the logical variables in D_j (in Figure 10.3, we again use the example of $D_j = x_1 \lor \bar{x}_2 \lor x_4$). We shall call a_j, b_j, and c_j the *triangular* vertices of G_j. Our model will require that v_i is in the edge cover if and only if x_i is true and $\bar{v}_i$ is in the edge cover if $\bar{x}_i$ is true. Note that to cover the edge $(v_i, \bar{v}_i)$, at least one of these two vertices must be in any edge cover.

The desired graph G consists of the subgraphs G_j plus the edges $(v_i, \bar{v}_i)$. The size of the desired edge cover will be $n + 2m$, where n is the number of logical variables and m the number of disjunctions. For the pair of disjunctions $(x_1 \lor$

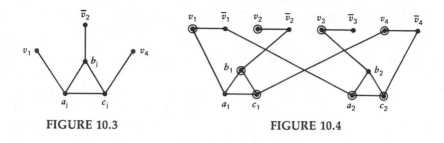

FIGURE 10.3   FIGURE 10.4

$\bar{x}_2 \vee x_4$), ($\bar{x}_1 \vee x_3 \vee \bar{x}_4$), we get the graph shown in Figure 10.4; the circled vertices form an edge cover.

Observe that to cover the three edges of the triangle (a_j, b_j, c_j) in G_j, at least two of the triangular vertices in G_j must be in any edge cover of G (any two will do). For all the G_j, we then require at least $2m$ vertices. To cover the edges ($v_i, \bar{v}_i$) requires n vertices. In total, we require at least $n + 2m$ vertices to cover the edges of G.

If an edge cover V' has exactly $n + 2m$ vertices, then exactly two of each triple a_j, b_j, c_j and exactly one of each pair v_i, $\bar{v}_i$ can be in V'. The problem is to be sure that V' covers the three edges in each G_j from the input vertices to the triangular vertices. Call these edges the *input edges* of G_j. Observe that for each G_j, one (or more) of the input vertices of G_j must be in V', since the two triangular vertices of G_j in V' can only cover two of the three input edges.

We conclude that G has an edge cover V' of size $n + 2m$ if and only if V' contains at least one of the input vertices in each G_j. This latter condition is equivalent to choices of x_i true or $\bar{x}_i$ true that satisfy all the disjunctions D_j, via the conversion mentioned above: x_i true if $v_i \in V'$ and $\bar{x}_i$ true if $v_i \in V'$. ∎

Example 6

Show that any satisfiability problem can be modeled by the question of whether there exists a complete subgraph K_q in some graph.

We will simply show that this problem is equivalent to the existence of an edge cover of a particular size in a related graph. Then modeling satisfiability with the existence of a K_q will follow from the result in Example 5. First observe that a set of vertices V' in a graph $G = (V, E)$ form a K_q if and only if in the complement $\bar{G}$, V' is an independent set (an independent set is a set of mutually nonadjacent vertices). Next we claim that V' is an independent set in a graph if and only if $V-V'$ is an edge cover: if V' is an independent set, no edge joins a pair of vertices in V' and so every edge is incident to at least one vertex in $V-V'$, and conversely.

Thus the vertices of V' form a K_q in the n-vertex graph G if and only if $V-V'$ is an edge cover (of size $n-q$) in the complement $\bar{G}$. Or in complementary form, V' is an edge cover of G if and only if $V-V'$ is a clique in G. For example, the graph G in Example 5 had a total of $2n + 3m$ vertices and required the existence of an edge cover of size $n + 2m$. This then is equivalent to the existence of a K_{n+m} in $\bar{G}$. ∎

The number of vertices and edges in the graphs in Example 4, 5, and 6 are linear functions of n and m, and so a polynomial algorithm for determining 3-colorability, existence of an edge cover, or existence of a K_q would mean that any satisfiability problem could be solved in polynomial time. This in turn means that any problem modeled (with polynomial effort) as a satisfiability problem would have a polynomial algorithm.

These graph models for satisfiability are very difficult to create, more like research problems than text exercises. We presented some examples of these models to illustrate the complexity of advanced graph modeling. Of course, the payoff from such a model, if found, is tremendous since it tells us that the graph problem cannot have an efficient solution ("cannot" assumes that NP-complete problems have no polynomial algorithms).

We close this section with a formal definition of NP-completeness. A decision problem will be in NP if a Turing machine can verify that a solution to it is valid in polynomial time, if a solution exists. Recall from the end of the previous section that a Turing machine is essentially a finite-state machine with a memory; Turing machines can generally do anything a modern digital computer can do. More precisely, a decision problem is in NP (Nondeteministic Polynomial) if a nondeterministic Turing machine, abbreviated NDTM, can be designed that recognizes in polynomial effort any instance of the decision problem with a positive answer. A NDTM can read in as input the statement of a problem, then correctly guess of the solution to problem (if a solution exists), and finally verify that the guess is indeed a valid solution. On verifying a solution, the NDTM prints out a "recognition" symbol and halts, indicating that it accepted this input string. If no solution exists, the NDTM will not have stopped after some predetermined length of time or will have halted with a nonrecognition symbol, and this fact will tell us that no solution exists.

A decision problem is **NP-complete** if any NP problem can be reformulated in polynomial effort as such a decision problem. The crucial result that got NP-completeness started was proved by S. Cook [1] in 1971.

Theorem (Cook)

The satisfiability problem is NP-complete.

With Cook's theorem, the results obtained in Examples 1, 2, and 3 are immediate. Graph coloring, Hamilton paths, and complete subgraphs can all be verified in polynomial time and, hence, by definition are in NP and, by Cook's theorem, can thus be modeled as satisfiability problems.

The proof of Cook's theorem consists of showing that the problem of recognizing any NP language (collection of input strings) with a NDTM can be modeled as a satisfiability problem. See Garey and Johnson [2] for the details of the proof and for a fuller presentation of the theory of NP-completeness.

R. Karp led the effort in the 1970s to make NP-completeness a practical tool by constructing graph models of satisfiability (like Examples 4–6 above). Today thousands of problems have been shown to be NP-complete. See Garey and Johnson [2] for a survey. We note in closing that there is one important problem—whether two graphs are isomorphic—that appears to require exponential effort but which has not yet been shown to be NP-complete.

EXERCISES

SUMMARY OF EXERCISES   The first three exercises involve concrete instances of the logical models in the first three examples of this section. The next three exercises involve concrete instances of the graph models in Examples 4–6. Exercises 7–18 involve increasingly difficult logical models of graph problems. Exercises 19–21 require one to show graph problems are NP-complete by modeling other NP-complete problems in terms of these problems—these exercises are not too hard. Exercises 22 and 23 require showing that graph problems are NP-complete by using them to model satisfiability—these problems are quite difficult.

1.   Write a collection of simple disjunctions whose satisfiability is equivalent to the existence of a 3-coloring for the following graph and find one choice of values for the logical variables that satisfy the collection.

2.   Write a collection of simple disjunctions whose satisfiability is equivalent to the existence of a Hamilton path for the graph in Exercise 1 and find one choice of values for the logical variables that satisfy the collection.

3.   Write a collection of propositions whose satisfiability is equivalent to the existence of a K_2 in the graph in Exercise 1 and find one choice of values for the logical variables that satisfy the collection.

4.   Model the satisfiability of the following collections of propositions in terms of the existence of a 3-coloring in some graph, as described in Example 4. Find a 3-coloring, if one exists, in each graph and use the 3-coloring to obtain an assignment of logical values that satisfy the propositions.
     (a) $x_1 \vee x_2 \vee x_3, x_1 \vee \bar{x}_2 \vee \bar{x}_4$
     (b) $x_1 \vee x_2 \vee x_3, x_1 \vee \bar{x}_3 \vee x_4, \bar{x}_2 \vee \bar{x}_3 \vee x_4$
     (c) $x_1 \vee x_2 \vee x_3, x_1 \vee x_2 \vee \bar{x}_4, \bar{x}_1 \vee x_3 \vee x_4, \bar{x}_1 \vee \bar{x}_2 \vee \bar{x}_3$

5. Model the satisfiability of the collections of propositions in Exercise 4 in terms of the existence of an edge cover as described in Example 5. Find the desired edge cover, if one exists, in each graph and use it to obtain an assignment of logical values that satisfy the propositions.

6. Model the satisfiability of the collections of propositions in Exercise 4 in terms of the existence of a K_q in some graph as described in Example 6. Find the desired K_q, if one exists, in each graph and use it to obtain an assignment of logical values that satisfy the propositions.

7. Reformulate the problem of deciding whether an n-vertex graph can be 2-colored as a satisfiability problem for a collection of simple disjunctive propositions.

8. Reformulate the problem of deciding whether an n-vertex graph has a Hamilton path (the starting vertex is not specified) as a satisfiability problem. (*Hint*: Use the result of Example 2 as a building block.)

9. Reformulate the problem of deciding whether an n-vertex graph has a Hamilton circuit as a satisfiability problem. (*Hint*: Modify the result of Example 2.)

10. (a) Another approach to modeling the existence of a Hamilton circuit would not worry about the order of vertices on the circuit but would simply seek a set C of n edges such that each vertex is adjacent to exactly two edges in C. Write a collection of simple disjunctive propositions to model the existence of such a set C.
    (b) What other constraint is needed to assure that C is a Hamilton circuit?

11. Reformate the problem of deciding whether an n-vertex graph contains the following subgraph:

as a satisfiability problem.

12. (a) Reformulate the problem of deciding whether an n-vertex graph has a path from vertex x to vertex y as a satisfiability problem.
    (b) Reformulate the problem of deciding whether an n-vertex graph is connected as a satisfiability problem.
    (c) Reformulate the problem of deciding whether an n-vertex graph has an Euler circuit as a satisfiability problem.

13. Reformulate the problem of deciding whether an n-vertex graph has a path of length $<k$ from vertex x to vertex y as a satisfiability problem.

14. Reformulate the problem of deciding whether an n-vertex graph has an edge cover of size k as a satisfiability problem.

15. Reformulate the problem of deciding whether an n-city Traveling Salesperson Problem has a tour of length r as a satisfiability problem. Let $d_{i,j}$ be the length of the edge from city i to city j. For simplicity, the tour can be a Hamilton path, not a Hamilton circuit. (*Hint*: Combine Example 2 with the Shortest Path Algorithm in Section 4.1; that is, let the second subscript in Example 2 represent the total distance traveled thus far on the tour).

16. Reformulate the problem of deciding whether a graph contains a circuit as a satisfiability problem.

17. (a) Reformulate the problem of deciding whether a directed graph is a rooted tree as a satisfiability problem. (*Hint*: Use Exercise 12(b).)
    (b) Reformulate the problem of deciding whether an undirected graph is a tree as a satisfiability problem.

18. Reformulate the problem of deciding whether a graph is an interval graph as a satisfiability problem (see Example 6 in Section 1.1). (*Hint*: The decisions to be made are whether the left (right) endpoint of the interval for the ith vertex is to the left or right of the left (right) endpoint of the interval for the jth vertex.)

19. Show that the question of whether a graph G contains a subgraph isomorphic to a graph H is NP-complete. (*Hint*: Use Example 6.)

20. Show that the question of whether a graph G contains a vertex cover is NP-complete. A *vertex cover* is a subset S of vertices of a given size such that every other vertex of G is adjacent to a vertex in S. (*Hint*: Show that any edge cover problem (Example 5) can be modeled as a vertex cover problem in a related graph.)

21. Let a collection C of subsets of a set S be given. A *hitting set* for C is a subset S' of S that contains at least one element of each subset in C. Show that the question of whether S has a hitting set for C of size k is NP-complete by modeling the edge cover problem as a hitting set problem.

22. Show that the following question is NP-complete by using it to model satisfiability. Given a collection C of subsets of a (finite) set S, is there a partition of S into two subsets S' and S'' such that no subset in C is entirely contained in either S' or S''.

23. Show that the following question is NP-complete by using it to model satisfiability. Given a directed graph G, does there exist a function f on the vertices of G such that f assumes nonnegative integer values (including 0) and for each x in G, $f(x)$ is the minimum nonnegative integer different from the set of values $\{f(y): (x,y) \text{ is an edge of } G\}$. Such a function is called a *Grundy function*; it is introduced in Section 11.2.

10.4 SUMMARY AND REFERENCES

This chapter introduced several topics from the foundations of computer science. The emphasis was on combinatorial analysis involving these topics rather than their role in computer science. Of course, it is hoped that readers learned some computer science theory along the way.

Formal languages and grammars were used to generate all outcomes in various combinatorial collections. Grammars formalize the process of generating outcomes that was implicitly used in Chapter 5 in counting. They also suggest recurrence relations for counting problems. In computer science, grammars systematize the construction of programming languages.

Finite-state machines are one of the simplest models of a computer imaginable. But they are powerful enough to recognize any regular language. This relationship between regular languages and finite-state machines is one of the fundamental results in theoretical computer science and has wide-reaching practical consequences.

NP-completeness involves a powerful marriage of combinatorial modeling and the theory of computation. Its results have great practical value to researchers trying to develop efficient algorithms for a wide spectrum of problems in computer science and operations research.

The topics in this section have only been studied extensively in the last two or three decades, although Turing machines were proposed by A. Turing in 1936 [6].

1. S. Cook, "The complexity of theorem-proving procedures," *Proceedings of 3rd ACM Symposium on Theory of Computing*, 1971, 151–158.

2. M. Garey and D. Johnson, *Computers and Intractability: A Guide to the Theory of NP-Completeness*, Freeman, San Francisco, 1979.

3. A. Gill, *Applied Algebra for the Computer Sciences*, Prentice-Hall, Englewood Cliffs N.J., 1976.

4. J. Hopcroft and J. Ullman, *Formal Languages and Their Relation to Automata*, Addison-Wesley, Reading, Mass., 1969.

5. R. Karp, "Reducibility among combinatorial problems," in *Complexity of Computer Computations*, Plenum Press, New York, 1972, 85–103.

6. A. Turing, "On computable numbers, with an application to the Entscheidungsproblem," *Procedings London Math. Soc.*, Series 2, Vol. 42, 1936, 230–265.

Chapter Eleven
Games with Graphs

11.1 GRAPH MODEL FOR INSTANT INSANITY

In this chapter we apply graphs to the analysis of some games. While clearly not an important use of graphs, this application to games is very interesting. Indeed, this chapter is meant as a sort of mathematical "dessert" following the more serious topics in preceding chapters. It is essential to study how graph models can be used to solve a variety of real-world problems, but there is a more personal reward in learning how graphs can permit one to solve a well-known difficult puzzle or to win at certain games. The first section is devoted to an ingenious graph model for the puzzle known as Instant Insanity. In the succeeding two sections, we develop the theory of progressively finite games and apply it to Nim-type games.

The Instant Insanity puzzle consists of four cubes whose faces are colored with one of the four colors: red(R), white (W), blue(B), and green (G). In Figure 11.1, we show four such cubes. The six faces on the ith cube are denoted: f_i—front face, l_i—left face, b_i—back face, r_i—right face, t_i—top face, and u_i—under face. The object of this puzzle is to place the four cubes in a pile (cube 1 on top of cube 2 on top of . . . etc.) so that each side of the pile has one face of each color. For example, if cube 1 had all red faces, cube 2 all blue faces, cube 3 all white faces, and cube 4 all green faces, then any pile made of the 4 cubes would automatically be a solution, that is, each side of the pile will have all 4 colors.

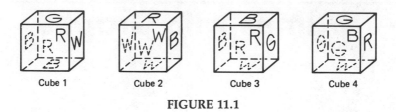

Cube 1 Cube 2 Cube 3 Cube 4

FIGURE 11.1

Determining how to arrange the 4 cubes in Figure 11.1 in such a pile is a very difficult task. In general, there are 24 symmetries of a cube, and, thus $24^4 = 331{,}776$ different piles that can be built. An enumeration tree search will be immense, although symmetries of the face colors and the constraint of no repeated colors on a side will eliminate many possibilities. A computer program to do this search is easily written but may require a lot of computer time (depending on the set of cubes). Fortunately, we can model this puzzle with a 4-vertex graph in such a fashion that the graph-theoretic restatement of the puzzle can be solved by inspection in a few minutes.

Before presenting the graph model, we need to discuss a simple decomposition principle for this puzzle. Arranging the cubes in a pile so that Left and Right sides of the pile have one face of each color is "independent" of arranging the Front and Back sides with one face of each color. By "independent," we mean that once cube 1 is arranged so that a given pair of (opposite) faces are on the Left and Right side of the pile, then any remaining pair of (opposite) faces on cube 1 can be on the Front and Back sides of the pile.

Let $l_i^*, r_i^*, f_i^*, b_i^*$ denote the colors of the respective faces of cube i on the four sides of the pile. Suppose that $l_1^* = t_1 = G$ and $r_1^* = u_1 = B$. Then by rotating cube 1 about the centers of these two faces, we can get $f_1^* = B$, $b_1^* = W$, or $f_1^* = R$, $b_1^* = R$, or $f_1^* = W$, $b_1^* = B$—all possible remaining choices for Front and Back sides. Since this same Left-Right and Front-Back "independence" holds for the other cubes, we see that the puzzle can be broken into two disjoint problems:

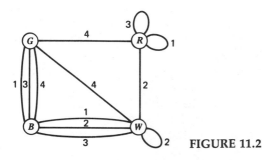

FIGURE 11.2

Decomposition Principle

(i)   Pick one pair of opposite faces on each cube for the Left and Right sides of the pile so that these two sides of the pile will have one face of each color; and

(ii) Pick a different pair of opposite faces on each cube for the Front and Back sides of the pile so that these two sides will have one face of each color.

Now we are ready to present our graph model (due F. de Carteblanche). Actually we use a multigraph. Make one vertex for each of the four colors. Draw an edge with label i between two vertices representing colors that are on a pair of opposite faces on cube i (the opposite pairs of faces are l_i, r_i and f_i, b_i and t_i, u_i). Figure 11.2 shows this multigraph for the cubes in Figure 11.1. For example, since $l_1 = B$, $r_1 = W$, we draw an edge labeled 1 between vertex B and vertex W; for $f_1 = R$, $b_1 = R$, we draw a self-loop labeled 1 at vertex R; and for $t_1 = G$, $u_1 = B$, we draw an edge labeled 1 between vertices G and B. The edges for the other cubes are drawn similarly.

We can now re-state the puzzle in graph-theoretic terms. By the decomposition principle, we can break the puzzle into a Left-Right part and a Front-Back part. We initially consider just the Left-Right part, that is, find one pair of opposite faces on each cube so that the Left and Right sides of the pile have one face of each color.

Let us simplify this Left-Right problem slightly by asking only for four pairs of opposite faces, one pair from each cube, such that among this total set of eight faces each color appears twice. Later we will show how to insure that each color appears once on the Left side and once on the Right side. Since a color corresponds to a vertex, a cube to an edge number, and a pair of opposite faces to an edge, this simplified Left-Right problem has the graph-theoretic re-statement: *find four edges, one with each number, such that the family of eight end vertices of these four edges contains each vertex twice.*

This condition on the end vertices of the four edges is equivalent to requiring that in the subgraph formed by these four edges, each vertex has degree 2 (a self-loop counts as degree 2 at its vertex). But a subgraph with all vertices of degree 2 is just a circuit or collection of disjoint circuits (a self-loop is a circuit of length 1). In a general n-vertex graph, a set of n edges forming disjoint simple circuits is called a **factor** of the graph. In our Instant Insanity model, let the term **labeled factor** denote a factor in which each edge number appears once.

The simplified Left-Right problem in our Instant Insanity graph model now reduces to: find a labeled factor. In Figure 11.3, we show three possible labeled factors for the multigraph in Figure 11.2.

We show how a labeled factor can be transformed into an arrangement of the cubes in which the Left and Right sides of the pile have one face of each

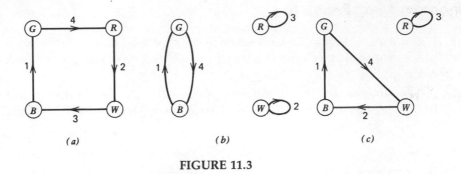

FIGURE 11.3

color. We direct the edges in each circuit in a consistent direction, say, in clockwise direction (see Figure 11.3). Consider the labeled factor in Figure 11.3a. As we go around this circuit, we arrange each cube with the color of the tail vertex of an edge as the color on the Left Side of the cube, and the color of the head vertex on the Right Side. Start with any edge on the circuit. Let us pick the edge labeled 1. This 1-edge in Figure 11.3a goes from B to G, and so, then we arrange cube 1 so that l_1^* (the left face of cube 1 in the pile) $= B$ and $r_1^* = G$. Continuing along the circuit from the 1-edge, we next encounter a 4-edge from G to R. Accordingly we arrange cube 4 so that $l_4^* = G$ and $r_4^* = R$. Next comes a 2-edge from R to W followed by a 3-edge from W to B. So we arrange cubes 2 and 3 with $l_2^* = R$, $r_2^* = W$ and $l_3^* = W$, $r_3^* = B$.

This process assures that each color appears once on each side, since each vertex (color) is at the head of one edge and at the tail of one edge. Figure 11.4 shows the Left and Right sides (only) of the pile made by the four cubes arranged as just described. If we had chosen the labeled factor in Figure 11.3b, we would use the clockwise traversal procedure for all three circuits, yielding $l_1^* = B$, $r_1^* = G$, $l_4^* = G$, $r_4^* = B$ for one circuit, and $l_3^* = r_3^* = R$ and $l_2^* = r_2^* = W$ for the two self-loops.

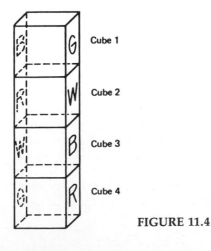

FIGURE 11.4

We are now ready to solve the Instant Insanity puzzle. Restating the decomposition principle in terms of labeled factors:

Graph-theoretic
Formulation of Instant Insanity

Find two edge-disjoint labeled factors, one for Left-Right sides and one for Front-Back sides; and then

Use the clockwise traversal procedure to determine the Left-Right and Front-Back arrangements of each cube.

It is not hard to find two disjoint labeled factors by inspection. The three labeled factors in Figure 11.3 all use the same 1-edge (between B and G). So no two of these three factors are disjoint. The easiest approach is to find one labeled factor, delete its edges, and look for a second labeled factor. If none is found, start with a different labeled factor. For example, suppose we use the labeled factor in Figure 11.3a as our first factor. After deleting its edges from the graph in Figure 11.2, we easily find a second labeled factor. Such a second labeled factor is shown in Figure 11.5 (the reader should be able to find another second factor).

Let us use the factor from Figure 11.3a to arrange the Left and Right sides, as shown in Figure 11.4. Next we rotate each cube about the centers of its Left and Right faces to arrange the Front and Back faces according to a clockwise traversal of the circuits in Figure 11.5. Starting with the 2-edge, we make $f_2^* = W$, $b_2^* = B$, $f_3^* = B$, $b_3^* = G$, $f_4^* = G$, $r_4^* = W$ and $f_1^* = b_1^* = R$. Figure 11.6 shows the resulting solution of the Instant Insanity puzzle.

Now go buy or borrow a set of Instant Insanity cubes and show your friends that you learned something really useful from this book!

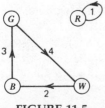

FIGURE 11.5

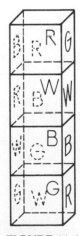

FIGURE 11.6

EXERCISES

1. Find all other labeled factors of the multigraph in Figure 11.2 besides the ones in Figure 11.3 and Figure 11.5. Give an argument in the process to show that there can be no other labeled factors.

2. Find all Instant Insanity solutions (a disjoint pair of labeled factors) to the game with associated graph (a).

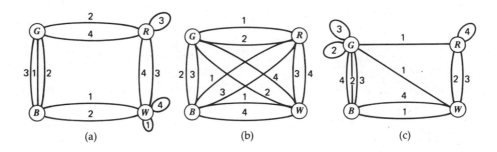

(a)　　　　　　(b)　　　　　　(c)

3. Find all Instant Insanity solutions to the game with the associated graph (b) shown above.

4. Show carefully that Instant Insanity graph (c) shown above does not possess a solution, that is, a pair of disjoint labeled factors.

5. (a) If a graph G has a Hamiltonian circuit, must it also have a factor? Prove true or give a counterexample.
   (b) Repeat part a for the case of G having an Euler circuit.

6. Find a factor in the following graphs, if possible:

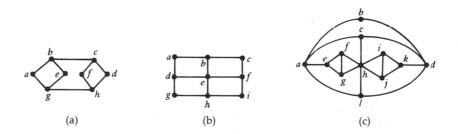

(a)　　　　　　(b)　　　　　　(c)

7. Show that the definition of a factor in a graph implies that each vertex is incident to exactly two edges (or one self-loop) in a factor.

8. A directed factor in an n-vertex directed graph is a set of n edges with one of these edges incident *from* and one incident *to* each vertex. Show how the problem of finding a directed factor in a directed graph can be formulated as a matching or network flow problem.

11.2 PROGRESSIVELY FINITE GAMES

In this section we develop the mathematical theory of progressively finite games. A game in which two players take turns making a move until one player wins (no ties are allowed) is called **progressively finite** if: (i) there is only a finite choice of moves possible at any particular position in the play of the game; and (ii) if the play of the game must end (with one player winning) after a finite number of moves.

Our objective in this section is winning: how to determine winning strategies for progressively finite games. We can model a progressively finite game by a directed graph with a vertex for each position that can occur in the play of the game and a directed edge for each possible move from one position to another.*

Observe that the graph of a progressively finite game cannot contain any directed circuits, since players could move around and around a circuit of positions forever (violating the constraint on finite play). Thus no positions can ever be repeated in the play of a game. Rather, the game moves inexorably towards some final position that is a win for one of the players. Games such as checkers and chess that permit ties and repetition of positions are not progressively finite. Most progressively finite games are "takeaway"-type games of the sort illustrated by the games in Examples 1 and 2.

Example 1

A set of 16 objects is placed on a table. Two players take turns removing 1, 2, 3, or 4 objects. The winner is the player who removes the last object. The graph of this game is shown in Figure 11.7 (all edges are directed from left to right). ■

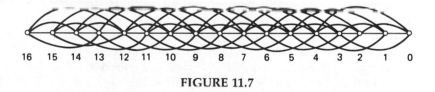

16 15 14 13 12 11 10 9 8 7 6 5 4 3 2 1 0

FIGURE 11.7

---

* There is one technical point here. This book's definition of a graph requires that it have a finite number of vertices, but the definition of progressively finite games does not explicitly require that there be a finite number of positions; it only requires a finite number of moves from any position and finite play. Fortunately, it can be proved that a progressively finite game has in total only a finite number of different possible positions. This result is known as Konig's Infinity Lemma.

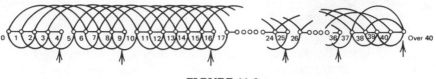

FIGURE 11.8

Example 2

Starting with an empty pile, two players add 1 penny or 2 pennies or 1 nickel to the pile until the value of the pile is the square of a positive integer (>1) or until the value exceeds 40. The player whose addition brings the value of the pile to one of these critical amounts takes all the money in the pile. The graph of this game is shown in Figure 11.8 (all edges are directed from left to right; arrows point to the game-winning positions). This game is an inverted form of a "takeaway" game in which 1, 2, or 5 objects are withdrawn from a pile until the pile is reduced to certain critical sizes. ■

The reader is encouraged to try playing these games with a friend. A winning strategy for the first player (who makes the first move) in the game in Example 1 can be found with a little thought. The game in Example 2 is harder. The details of this second game only serve to confuse systematic attempts to find a winning strategy. It is easier to develop a general theory of winning strategies in progressively finite games and then apply the theory to Example 2.

A **winning position** in a progressively finite game is a position at which play stops, and the player who moved to this position is declared the winner. Every progressively finite game must have at least one winning position or else play would go on without end. A **winning strategy** for a player is a rule that tells the player which move to make at each stage in a game so as to insure that the player will eventually win, that is, finally move to a winning position. Obviously only one player can have a winning strategy.

In the graph of a progressively finite game, vertices with 0 out-degree must represent winning positions, for if no edges leave a vertex, the game must stop at that vertex. We call such vertices **winning vertices**. The graph of the game in Example 1 has just one winning vertex, numbered 0, while the graph of the game in Example 2 has six winning vertices, numbered 4, 9, 16, 25, 36, and "over 40". Any vertex adjacent to a winning vertex is a "losing" vertex in the sense that if one player moves to a "losing" vertex, the other player can then move to a winning vertex (and win the game). The vertices numbered 1, 2, 3, and 4 are losing vertices in the game in Example 1. The vertices numbered 2, 3, 7, 8, 11, 14, 15, 20, 23, 24, 31, 34, 35, 37, 38, 39, and 40 are losing vertices in the game in Example 2.

Stepping one more move from the end of the game, we see that if all edges from a vertex x go to losing vertices, then x is a "pre-winning" vertex. Whenever player A moves to such a pre-winning vertex, then player B's next move must be to a losing vertex and now player A can move to a winning vertex. Vertex 5 in the game in Example 1 is a pre-winning vertex, since all edges from 5 go to losing vertices (1, 2, 3, and 4). Vertices 6 and 33 are the pre-winning vertices in the game in Example 2; all other vertices adjacent to losing vertices are also adjacent to nonlosing vertices.

The general theory of winning strategies in progressively finite games is based on a recursive extension of the preceding reasoning. We seek good vertices, such as winning and pre-winning vertices, that win or lead only to bad vertices. When our opponent is forced to move to a bad vertex, such as losing vertices, then we will always be able to move to another good vertex. A winning strategy for the first player will tell the player how to find a good vertex from any bad vertex. This pattern of play, with the first player moving to successive good vertices and the second player forced to move to bad vertices, will continue until finally the first player reaches a good vertex that is a winning vertex. If the game starts at a good vertex and the first player's initial move must be to a bad vertex, then roles are reversed and it is the second player who now has a winning strategy using the good vertices.

Example 1 (continued)

The winning strategy for the first player in this game is to move to a vertex whose number is a multiple of 5. These are the good vertices. Thus the first move from the starting vertex 16 is to vertex 15 (i.e., the first player removes one object). From 15, the second player must move to one of the bad vertices 11, 12, 13, or 14. From any of these bad vertices the first player now moves to vertex 10. Whatever the second player's next move, the first player will always be able to move to the pre-winning vertex 5, and one round later the first player will win. Note that if the game started with only 15 objects, then the roles would be reversed and the second player would be able to use this winning strategy. ∎

We formalize the concept of good vertices with the following definition. A **kernel** in a graph is a set of vertices such that:

(i) there is no edge joining any two vertices in the kernel; and
(ii) there is an edge from every nonkernel vertex to some kernel vertex.

The kernel of the graph in Example 1 is the set of vertices numbered 0, 5, 10, and 15.

Theorem 1

If the graph of a progressively finite game has a kernel K, then a winning strategy for the first player is to move to a kernel vertex on every turn.

However, if the starting vertex is in the kernel, then the second player can use this winning strategy.

Proof

First we show that all winning vertices must be in K. The reason is that vertices not in a kernel must have an edge directed to a vertex in the kernel. But winning vertices have 0 out-degree. Thus they must be in any kernel.

Next we show that moving to a kernel vertex on every turn is a winning strategy for the first player. By property (i) of a kernel, when the first player moves to a kernel vertex, the second player must then move to a nonkernel vertex. Then by property (ii), the first player can always move from the nonkernel vertex to a kernel vertex. The play proceeds in this fashion with the first player always moving to a kernel vertex and the second player always moving to a nonkernel vertex. Since the game is progressively finite, the play must eventually end. Since all winning vertices are in the kernel, the first player must win.

If the starting vertex is in the kernel, then the roles of the first and second players are reversed and the second player can always move to a kernel vertex for an eventual win. ∎

The preceding proof does not show explicitly how successively moving to kernel vertices leads to a winning vertex. The proof is existential. It only shows that by moving to kernel vertices the first player must eventually arrive at a winning vertex.

A more immediate problem is to prove that the graph of any progressively finite game has a kernel and to show how to find this kernel. Not all graphs have kernels. For example, the graph in Figure 11.9 has no kernel. To show this we argue: If there were a kernel, we could assume by the symmetry in this graph that vertex a is in the kernel. Then vertex e, which has an edge to a, cannot be in the kernel. Vertex d's only outward edge goes to nonkernel vertex e, and so d must be in the kernel. Similarly, c cannot be in the kernel and then b must be in the kernel. But now the edge (a,b) joins two kernel vertices—a contradiction.

Fortunately, graphs of progressively finite games do have kernels. To demonstrate the existence of kernels in such graphs, we need to organize the vertices in a graph into levels based on the "distance" of the vertices from a

FIGURE 11.9

winning vertex. We recursively define the level $l(x)$ of vertex x in a directed graph and the sets L_k of vertices at level $\leq k$ as follows. If $s(x) = \{y | x$ has an edge to $y\}$ is the set of successors of x, then

$$l(x) = 0 \Leftrightarrow s(x) \text{ is empty} \qquad \text{and} \qquad L_0 = \{x | l(x) = 0\}$$

$$l(x) = 1 \Leftrightarrow x \notin L_0 \text{ and } s(x) \subseteq L_0, \qquad \text{and} \qquad L_1 = L_0 \cup \{x | l(x) = 1\}$$

and in general,

$$l(x) = k \Leftrightarrow x \notin L_{k-1} \text{ and } s(x) \subseteq L_{k-1} \qquad \text{and} \qquad L_k = L_{k-1} \cup \{x | l(x) = k\}$$

Observe that $L_k - L_{k-1}$ is the set of vertices at level k.

It follows from this definition of level that every vertex at level 0 is a winning vertex and that every vertex at level 1 is a losing vertex. A vertex at level 2 is also a losing vertex if it is adjacent to a vertex at level 0; it is a pre-winning vertex only if all its successors are at level 1. Note that every vertex at level $k(k > 0)$ must be adjacent to a vertex at level $k - 1$ but cannot be adjacent to any other vertex at level k (or greater). Now we can prove the fundamental theorem for progressively finite games.

Theorem 2

Every progressively finite game has a unique winning strategy. That is, the graph of any progressively finite game has a unique kernel.

Proof

The proof is by induction on the levels, or more precisely, on the sets L_k. Let K_k be the set of kernel vertices in L_k. First consider the set L_0. As noted in the proof of Theorem 1, all winning vertices (vertices in L_0) must be in the kernel. Thus $K_0 = L_0$.

Next let us inductively assume that K_{n-1} is the unique, well-defined set of kernel vertices in L_{n-1}. We must show that we can find a unique set of level-n vertices that must be added to K_{n-1} to form the kernel K_n for L_n. By the way that level numbers were defined, $l(x) = n$ means that $s(x) \subseteq L_{n-1}$. If a level-n vertex x is adjacent to no kernel vertex of K_{n-1}, this x must be in K_n; conversely, if x is adjacent to a kernel vertex of K_{n-1}, x cannot be in the kernel. Hence $K_n = K_{n-1} \cup \{x | l(x) = n \text{ and } s(x) \cap K_{n-1} = \emptyset\}$ is the unique, well-defined set of kernel vertices in L_n. It follows by mathematical induction that the graph has a unique kernel. By Theorem 1, this kernel is the unique winning strategy. ∎

The proof of Theorem 2 tells us how to build a kernel. First put the winning vertices in the kernel. Then recursively add the vertices at successive levels not adjacent to the current set of kernel vertices. We implement this procedure for finding kernels using a labeling rule called a Grundy function. A **Grundy function** $g(x)$ for the vertices of a directed graph is defined as follows:

For each vertex x, $g(x)$ is the smallest nonnegative integer that has not been assigned to any of x's successors in $s(x)$.

We shall prove shortly that vertices with Grundy number 0 are exactly the set of kernel vertices. In the next section, Grundy numbers will be seen to play a fundamental role in more complex games.

In the graph of a progressively finite game, Grundy values are easily determined using a level-by-level approach. The vertices on level 0 have no successors and so their Grundy number will be 0 (the smallest nonnegative integer). Next we determine $g(x)$ for vertices x at level 1, then vertices at level 2, and so forth. In this way, all of a vertex x's successors are assigned Grundy numbers before it is time to determine the Grundy number of x, since x's successors are at lower levels. Actually we do not need to proceed in a totally level-by-level fashion. We can use any method that does not try to assign a vertex a Grundy number until all the vertex's successors have Grundy numbers.

No Grundy function can be defined for the graph in Figure 11.9. Each vertex x in Figure 11.9 has a successor whose Grundy number must be defined before $g(x)$ can be determined. Even if we try to invent simultaneously a $g(x)$ for all vertices in Figure 11.9 at once, no set of values will satisfy the definition of a Grundy function (details are left to the reader).

The Grundy numbers for the vertices in Figures 11.7 and 11.8 are given in the tables on p. 401 (winning vertices are indicated by boxes). Note that all the vertices in Figure 11.7 that are in the kernel,—namely, 0, 5, 10, 15, all have Grundy number 0. This is no accident.

Theorem 3

The graph of a progressively finite game has a unique Grundy function. Further, the vertices with Grundy number 0 are the vertices in the kernel.

Proof

The recursive level-by-level construction of a Grundy function for the graph of a progressively finite game gives each vertex a unique Grundy number, as described above. By the definition of a Grundy function, a vertex x with Grundy number 0 cannot have an edge to another vertex with Grundy num-

TABLE OF GRUNDY NUMBERS FOR FIGURE 11.7

Vertex x	16	15	14	13	12	11	10	9	8	7	6	5	4	3	2	1	0
$g(x)$	1	0	4	3	2	1	0	4	3	2	1	0	4	3	2	1	0

TABLE FOR GRUNDY NUMBERS FOR FIGURE 11.8

Vertex x	0	1	2	3	4	5	6	7	8	9	10	11	12	13	14	15
$g(x)$	0	3	1	2	0	1	0	2	1	0	3	1	0	3	1	2
	16	17	18	19	20	21	22	23	24	25	26	27	28	29	30	31
	0	1	0	2	1	0	3	2	1	0	2	0	1	2	0	1
	32	33	34	35	36	37	38	39	40	"over 40"						
	2	0	1	2	0	1	3	2	1	0						

ber 0 (or else x would have to have a different number). Similarly any vertex y with Grundy number $k > 0$ must have an edge to some vertex with Grundy number 0 (or else y's number would be 0). Thus the set of vertices with Grundy number 0 satisfy the two defining properties of a kernel. ■

Example 2 (continued)

By Theorem 3 and the above table of Grundy numbers for Figure 11.8, we see that the kernel is the set 0, 4, 6, 9, 12, 16, 18, 21, 25, 27, 30, 33, 36, "over 40". Since the starting vertex is in the kernel, the second player has the winning strategy in this game. A play of the game might proceed as follows (let player A be the first player and B be the second player): first A moves to 1, then B moves to kernel vertex 6 (adding a nickel to the pile), then A must move to 11 (a move to 7 or 8 lets B win at 9), then B moves to kernel vertex 12, then A must move to 17, then B moves to kernel vertex 18, then A moves to 20, then B to 21, then A to 22, then B to 27, then A to 32, then B to 33, then A to 35 (A senses defeat and gives up), and then B moves to the winning vertex 36 (and collects the 36 cents). ■

EXERCISES

1. Find a kernel in the following graphs or show why none can exist:

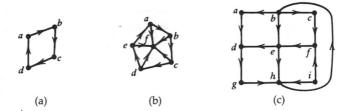

(a)   (b)   (c)

2. Suppose there are 25 sticks and two players take turns removing up to 5 sticks. The winner is the player who removes the last stick. Which player has a winning strategy? Describe this strategy.

3. Repeat Example 2 with 2, 3, or 7 cents added each time. Find the set of positions in the kernel.

4. Show that in Example 2 the second player can always win by the second move.

5. Suppose in Example 2 that the first player A knows he/she will lose and wants to minimize his/her loss. What is the smallest winning amount A can force B (the second player) to accept (such that if B tried to make the game no longer, then A could get to the kernel and win).

6. (a) Suppose we have a pile of 7 red sticks and 10 blue sticks. A player can remove any number of red sticks or any number of blue sticks

or equal numbers of red and blue sticks. The winner is the player who removes the last stick. Find the set of positions in the kernel.

(b) Repeat part a with a limit of removing at most 5 sticks of one color (or both colors) on a move.

7. Repeat Example 1 but now the player to remove the last object loses. Describe the winning strategy for this game.

8. Find the Grundy function for graphs or games in:
(a) Exercise 1(a)   (b) Exercise 1(b)   (c) Exercise 3

9. Show that there is no Grundy function for the graph in Figure 11.9.

10. If $W(S)$ is the set of vertices without an edge directed to any vertex in the set S, show that a set S is a kernel if and only if $S = W(S)$.

11. (a) Show that if $l(x) = k$ for a vertex x in the progressively finite graph G, then there is a path of length k starting at x in G.
(b) Show that if $g(x) = k$, there is a path of length k starting at x in G.

12. Show that for any vertex x in a progressively finite graph, $g(x) \leq l(x)$.

13. Find a directed graph possessing a kernel but no Grundy function.

14. Show that if a directed graph G and all subgraphs (obtained by deleting various vertices) have kernels, then G has a Grundy function.

15. Show that both the level numbers and the Grundy function in a progressively finite graph G constitute proper colorings of G.

16. Show that no matter how the edges are directed in a bipartite graph, it will always have a Grundy function.

17. Show that the graph of a progressively finite game can have only a finite number of vertices. (*Hint:* Show that if there are an infinite number of vertices, then there must be an infinite path (infinite play).)

18. Consider the following graph game. Player A tries to make a path with a set of vertices from the left to the right side of this graph. Player B tries to make a path from top to bottom. The players take turns picking vertices until one player gets the desired path.

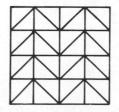

(a) Show that this is a progressively finite game.
(b) Find a winning strategy for the first player to move.

19. The game of *kayles* has a row of n equally spaced stones. Two players alternate turns of removing one or two consecutive stones (with no intervening spaces). The player to remove the last stone wins.

(a) Draw the graph and find its Grundy function for a four-stone game.

(b) Repeat part a for a five-stone game.

11.3 NIM-TYPE GAMES

In this section we extend the theory of progressively finite games to takeaway games involving several piles of objects. The simplest game of this form is called **Nim**, a game in which two players take turns removing any amount they wish from one of the piles. The winner is the player who removes the last object from the last remaining (nonempty) pile. While the positions in the two games in Examples 1 and 2 in the previous section could be described with a single nonnegative integer representing the size or monetary value of the single pile, the position in a Nim game requires a vector of nonnegative integers $(p_1, p_2, \ldots, p_m)$, the kth number p_k representing the current size of the kth pile.

Example 1

Consider the Nim game with four piles of sticks: one stick in the first pile, two sticks in the second pile, three in the third, and four in the fourth. See Figure 11.10. We represent the initial position of this game with the vector (1,2,3,4).

Let the first and second players be named A and B, respectively. A sample play of the game might go as follows. First A removes all four sticks from the fourth pile. The new position is (1,2,3,0). Next B removes one stick from the third pile to produce position (1,2,2,0). Now A removes the one stick in the first pile to produce position (0,2,2,0). A has been moving into kernel positions (playing a winning strategy) and is now about to win: If B removes all of the second or third pile, A will remove the other pile; or if B removes just one stick, A will remove one stick from the other pile and A will win on the next round. The reader is encouraged to play this Nim game with a friend. ∎

The game in Example 1 has $2 \cdot 3 \cdot 4 \cdot 5 = 120$ different positions (the ith pile has $i + 1$ possible sizes, $0, 1, \ldots, i$). Thus it would be very cumbersome to draw the graph of this game and compute its kernel with a Grundy function, as in the previous section. In general, a Nim game with m piles and n_i objects in the ith pile will have $(n_1 + 1)(n_2 + 1) \cdots (n_m + 1)$ different positions. The only way that we could ever play winning Nim without a computer is if we could determine the Grundy number of a position (vertex) directly from the position vector $(p_1, p_2, \ldots, p_m)$. Fortunately, such direct computation is possible.

First we need to examine the Grundy function for a single-pile Nim game. Consider the Nim game with one pile of four sticks. See Figure 11.11. The edges are directed from right to left with an edge (i, j) for all $0 \leqslant j \leqslant i \leqslant 4$.

|
| |
| | |
| | | |

FIGURE 11.10

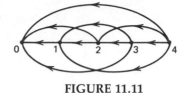

FIGURE 11.11

The Grundy number of vertex 0, the winning vertex, is 0; the Grundy number of vertex 1 is 1; and so on. In any one-pile Nim game, vertex i will have a Grundy number of i, since vertex i has edges to all lower-numbered vertices. Of course, the strategy of any one-pile Nim game is trivial: the first player removes the whole pile and wins. The nice form of this single-pile Grundy function will simplify the computation of Grundy functions for multi-pile Nim games.

Although the graph of a multi-pile Nim game is too complex to draw, we can still describe it symbolically. We have a vertex for each position $(p_1, p_2, \ldots, p_m)$, where $0 \le p_i \le n_i$ (n_i is the initial size of the ith pile). Since the only permissible moves are removing some amount from one pile, the associated graph has edges from vertex $(p_1, p_2, \ldots, p_m)$ to vertex $(q_1, q_2, \ldots, q_m)$ for each pair of vertices such that for one j, $q_j < p_j$, and for all $i \ne j$, $q_i = p_i$. This graph is in some sense a "vector" of the graphs for each pile, for if we fix all p_i except one, say p_3, then the subgraph of vertices $(p_1, p_2, 0, p_4, \ldots, p_m)$, $(p_1, p_2, 1, p_4, \ldots, p_m), \ldots, (p_1, p_2, n_3, p_4, \ldots, p_m)$ is exactly the graph for pile three alone.

This type of vector of graphs can be formalized as follows. The **direct sum** $H = H_1 + H_2 + \cdots + H_m$ of graphs $H_1, H_2, \ldots, H_m$ with vertex sets $X_1, X_2, \ldots, X_m$, respectively, has vertex set $X = \{(x_1, x_2, \ldots, x_m) | x_i \in X_i, 1 \le i \le m\}$ and edges defined by the successor sets

$$s((x_1, x_2, \ldots, x_m)) = \{(y, x_2, x_3, \ldots, x_m) | y \in s(x_1)\}$$
$$\cup \{(x_1, y, x_3, \ldots, x_m) | y \in s(x_2)\}$$
$$\vdots$$
$$\cup \{(x_1, x_2, \ldots, x_{m-1}, y) | y \in s(x_m)\}$$

It follows from this definition that the graph G of an m-pile Nim game is the direct sum of the graphs G_i of the ith piles: $G = G_1 + G_2 + \cdots + G_m$.

Example 2

Consider the simple Nim game with two piles of two objects each. Figure 11.12a shows the graph $G_1 = G_2$ of one pile alone. Figure 11.12b shows the

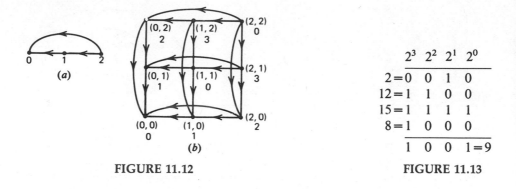

FIGURE 11.12

$$\begin{array}{cccc} 2^3 & 2^2 & 2^1 & 2^0 \\ \hline 2=0 & 0 & 1 & 0 \\ 12=1 & 1 & 0 & 0 \\ 15=1 & 1 & 1 & 1 \\ 8=1 & 0 & 0 & 0 \\ \hline 1 & 0 & 0 & 1=9 \end{array}$$

FIGURE 11.13

graph $G = G_1 + G_2$ of the two pile game. Find the Grundy function and kernel of this graph.

The Grundy numbers for the vertices are shown in Figure 11.12b. The vertices with Grundy number 0 are in the kernel. Since $g((2,2)) = 0$, the second player has the winning strategy (according to Theorem 3 in the previous section). ∎

Next we show how to compute the Grundy number of $(x_1, x_2, \ldots, x_m)$, a vertex in a direct sum of graphs, from the Grundy numbers $g(x_i)$. The computation to be performed on these Grundy numbers is called a digital sum. The **digital sum** c of nonnegative integers $c_1, c_2, \ldots, c_m$, written

$$c = c_1 \dotplus c_2 \dotplus c_3 \dotplus \cdots \dotplus c_m$$

is computed in the following manner. Let $c_i^{(k)}$ be the kth binary digit in a binary expansion of c_i, that is, $c_i = c_i^{(0)} + c_i^{(1)}2 + c_i^{(2)}2^2 + \cdots$, and similarly let $c^{(k)}$ be the kth digit in a binary expansion of c. Then $c^{(k)} \equiv c_1^{(k)} + c_2^{(k)} + \cdots + c_m^{(k)}$ (modulo 2). That is, the kth binary digit of c is 1 if the sum of the kth binary digits of the c_i's is odd, and the kth binary digit of c is 0 if the sum of the kth binary digits of the c_i's is even.

Example 3

Compute the value of the digital sum $2 \dotplus 12 \dotplus 15 \dotplus 8$. We write these numbers in binary form and sum (modulo 2) the digits as shown in Figure 11.13. Translating the sum back into an integer, we see that this digital sum equals 9. ∎

Observe that $c \dotplus b_1 = c \dotplus b_2$ if and only if $b_1 = b_2$. We now present a remarkable theorem. In essence, it says that digital sums are the only way to win at Nim.

Theorem

If the graphs $G_1, G_2, \ldots, G_m$ possess Grundy functions $g_i(x)$, then the direct sum $G = G_1 + G_2 + \cdots + G_m$ possesses a Grundy function $g(x)$ defined by

$$g(x) = g((x_1, x_2, \ldots, x_m)) = g_1(x_1) \dotplus g_2(x_2) \dotplus \cdots \dotplus g_m(x_m)$$

Proof

We must show that $g(x)$ is the smallest nonnegative integer that is not equal to any number in the set $\{g(y)|y \in s(x)\}$. This definition of a Grundy function can be broken into two parts: (i) If $y \in s(x)$, then $g(x) \neq g(y)$: and (ii) for each nonnegative integer $b < g(x)$, there exists some $y \in s(x)$ with $g(y) = b$.

Part (i)—Show that if $y \in s(x)$, then $g(x) \neq g(y)$. Since $y = (y_1, y_2, \ldots, y_m) \in s((x_1, x_2, \ldots, x_m))$, then by the definition of direct sum, for some j, $y_j \in s(x_j)$, and for all $i \neq j$, $y_i = x_i$. If $g_k(x_k) = c_k$ and $g_k(y_k) = d_k$, then $d_j \neq c_j$ and $d_i = c_i$. So

$$g(x) = c_1 \dotplus c_2 \dotplus \cdots \dotplus c_j \dotplus \cdots \dotplus c_m = c' \dotplus c_j$$

and

$$g(y) = d_1 \dotplus d_2 \dotplus \cdots \dotplus d_j \dotplus \cdots \dotplus d_m = c' \dotplus d_j$$

where c' is the digital sum of all c_i's except c_j. As noted above $c' \dotplus c_j = c' \dotplus d_j$ if and only if $c_j = d_j$. Since $c_j \neq d_j$, we conclude that $g(x) \neq g(y)$, as required.

Part (ii)—Show that for each nonnegative integer $b < g(x)$, there exists some $y \in s(x)$ with $g(y) = b$. A general proof of this part is fairly technical (see Berge [2], p. 25 for details). The practical side of this proof, discussed below, is finding a y with $g(y) = 0$ (a kernel vertex) and moving to it. The proof of part (ii) is a generalization of this discussion. ∎

Corollary

The vertex $(p_1, p_2, \ldots, p_m)$ in the graph of an m-pile Nim game has the Grundy number $p_1 \dotplus p_2 \dotplus \cdots \dotplus p_m$. Thus $(p_1, p_2, \ldots, p_m)$ is a kernel vertex if and only if $p_1 \dotplus p_2 \dotplus \cdots \dotplus p_m = 0$. ∎

The reader should go back to the Nim game of two piles of two objects each in Example 2 and check that the Grundy numbers obtained for the vertices of the graph of that game are the digital sums of the pile sizes.

Example 1 (continued)

The Grundy number of the starting vertex of the Nim game in Figure 11.10 is the digital sum $1 \dotplus 2 \dotplus 3 \dotplus 4 = 4$, as computed in Figure 11.14. The first player A wants to decrease the size of one pile so that the new digital sum is 0 (a kernel position). That is, A should alter the binary digits in one row of Figure 11.14 so that the sum (bottom) row is 0   0   0. For the sum row to become all 0's, every column now having an odd number of 1's should have one of its digits changed (either a 1 to a 0 or a 0 to a 1) to make the number of 1's even. Since the sum row has a 1 in just the 2^2 column, we can change the (single) 1 in that column to a 0. That is, pile 4's binary expansion should be 0   0   0, and so player A should remove all sticks in the fourth pile. This new position (1,2,3,0) has Grundy number $1 \dotplus 2 \dotplus 3 \dotplus 0 = 0$. Note that to make the number of 1's even in the 2^2 column, we could not change any 0 to a 1, for the new binary expansion in the altered row would have to be a larger number—an impossible move in Nim.   ∎

We now generalize the method of finding a vertex with Grundy number 0 given in the preceding example. Form a digital sum table as in Figure 11.14 for the current game position. *Pick a row e with a 1 in the leftmost column having a 1 in the sum row. In row e, change the digit in every column having a 1 in the sum row, that is, in every column with an odd number of 1's.* After this change of digits, every column will have an even number of 1's and the sum row will be all 0's. So this new position will be a kernel vertex. Note that since the leftmost digit that is changed in row e is a 1 (this is how row e was chosen), changing digits in row e will yield a smaller number, call it h. The first player should thus decrease the size of the eth pile to a size of h objects.

Example 4

Consider the Nim game of four piles with 2, 3, 4, and 6 sticks shown in Figure 11.15a. The digital sum table for the initial position is shown in Figure 11.15b. The 2^1 column is the leftmost column with a 1 in the sum row, and so we must change a row with a 1 in the 2^1 column. We can use the first, second, or fourth row. Suppose that we choose the first row. Then since the 2^1 and 2^0 columns in the sum row have 1's, we change the digits in these columns in the first row. The new first row is 0   0   1. Thus the first player should reduce the size of the first pile to 1. The reader should continue the play of this Nim game with a friend (or against oneself) to practice this kernel-finding rule.   ∎

	2^2	2^1	2^0
1 =	0	0	1
2 =	0	1	0
3 =	0	1	1
4 =	1	0	0
	1	0	0 = 4

FIGURE 11.14

		2^2	2^1	2^0
11	2 =	0	1	0
111	3 =	0	1	1
1111	4 =	1	0	0
111111	6 =	1	1	0
(a)				
		0	1	1 = 3

(b)

FIGURE 11.15

EXERCISES

1. Find the Grundy number of the initial position and the correct first move in a winning strategy for the following Nim games:

||	|	||	||
|||	|||	||||	|||
|||	|||||	||||	||||
|||||	|||||||	||||||	|||||
(a)	(b)	(c)	(d)

2. Suppose that no more than two sticks can be removed at a time from any pile. Repeat Exercise 1 with this additional condition.

3. Suppose that no more than i sticks can be removed at a time from the ith pile (piles are numbered from top to bottom). Repeat Exercise 1 with this additional condition.

4. For the Nim game in Exercise 1(c), what moves yield positions with Grundy number equal to:
   (a) 1   (b) 2   (c) 3.

5. Suppose three copies of the game in Example 2 of Section 11.2 are played simultaneously. Players stop adding money to a pile when the value of the pile is a square or exceeds 40. The player who adds the last amount to the last pile wins the money in all three piles.
   (a) If initially there are 2¢ in two piles and 1¢ in the third pile, what is the Grundy number of this position and what is a correct winning move?
   (b) Find all kernel positions in which the sum of the money in the first two piles is less than 10¢ and the third pile is empty.
   (c) Using the table of Grundy numbers for the one-pile game near the end of Section 11.2, write a computer program to compute the next winning move in this three-pile game.

6. Draw the direct sums involving the graphs shown below.
   (a) $G_1 + G_1$ (b) $G_1 + G_2 + G_3$

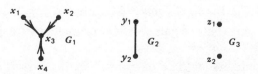

7. Generalize the argument for finding kernel vertices in Nim (preceding Example 4) to obtain part (ii) of the proof of the theorem.

8. Consider the variation of Nim in which the last player to move *loses*. Show that the winning strategy is to play the regular last-player-wins

Nim strategy but when only one pile has more than one stick now decrease that pile's size to 1 instead of 0 (or to 0 instead of 1).

9. Write a computer program to play winning Nim.

10. Consider the following variation on Nim. Place $C(n + 1,2)$ identical balls in a triangle, like bowling pins. Two players take turns removing any number of balls in a set that all lie on a straight line. The player to remove the last ball wins.

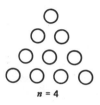

$n = 4$

(a) Find a winning first move for this game when $n = 3$.
(b) Repeat for $n = 4$.

11.4 SUMMARY AND REFERENCES

The graph model for Instant Insanity is due to de Carteblanche [4] (reputedly a pseudonym for the famous graph theorist W. Tutte). The first published analysis of Nim appeared in 1902 by C. Bouton [3]. The Grundy function for progressively finite games was presented by P. Grundy [7] in 1939. It was independently discovered a few years earlier by the German mathematician Sprague.

This chapter has only scratched the surface of the theory of finitely progressive games. Interested readers should turn to *On Numbers and Games*, by J. Conway [5], the world's acknowledged master gamesman, and to *Winning Ways* by Berlekamp, Conway, and Guy [1]. The reader should also see the classic book about mathematics and games, *Mathematical Recreations and Essays* by Coxeter and Ball [6].

1. E. Berlekamp, J. Conway, and R. Guy, *On Winning Ways*, 2 volumes, Academic Press, New York, 1982.
2. C. Berge, *The Theory of Graphs*, Methuen-Wiley, New York, 1962.
3. C. L. Bouton, "Nim, a game with a complete mathematical theory," *Ann. Math.* 3 (1902), 35–39.
4. F. de Carteblanche, "Pile of cubes," *Eureka*, April, 1947.
5. J. H. Conway, *On Numbers and Games*, Academic Press, New York, 1976.
6. H. Coxeter and W. Ball, *Mathematical Recreations and Essays*, Univ. of Toronto Press, 1972.
7. P. M. Grundy, "Mathematics and games," *Eureka*, January, 1939.

Appendix

A.1 SET THEORY AND LOGIC

A set is a collection of distinct objects. In contrast to a sequence of objects, a set is unordered. Usually we refer to the objects in a set as the **elements,** or **members,** of the set. These elements may themselves be sets, as in the set of all 5-card hands; each hand is a set of 5 cards.

A set S is a **subset** of set T if every element of S is also an element of T. In the case of a 5-card hand, we are really working with a 5-card subset of the set of all 52 cards. Any set is trivially a subset of itself. A **proper** subset is a nonempty subset with fewer elements than the whole set. Two sets with no common members are called **disjoint** sets. A **family** is a collection in which multiple appearances of objects is allowed; for example, $\{a,a,a,b,b,c\}$.

This text uses capital letters to denote sets and lower-case letters to denote elements (unless the elements are themselves sets). The number of elements in a set S is denoted by $N(S)$. The symbol $\in$ represents set membership, for example, $x \in S$ means that x is an element of S; and $x \notin S$ means that x is not a member of S. The symbol $\subseteq$ represents subset containment, for example, $T \subseteq S$ means that T is a subset of S.

There are three ways to define a set with formal mathematical notation: first, by listing its elements, as in $S = \{x_1, x_2, x_3, x_4\}$ or $T = \{\{1,-1\}, \{2,-2\}, \{3,-3\}, \{4,-4\}, \{5,-5\}, \ldots\}$—implicitly, T is the set of all pairs of a positive integer and its negative; second, by a defining property, as in $P = \{p \mid p$ is a person taking this course$\}$ or $R = \{r \mid$ there exist integers s and t with $t \neq 0$ and

$r = s/t$}—R is the set of all rational numbers; third, as the result of some operation(s) on other sets (see below).

There are two special sets we often use: the **empty**, or **null**, set, written $\emptyset$; and the **universal** set of all objects currently under consideration, written $\mathcal{U}$.

It is important to bear in mind that in many real-world problems, sets cannot be precisely defined or enumerated. For example, the set of all ways in which a certain computer program can fail is ill defined. A census of the population of the United States involves substantial error for several important subcategories of the populace, yet the official United States population is given as 226,504,825. Beware of any calculations based on imprecise sets!

The three basic operations on sets that we will use are:

1. The **intersection** of S and T, $S \cap T = \{x \in \mathcal{U} | x \in S \text{ and } x \in T\}$.
2. The **union** of S and T, $S \cup T = \{x \in \mathcal{U} | x \in S \text{ or } x \in T\}$.
3. The **complement** of S, $\bar{S} = \{x \in \mathcal{U} | x \notin S\}$.

A fourth operation that is sometimes useful is:

4. The **difference** of S of minus T, $S - T = \{x \in \mathcal{U} | x \in S \text{ and } x \notin T\}$.

Observe that $S - T$ can be expressed in terms of the preceding operations as $S - T = S \cap \bar{T}$.

Example 1

A sample of eight brands of calculators consists of four machines that have rechargeable batteries and four that do not. Also four out of the eight have memory. We want to select four calculators to be taken apart for thorough analysis. Half of this set of four should be rechargeable and half should have memory. How many ways can such a set of four machines be chosen from the eight brands?

To answer this question, we must know how many machines are rechargeable and have memory, how many are rechargeable and have no memory, and so on. That is, if R is the set of rechargeable machines and M the set of machines with memory, then we need information about the sizes of the intersections $N(R \cap M)$, $N(R \cap \bar{M})$, $N(\bar{R} \cap M)$, and $N(\bar{R} \cap \bar{M})$. Assume that there are two machines in each of these four subsets. See Figure A1.1. Then

	Memory	No Memory
Rechargeable	2	2
Not Rechargeable	2	2

FIGURE A1.1

one strategy to get the desired mix of four machines would be to pick one machine from each category in Figure A1.1 ($2 \cdot 2 \cdot 2 \cdot 2 = 16$ choices). There are two other strategies for getting the four machines from Figure A1.1 (see Exercise 4 at the end of this section). ∎

The study of set expressions involving the above set operations and associated laws is called **Boolean algebra.** Three of the most important laws of Boolean algebra are:

BA1. $S = \bar{\bar{S}}$.

BA2. $\overline{S \cap T} = \bar{S} \cup \bar{T}$.

BA3. $\overline{S \cup T} = \bar{S} \cap \bar{T}$.

To visualize set expressions, we use a picture called a **Venn diagram.** Figure A1.2a shows a Venn diagram for the sets S and T. The whole rectangle represents the universe $\mathcal{U}$ of all elements under consideration. The circles represent the sets S and T. In Figure A1.2b, the darkened region represents $S \cap T$ and the striped area $(S \cup T) - (S \cap T)$. It is possible that certain regions in a Venn diagram may actually be empty sets. Laws BA2 and BA3 can often simplify a counting problem.

Example 2

Consider the universe $\mathcal{U}$ of all 52^2 outcomes obtained by picking one card from a deck of 52 playing cards, replacing that card, and then again picking a card (possibly the same card). Suppose that we want to compute the size of the set Q of outcomes with at least one spade or at least one heart.

Let S be the set of outcomes with a void in spades, that is, in which no spade is chosen on either pick. Let H be the set of outcomes with a void in hearts. Write the set Q as an expression in terms of S and H, and calculate $N(Q)$.

The set Q equals $\bar{S} \cup \bar{H}$, the union of the set of outcomes with one or more spades (not a void in spades) and the set of outcomes with one or more hearts. The set $\bar{S} \cup \bar{H}$ is the shaded area in Figure A1.3. By BA3, $\bar{S} \cup \bar{H} = \overline{S \cap H}$,

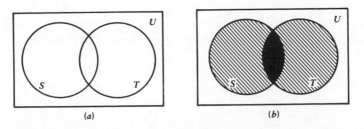

(a) (b)

FIGURE A1.2

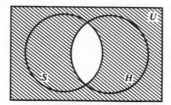

FIGURE A1.3

where $S \cap H$ is the set of outcomes with no spades and no hearts ($S \cap H$ is the unshaded region in Figure A1.3). That is, $S \cap H$ is the set of outcomes where each pick is one of the 26 diamonds or clubs. Thus $N(S \cap H) = 26^2$. By Figure A1.2, $N(\overline{S} \cap \overline{H}) = N(\mathcal{U}) - N(S \cap H) = 52^2 - 26^2 = 2704 - 676 = 2028$, and so $N(\overline{S} \cup \overline{H}) = N(\overline{S \cap H}) = 2028$. ∎

At the elementary level, mathematical logic is very similar to set theory. Instead of sets, logic deals with **statements** or **propositions**. Statements are usually represented by lower-case letters. They have the value true (T) or false (F). The three principal operations on statements are as follows.

1. The **conjunction** of p and q, $p \wedge q$.
2. The **disjunction** of p and q, $p \vee q$.
3. The **negation** of p, $\sim p$.

If p is "this integer is even" and q is "this integer is less than 25," then $p \wedge \sim q$ is "this integer is even and is not less than 25." A truth table is used to enumerate the value (T or F) of the above operations for each possible pair of values of p and q.

p	q	$p \wedge q$	$p \vee q$	$\sim p$
T	T	T	T	F
T	F	F	T	F
F	T	F	T	T
F	F	F	F	T

The study of expressions built from such logical operations is called **propositional calculus**. The three axioms of propositional calculus corresponding to laws BA1, BA2, BA3 above are:

PC1. $\sim \sim p = p$.

PC2. $\sim (p \wedge q) = \sim p \vee \sim q$.

PC3. $\sim (p \vee q) = \sim p \wedge \sim q$.

With each statement p, one can associate a set S_p of all elements (in a given universe) that satisfy statement p. Conversely, to each set S, one can associate a statement p_s, "this object is an element in S."

Note that if the values of T and F are replaced by 1 and 0, respectively, then computer logic design, which uses "AND," "OR," and "NOT," gates, is seen to be a problem in propositional calculus.

Set theory, and more generally, nonnumerical mathematics, were studied little until the nineteenth century. G. Peacock's *Treatise on Algebra*, published in 1830, first suggested that the symbols for objects in algebra could represent nonnumeric entities. A. De Morgan discussed a similar generalization for algebraic operations a few years later. G. Boole's *Investigation of the Laws of Thought* (1854) extended and formalized Peacock's and De Morgan's work to present a formal algebra of sets and logic. The great philosopher mathematician Betrand Russell has written "Pure Mathematics was discovered by Boole." Subsequently it was found that numerical mathematics also needs to be defined in terms of set theory in order to have a proper mathematical foundation (see Halmos's *Naive Set Theory*). See Chapter 26 of Boyer's *History of Mathematics* for more details about the history of set theory.

EXERCISES

1. Let A be the set of all positive integers less than 30. Let B be the set of all positive integers that end in a 7 or 2. Let C be the set of all multiples of 3. List the numbers in the following sets:
    (a) $A \cap (B \cap C)$ (c) $A \cap (B \cup C)$
    (b) $A \cap (B \cup C)$ (d) $A - (B \cap C)$

2. If $A = \{1,2,4,7,8\}$, $B = \{1,5,7,9\}$, and $C = \{3,7,8,9\}$ and $\mathcal{U} = \{1,2,3,4,5,6,7,8,9,10\}$, then find set expressions (if possible) equal to:
    (a) $\{7\}$ (c) $\{1\}$ (e) $\{4,8,10\}$
    (b) $\{6,10\}$ (d) $\{2,7,9\}$ (f) $\{3,5,6,7,9,10\}$

3. Suppose that in Example 1 we were only given the additional information that two calculators have memory but are not rechargeable. Show that we can now deduce how many machines are in each of the other three categories.

4. What are the other two strategies in Example 1 to pick a subset of four calculators, two of which are rechargeable and two of which have memory? How many choices are there for each of these ways?

5. Suppose we are given a group of 20 people, 13 of whom are women. Suppose 8 of the women are married. In each of the following cases, tell whether the additional piece of information is sufficient to determine the number of married men. If it is, give this number.
    (a) There are 12 people who are either married or male (or both).
    (b) There are 8 people who are unmarried.
    (c) There are 15 people who are female or unmarried.

6. Label each region of the Venn diagram in Figure A1.2a with the set expression it represents (e.g., the region where both sets intersect would be labeled $A \cap B$).

7. Draw Venn diagrams and shade the area representing the following sets:
   (a) $\overline{A - B}$
   (b) $\overline{A \cup B}$
   (c) $\overline{(A \cup B)} \cap \overline{(A \cap B)}$
   (d) $A - (B - A)$

8. Write as simple a set expression as you can for the shaded areas in the following Venn diagrams.

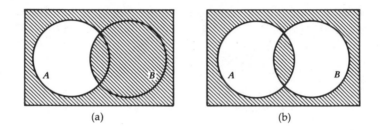

    (a)   (b)

9. Create a Venn diagram for representing sets A, B, and C. The diagram should have regions for all possible intersections of A, B, and C. How many different regions are there? Label each region with the set expression it represents (e.g., the region where all three sets intersect would be labeled $A \cap B \cap C$).

10. Draw Venn diagrams for sets A, B, and C and shade the areas representing the sets in Exercise 1.

11. Verify with Venn diagrams the following laws of Boolean algebra:
    (a) $(A \cup B) \cap A = A$
    (b) $(A \cap B) \cup A = A$
    (c) $A \cup \overline{A} = \mathcal{U}$
    (d) $(A \cup B) \cap C = (A \cap C) \cup (B \cap C)$
    (e) $(A \cap B) \cup C = (A \cup C) \cap (B \cup C)$

12. Use Venn diagrams to determine which pairs of the following set expressions are equivalent.
    (a) $(A \cap B) \cup (\overline{A} \cap B)$
    (b) $A - (B - (A - B))$
    (c) $(\overline{A} - \overline{B}) \cap (\overline{A} \cup \overline{B})$
    (d) $(\overline{A} \cap (B - \overline{A}))$

13. Use Law BA2 to prove that $\overline{A \cap B \cap C} = \overline{A} \cup \overline{B} \cup \overline{C}$ [note that $A \cap B \cap C = (A \cap B) \cap C = A \cap (B \cap C)$].

14. Use Law BA3 to prove that $\overline{A \cup B \cup C} = \overline{A} \cap \overline{B} \cap \overline{C}$ [note that $A \cup B \cup C = (A \cup B) \cup C = A \cup (B \cup C)$].

15. Use Laws BA1, BA2, and BA3 plus the identities $A \cap A = A$, $A \cup A = A$, $A \cap B = B \cap A$, and $A \cup B = B \cup A$ to simplify the following set expressions:
    (a) $(A \cup B) \cap (A \cap B)$
    (b) $A - (\overline{B} \cup \overline{A})$
    (c) $(\overline{A \cup C}) \cap (\overline{A \cup B}) \cap (\overline{B \cup C})$
    (d) $(A \cup B) \cap ((C \cup B) - (\overline{A} \cap \overline{B}))$

16. Suppose two dice are rolled. How many outcomes have a 1 or 2 showing on at least one of the dice?

17. Suppose that a card is drawn from a deck of 52 cards (and replaced before the next draw) three times. Let S be the set of outcomes where the first card is a spade, H be the set where the second card is a heart, and C be the set where the third card is a club. For each of the following sets E, write E as a set expression in terms of S, H, C, and determine $N(E)$.

    (a) Let E be the set of outcomes contained in at least one of the sets S, H, C.

    (b) Let E be the set of outcomes contained in at least one but not all three of the sets S, H, C.

    (c) Let E be the set of outcomes contained in exactly two of the sets S, H, C.

18. For what values of propositions p and q are each of the following statements true?

    (a) $\sim(p \vee \sim q)$ (c) $(p \vee \sim q) \wedge (\sim p \wedge q)$

    (b) $(\sim(p \vee q)) \vee (\sim(p \wedge q))$ (d) $\sim(q \wedge q)$

19. For what values of propositions p, q, and r are each of the following statements true?

    (a) $(p \wedge q) \vee \sim r$ (c) $(\sim(p \vee q)) \wedge (\sim(q \vee r))$

    (b) $(\sim p \wedge \sim q) \vee \sim r$ (d) $((p \wedge q) \vee (p \vee r)) \wedge (\sim q \vee \sim r)$

20. Build a statement using propositions p, q, and r that is true only when exactly two of the three propositions are true.

21. Two statements involving propositions $p_1, p_2, \ldots, p_k$ are equivalent if they are true for exactly the same sets of values of the propositions. How many different (nonequivalent) statements involving k propositions are there?

22. Let $a_1 a_2$ be the binary digits in the binary representation of a number a, $0 \leqslant a \leqslant 3$ (e.g., if $a = 2$, then $a_1 = 1$ and $a_2 = 0$). Let $b_1 b_2$ be similar binary digits for b, $0 \leqslant b \leqslant 3$. Now let $c = a + b$, where c's binary digits are $c_1 c_2 c_3$. Treating 1 as true and 0 as false, express c_1, c_2, and c_3 as propositional statements in terms of a_1, a_2, b_1, and b_2.

A.2 MATHEMATICAL INDUCTION

The most useful, and simplest, proof technique in combinatorial mathematics and computer science is **mathematical induction.** Let p_n denote a statement involving n objects. Then an induction proof that p_n is true for all $n \geqslant 0$ requires two steps:

1. Initial step: Verify that p_0 is true.

2. Induction step: Show that if $p_0, p_1, p_2, \ldots, p_{n-1}$ are true, then p_n must be true.

Sometimes p_n will only be true for $k \leqslant n$. Then the initial step is to verify that p_k is true, and in the induction step we assume only that $p_k, p_{k+1}, \ldots, p_{n-1}$ are true.

Example 1

Let s_n denote the sum of the integers 1 through n, that is, $s_n = 1 + 2 + 3 + \cdots + n$. Show that $s_n = \frac{1}{2}n(n + 1)$.

We use mathematical induction to verify this formula. Since s_0 is not defined, the initial step is to verify the formula for s_1. The formula says $s_1 = \frac{1}{2}1(1 + 1) = 1$—obviously correct. For the induction step, we assume that the formula is true for $s_1, s_2, \ldots, s_{n-1}$. In this problem (as in most induction problems), we only need to assume that the formula is true for s_{n-1}:

$$s_{n-1} = 1 + 2 + 3 + \cdots + n - 1 = \tfrac{1}{2}(n - 1)((n - 1) + 1) = \tfrac{1}{2}(n - 1)n$$

We now use this expression for s_{n-1} to prove that the formula is true for s_n.

$$s_n = 1 + 2 + \cdots + (n - 1) + n = s_{n-1} + n$$
$$= \tfrac{1}{2}(n - 1)n + n$$
$$= \tfrac{1}{2}[(n - 1)n + 2n] = \tfrac{1}{2}n(n + 1)$$

This completes the induction proof that $s_n = \frac{1}{2}n(n + 1)$ for all positive n. ■

Example 2

Suppose that the population of a colony of ants doubles in each successive year. A colony is established with an initial population of 10 ants. How many ants will this colony have after n years?

Let a_n denote the number of ants in the colony after n years. We are given that $a_0 = 10$. Since the colony's population doubles annually, then $a_1 = 20$, $a_2 = 40$, and $a_3 = 80$. Looking at the first few values of a_n, we are led to conjecture that $a_n = 2^n \cdot 10$. For the initial step, we check that $a_0 = 10 = 2^0 \cdot 10$, as required. For the induction step, we assume that $a_{n-1} = 2^{n-1} \cdot 10$. Now we use the annual doubling property of the colony:

$$a_n = 2a_{n-1} = 2(2^{n-1} \cdot 10) = 2^n \cdot 10$$

■

Example 3

A prime number is an integer $p > 1$ that is divisible by no other integer besides 1 and p. Show that any integer $n > 1$ can be written as a product of prime numbers.

We prove this fact by mathematical induction. Since the assertion concerns integers $n > 1$, the initial step is to verify that 2 can be written as a product of prime numbers. But 2 is itself a prime. Thus 2 is trivially the product of primes (i.e., actually of a single prime), itself. Next assume that the numbers 2, 3, . . . , $n - 1$ can be written as a product of primes, and use this assumption to prove that n can also be written as a product of primes.

If n is itself a prime, then as in the case of 2, there is nothing more to do. Suppose n is not a prime, and so there is an integer m that divides n and for some integer k, $n = km$. Since k and m must be less than n, they can each be written as a product of primes. Multiplying these two prime products for k and m together yields the desired representation of n as a product of primes. ∎

In his classic work *Fundamental Algorithms, The Art of Computer Programming*, Vol. I (Addison-Wesley, Reading, Mass., 1975), D. Knuth refers to mathematical induction as the "algorithmic proof procedure." The simplest way to verify the formulas in Examples 1 or 2 (or the assertion in Example 3) with a computer for a particular value of n, say, $n = 50$, would be to write a program that recursively computed values for successive n up to $n = 50$. For Example 1, the program would be

$$s \leftarrow 0;$$
$$\text{FOR } n = 1 \text{ TO } 50 \text{ DO } s \leftarrow s + n;$$

A program to determine the factorization into primes of each successive integer up to 50 would similarly mimic mathematical induction. Thus induction represents a generalized case-by-case verification of a formula or assertion.

The relationship between mathematical induction and algorithms is a two-way street. To verify that an algorithm or program works correctly, one usually uses some type of induction argument, based on the value of an input variable or on the number of iterations of a loop.

While the Greeks used certain iterative arguments in geometric calculations and the principle of reductio ad absurdum, mathematical induction was first used explicitly by Maurolycus around 1550. Pascal used an induction argument in 1654 to verify the additive property of binomial coefficients in the array now known as Pascal's triangle (this property is identity 2 in Section 5.5). The actual term induction was coined by De Morgan 200 years later. See Bussey, "Origin of mathematical induction," *American Math. Monthly*, 1917, for a fuller discussion of the history of mathematical induction.

EXERCISES

1. Write a computer program to compute for Example 2 the number of ants after 20 years.

2. Prove by induction that $1 + 3 + \cdots + (2n + 1) = (n + 1)^2$.

3. Prove by induction that $1^2 + 2^2 + \cdots + n^2 = \dfrac{n(n + 1)(2n + 1)}{6}$.

4. Prove by induction that $1 - 2^2 + 3^2 - \cdots + (-1)^n n^2 = (-1)^n \dfrac{n(n + 1)}{2}$.

5. Prove by induction that $1 \cdot 2 + 2 \cdot 3 + \cdots + n(n + 1) = \dfrac{n(n + 1)(n + 2)}{3}$.

6. Prove by induction that $1^3 + 2^3 + \cdots + n^3 = \dfrac{n^2(n + 1)^2}{4}$.

7. Prove by induction that $(1 + 2 + \cdots + n)^2 = 1^3 + 2^3 + \cdots + n^3$ (assume the result in Example 1).

8. Prove by induction that $\dfrac{1}{1 \cdot 2} + \dfrac{1}{2 \cdot 3} + \cdots + \dfrac{1}{n(n + 1)} = \dfrac{n}{n + 1}$.

9. Prove by induction that $\dfrac{1}{1 \cdot 4} + \dfrac{1}{4 \cdot 7} + \dfrac{1}{7 \cdot 10} + \cdots +$

   $\dfrac{1}{(3n - 2)(3n + 1)} = \dfrac{n}{3n + 1}$.

10. Prove by induction that $1 \cdot 1! + 2 \cdot 2! + \cdots + n \cdot n! = (n + 1)! - 1$.

11. (a) Prove by induction that for $a \neq 1$, $\dfrac{1 - a^{n+1}}{1 - a} = 1 + a + a^2 + \cdots + a^n$.

    (b) Guess a value, and confirm by induction, for the sum $1 + 2 + 2^2 + \cdots + 2^n$.

12. Prove by induction that $1^2 + 3^2 + 5^2 + \cdots + (2n - 1)^2 = \dfrac{n(2n - 1)(2n + 1)}{3}$.

13. If the number a_n of calf births at Dr. Smith's farm after n years obeys the law $a_n = 3a_{n-1} - 2a_{n-2}$ and in the first two years $a_1 = 3$ and $a_2 = 7$, then prove by induction that $a_n = 2^{n+1} - 1$.

14. Prove that any positive integer has a *unique* factorization into primes. You may assume the result of Example 3; you need to use the fact that if a prime p divides a product of positive integers then p divides one of the integers.

15. Write a computer program to find and print the prime factorizations of the first 50 integers.

16. Prove by induction that for any integer $m > 0$, $m \cdot n = n \cdot m$. By $r \cdot s$, we mean the sum of r copies of s.

17. Prove by induction that for integers $n \geq 5$, $2^n > n^2$.

18. Prove that the number of different subsets (including the null set and full set) of a set of n objects is 2^n.

19. Prove by induction that $\overline{A_1 \cap A_2 \cap \cdots \cap A_n} = \bar{A}_1 \cup \bar{A}_2 \cup \cdots \cup \bar{A}_n$ (see Exercise 13 in Appendix A1).

20. Prove that if n distinct dice are rolled, the number of outcomes where the sum of the faces is an even integer equals the number of outcomes with an odd sum.

21. Prove that the number of n-digit binary sequences with an even number of 1's equals the number of n-digit binary sequences with an odd number of 1's.

22. Prove by induction that the sum of the cubes of three successive position integers is divisible by 9.

23. The cat population in a dormitory has the property that the number of cats in one year is equal to the sum of the number of cats in the two previous years. If in the first year there was one cat and if in the second year there were two cats, then prove that the number of cats in the nth year is equal to

$$\frac{1}{\sqrt{5}}\left[\left(\frac{1}{2} + \frac{1}{2}\sqrt{5}\right)^{n+1} - \left(\frac{1}{2} - \frac{1}{2}\sqrt{5}\right)^{n+1}\right].$$

24. Why cannot one prove by induction that the number of binary sequences of all finite lengths is finite?

25. What is wrong with the following induction proof that all elements x_1, . . . , x_n in a set of n elements are equal? (1) Initial step ($n = 1$): the set has one element x_1 equal to itself. (2) Induction step: Assume $x_1 = x_2 = \cdots = x_{n-1}$; since also $x_{n-1} = x_n$ by induction assumption (when x_{n-1}, x_n are considered alone as a two-element set), then $x_1 = x_2 = \cdots = x_{n-1} = x_n$.

26. What is wrong with the following induction proof that for $a \neq 0$, $a^n = 1$? (1) Initial step ($n = 0$): $a^0 = 1$—always true. (3) Induction step: Assume $a^{n-1} = 1$ and now $a^n = \dfrac{a^{n-1}a^{n-1}}{a^{n-2}} = \dfrac{1 \cdot 1}{1} = 1$.

A.3 A LITTLE PROBABILITY

Historically, counting problems have been closely associated with probability. Indeed, any problem of the form "how many hobbits are there who . . ." has the closely related form "what fraction of all hobbits . . . ," which in turn can be posed probabilistically as "what is the probability that a randomly chosen hobbit . . . ?" The famous Pascal's triangle for binomial coefficients was developed by Pascal around 1650 in the process of analyzing some gambling probabilities. The probability of getting at least 7 heads on 10 flips of a fair coin, the probability of being dealt a five-card poker hand (from a well-shuffled deck) with a pair or better, and the probability of finding a faulty calculator in a sample of 20 machines if 5 percent of the machines from which the sample was drawn are faulty—all these probabilities are essentially counting problems.

Two hundred years ago, the French mathematician Laplace first defined probability as follows.

$$\text{Probability} = \frac{\text{number of favorable cases}}{\text{total number of cases}}$$

This definition corresponds to the "person in the street's" intuitive idea of what probability is. In this text, we only treat probability problems where Laplace's definition of probability applies. Implicit in this definition is the assumption that each case is equally likely. If the total number of cases is infinite, for example, all real numbers between 0 and 1, then we would not be able to use Laplace's definition. A more subtle difficulty is that in some probability problems, "cases" have to be carefully defined or else they may not all be equally likely, as, for example, the possible numbers of heads observed when a coin is flipped 10 times. To clarify this difficulty, we need to introduce a little of the basic terminology of probability theory.

An **experiment** is a clearly defined procedure that produces one of a given set of outcomes. These outcomes are called **elementary events** and the set of all elementary events is called the **sample space** of the experiment. An event that is a subset of several elementary events is sometimes called a **compound event**.

For example, when a single die is rolled, then obtaining a specific number, such as 5, is an elementary event, while obtaining an even number is a compound event. If S is the sample space of the experiment and E is an event in S, then Laplace's definition of probability says that the probability of event E, prob(E), is

$$\text{prob}(E) = \frac{N(E)}{N(S)}$$

In most instances, the size of S is easily determined, and so the problem of determining prob(E) reduces to counting the number of outcomes (elementary events) in event E. Returning to the die roll, the sample space of outcomes is $S = \{1,2,3,4,5,6\}$. If the event E is obtaining an even number, then $E = \{2,4,6\}$, and prob(E) = $\frac{3}{6}$ = $\frac{1}{2}$.

Many experiments we discuss involve a repeated (or simultaneous) series of simple experiments. Each round of the simple experiment is called a **trial**. For example, the experiment of flipping a coin three times involves three successive trials of the simple experiment of flipping a coin. Rolling two dice and recording the sum of the two values rolled involves the two simultaneous simple experiments of rolling a single die. In any experiment involving multiple trials, the elementary events are the sequences of outcomes of the simple experiments. If the simple experiments are performed simultaneously, we number the simple experiments and list their outcomes in order of the experiments' numbers.

The sample space of elementary events for the experiment of flipping a coin three times is

$$\{HHH, HHT, HTH, HTT, THII, THT, TTH, TTT\}$$

The sample space for the experiment of rolling two dice is

$$S = \{(i,j) | 1 \leqslant i \leqslant 6,\ 1 \leqslant j \leqslant 6\}$$

Let us consider more closely the experiment of rolling two dice and recording the sum of the two values on each die (the sample space is the set S listed above).

In terms of our formal definition of events, the sum of the two values equaling k is not in general an elementary event but rather a compound event. The event, the sum of the two dice equals 7, is the subset of elementary events

$$S_7 = \{(1,6),(2,5),(3,4),(4,3),(5,2),(6,1)\}$$

Thus, prob(sum = 7) = $N(S_7)/N(S) = \frac{6}{36}$. Similarly, $S_2 = \{(1,1)\}$ and so prob(sum = 2) = $N(S_2)/N(S) = 1/36$. Observe that a sum of 7 is six times as likely as a sum of 2. If we had considered the possible sums of the two dice as the elementary events, we would have had a sample space $S^* = \{2,3,4,5,6,7,8,9,10,11,12\}$ and each sum would have had probability $1/N(S^*) = 1/11$—clearly a mistake. The same type of error would have been made if we were to regard the number of heads as the elementary events in the experiment of flipping three coins (see the correct sample space listed above).

Suppose we have a box containing 5 identical red balls and 20 identical black balls. Our experiment is to draw a ball. What is the sample space of elementary events? All we can record as an outcome is the color of the ball, leading to the sample space $S = \{\text{red, black}\}$. But then by Laplace's definition of probability, picking a red ball will have probability $\frac{1}{2}$. Although this experiment consists of just a single trial, we are clearly making the same sort of mistake that occurs when the sums of dice are treated as elementary events. You are encouraged to pause a moment and try to supply your own explanation for resolving this red–black ball paradox before reading the next paragraph.

The logical solution is to consider not what color ball we withdraw from the box but where in the box we direct our hand to grasp a ball. There are 25 different spatial positions where balls are located in the box. So, in this sense, there are 25 different balls. We resolve this complication about "identical" objects with the following rule.

Identical Objects Rule

In probability problems, there are no collections of identical objects; all objects are distinguishable.

In complex experiments one should always take a moment to be sure that the elementary events are properly identified.

The original motivation for studying probability was the same as used in this book, games of chance—rolling dice, flipping coins, and card games. Cardano calculated the odds for different sums of two dice in the 1500s. Around 1600, Galileo calculated similar odds for three dice. A series of letters about gambling probabilities between Pascal and Fermat, written around 1650, constitute the real beginning of probability theory. Jacques Bernoulli's *Ars Conjectandi* (1713) was the first systematic treatment of probability (and associated combinatorial methods). One hundred years later, S. Laplace published his epoch treatise *Theorie Analytique des Probabilities,* a work containing both the definition of probability in a finite sample space used in this book and also advanced-calculus-based derivations of modern probability theory. See David [*Games, Gods, and Gaming* (Hafner)] for an excellent history of probability theory.

EXERCISES

1. An integer between 5 and 12 inclusive is chosen at random. What is the probability it is even?

2. In the experiment of flipping a coin three times, let E_k be the event that the number of heads equals k. Determine $\text{prob}(E_k)$, for $k = 0, 1, 2, 3$.

3. In the experiment of rolling two distinct dice, find the probability of the following events.
   (a) Both dice show the same value.
   (b) The sum of the dice is even.
   (c) The sum of the dice is the square of one of the dice's value.

4. A die is rolled three times. What is the probability of having a "5" appear at least two times?

5. Find the probability that in a randomly chosen arrangement of the letters h, a, t the following occur:
   (a) The letters are in alphabetical order.
   (b) The letter h occurs somewhere after the letter t in the arrangement.

6. Find the probability that in a randomly chosen (unordered) subset of two numbers from the set 1, 2, 3, 4, 5 the following occur:
   (a) The subset is {1,2}.   (b) 1 is not in the subset.

7. An urn contains six red balls and three black balls. If a ball is chosen, then returned, and a second ball chosen, what is the probability that one of the following is true?
   (a) Both balls are black   (b) One ball is black and one is red.

8. An urn contains two black balls and three red balls. If two different balls are successively removed, what is the probability that both balls are of the same color?

9. Two boys and two girls are lined up randomly in a row. What is the probability that the girls and boys alternate?

10. Five distinct dice are rolled. What is the probability of getting at least one "6"?

11. If a school has 100 students of whom 50 take French, 40 take Spanish, and 20 take French and Spanish, then what is the probability that a randomly chosen student takes no language?

12. What is the probability that an integer between 1 and 50 inclusive is divisible by 3 or 4?

13. There are three urns each with two balls: one urn has two black balls, the second has two red balls, and the third a black and a red ball. If an urn is randomly chosen and a randomly chosen ball in that urn is red, what is the probability that the other ball in this urn is also red? (*Hint:* Let the sample space be the set of all experimental outcomes where a red ball is chosen first.)

14. What should the sample space be for the following problem: An urn has 10 black balls and 5 red balls; 4 balls are removed *but not seen* (our eyes are shut) and then a fifth is removed and observed; now what is the probability that the fifth ball is red?

15. Are the elementary events in the following experiments equally likely?
    (a) A fair coin is tossed until two heads occur.
    (b) A positive integer is chosen at random.
    (c) Two positive integers are chosen at random and their difference is recorded.
    (d) A ball is randomly chosen from an urn with two red and two black balls and this ball is replaced and a ball chosen again until a red ball is obtained.

A.4 THE PIGEONHOLE PRINCIPLE

The pigeonhole principle is one of the most simpleminded ideas imaginable, and yet its generalizations involve some of the most profound and difficult results in all of combinatorial theory. This topic is included in these appendices because it does not fit naturally into any chapter of this book. More mathematical texts on combinatorics devote a whole chapter to Ramsey theory—for example, see Cohen's *Basic Techniques of Combinatorial Theory* (John Wiley) (Chapter 5). Also see Graham, Rothchild, and Spencer, *Ramsey Theory* (John Wiley).

Pigeonhole Principle

If there are more pigeons than pigeonholes, then some pigeonhole must contain two or more pigeons. More generally, if there are more than k times as many pigeons as pigeonholes, then some pigeonhole must contain at least $k + 1$ pigeons.

This principle is also called the Dirichlet drawer principle. An application of it is the observation that two people in New York City must have the same number of hairs on their heads. New York City has over 7,500,000 people and the average scalp contains 100,000 hairs. Indeed, the pigeonhole principle allows us to assert that it is theoretically possible to fill a subway car with about 75 New Yorkers all of whom have the same number of hairs on their head.

The following three problems suggest some of the diverse generalizations of the pigeonhole principle.

1. How large a set of distinct numbers between 1 and 200 is needed to assure that two numbers in the set have a common divisor?

2. How large a set of distinct numbers between 1 and n is needed to assure that the set contains a subset of five equally spaced numbers a_1, a_2, a_3, a_4, a_5, that is, $a_2 - a_1 = a_3 - a_2 = a_4 - a_3 = a_5 - a_4$?

3. Given a positive integer k, how large a group of people is needed to assure that either there exists a subset of k people in the group who all know each other or there exists a subset of k people none of whom know each other?

In Problem 3, it might be possible that for large values of k, there are groups of arbitrarily large size that do not meet the k mutual friends or k mutual strangers property. Such is not the case. There is a theorem stating that there exists a finite lower bound r_k such that any group with at least r_k people must satisfy this property. The theorem does not say what number r_k is. Actually, r_k has been determined for only a few values of k.

Most generalizations of the pigeonhole principle require special, research-level skills rather than the combinatorial problem-solving logic and techniques that this text seeks to develop. However, we include the basic pigeonhole principle in this text because it can occasionally be very helpful. The following examples illustrate an obvious and two not so obvious uses of the principle.

Example 1

There are 20 small towns in a region of west Texas. We want to get three people from one of these towns to help us with a survey of their town. If we go to any particular town and advertise for helpers, we know from past experience that the chances of getting three respondents is poor. Instead, we advertise in a regional newspaper that reaches all 20 towns. How many responses to our ad do we need to assure that the set of respondents will contain three people from the same town?

By the pigeonhole principle, we need more than $2 \cdot 20 = 40$ responses.

■

Example 2

We have 15 minicomputers and 10 printers. Every five minutes, some subset of the computers requests printers. How many different connections between various computers and printers are necessary to guarantee that if 10 (or less) computers want a printer there will always be connections to permit each of these computers to use a different printer?

Using a variant of the pigeonhole principle, we see that there must be at least 60 connections. Otherwise one of the 10 printers, call it printer A, would be connected to 5 (or less) computers and if none of the 5 computers connected to A were in the subset of 10 computers seeking printers, then printer A could not be used by any of these 10 computers. It happens that exactly 60 connections, if property made, will solve the connection problem (see Exercise 13).

■

Example 3

Show that any collection of 8 positive integers whose sum is 20 has a subset summing to 4.

We show that the collection must have one of the following four subsets with a sum of 4: (i) four 1's; (ii) two 2's; (iii) two 1's and a 2; (iv) a 1 and a 3. That one of these possibilities must hold is a Pigeonhole-type argument.

If S contained no 1 or 2, then its sum would be at least $3 \cdot 8 = 24$. So by the Pigeonhole Principle, S must contain a 1 or a 2.

Suppose S contains a 2. We are finished if there is a second 2 or if there are also two 1's. Possibly there is one 1, but all other integers are at least 3. If there is no 1, the other 7 integers, each at least 3, must sum to $20 - 2 = 18$; if one 1, the other 6 integers, each at least 3, must sum to $20 - 2 - 1 = 17$. By the Pigeonhole Principle, neither sum is possible.

Suppose S contains at least one 1 but no 2. If there is a 3 in S, case (iv) applies. If there are four 1's in S, case (i) applies. The alternative is no 2's and no 3's in S and at most three 1's. Then there are at least five other integers in S of size at least 4 whose sum must be less than 20—impossible. ∎

EXERCISES

1. Given a group of n women and their husbands, how many people must be chosen from this group of $2n$ people to guarantee the set contains a married couple?

2. Show that at a party of 20 people, there are 2 people who have the same number of friends.

3. In a round-robin tournament, show that there must be 2 players with the same number of wins if no player loses all matches.

4. Given 10 French books, 20 Spanish books, 8 German books, 15 Russian books, and 25 Italian books, how many books must be chosen to guarantee there are 12 books of the same language?

5. If there are 48 different pairs of people who know each other at a party of 20 people, then show that some person has 4 or fewer acquaintances.

6. A professor tells three jokes in his ethics course each year. How large a set of jokes does the professor need in order never to repeat the exact same triple of jokes over a period of 12 years?

7. Show that given any set of seven distinct integers, there must exist two integers in this set whose sum or difference is a multiple of 10.

8. Suppose the numbers 1 through 10 are randomly positioned around a circle. Show that the sum of some set of three consecutive numbers must be at least 17.

9. A computer is used for 99 hours over a period of 12 days, an integral number of hours each day. Show that on some pair of 2 consecutive days, the computer was used at least 17 hours.

10. Show that any subset of eight distinct integers between 1 and 14 contains a pair of integers k, ℓ such that k divides l.

11. Show that any subset of n distinct integers between 2 and $2n$ ($n \geqslant 2$) always contains a pair of integers with no common divisor.

12. Show that any set of 16 positive integers (not all distinct) summing to 30 has a subset summing to n, for $n = 1, 2, \ldots , 29$.

13. In Example 2, this section, find the required set of 60 computer-printer connections.

14. There used to be 6 computers and 10 printers in a large computing center (each computer used at most one printer at a time but the printers

often broke down). Now the 10 old printers are being replaced by 6 more reliable printers, but temporarily the computers will still be allowed to believe that there are 10 printers. Four dummy printers will pass on computer requests to the 6 other real printers. How many connections between the 4 dummy printers and the 6 real printers are needed to handle any set of 6 printing requests?

15.   Two circular disks each have 10 0's and 10 1's spaced equally around their edges in different orders. Show that the disks can always be superimposed on top of each other so that at least 10 positions have the same digit.

16.   A student will study basketweaving at least an hour a day for seven weeks, but never more than 11 hours in any one week. Show that there is some period of successive days during which the student studies a total of exactly 20 hours.

17.   Show that any sequence of $n^2 + 1$ distinct numbers contains an increasing subsequence of $n + 1$ numbers or a decreasing subsequence of $n + 1$ numbers.

18.   Show that at any party with at least six people, there either exists a set of three mutual friends or a set of three mutual strangers.

A.5   MASTERMIND

This appendix discusses the most popular logical-puzzle game of the 1970s — Mastermind. The counting problems associated with Mastermind are interesting applications of the formulas and counting principles introduced in Chapter 5.

Mastermind is a game manufactured by Invicta Ltd. of England. The game has attained such popularity in England that in 1975 a British National Mastermind Championship was held with overflow crowds. The objective of the game is to guess a secret code consisting of colored pegs. The secret code is a row of four pegs that may be chosen (with repeats) from the colors red (R), white (W), yellow (Y), green (G), blue (Bu), and black (Bk).

Each guess of a possible secret code is scored to give some information about how close the guess is to the real secret code. Specifically, the player who chose the secret code indicates (1) how many of the code pegs in the guess are of the right color *and* in the right position in the row and (2) how many of the code pegs are of the right color (occur somewhere in the secret code) but in the wrong position. These two pieces of information are recorded in the form of black keys (markers) for item 1 and white keys for item 2. The game ends when the secret code has been correctly guessed, that is, a score of four black keys is given to a guess. The following examples indicate how this scoring procedure works.

Example 1

Secret Code	R	Bu	Y	Y	Score	Comments
First guess	Bu	W	G	Y	●○	A code peg can receive at most one key; so the Y earns one black key.
Second guess	Y	G	G	Y	●○	
Third guess	Bk	G	G	W		A null score is actually very helpful; it eliminates three colors.
Fourth guess	Bu	Bu	R	R	●○	There is only one blue peg in the secret code, and so only one blue peg earns a key; because one blue peg is in the right position, we score one black key.

This game has many good associated counting problems. How many secret codes are possible, how many different sets of black and white scoring keys, how many secret codes are possible given a first guess and its scoring?

This game is conceptually similar to several substantive problems in information classification and retrieval. In computer recognition of human speech, chemical compounds, or other complex data sets, a number of cleverly planned tests must be performed on the data. Compilers perform a variety of sequential tests, usually in the form of binary search trees (discussed in Chapter 3), to identify commands in the text they are compiling. Many of the counting problems associated with efficient recognition require the same reasoning and techniques as the Mastermind counting problems.

Although we justify our discussion of Mastermind in terms of related counting problems, the game itself provides enjoyable recreation and we encourage readers to play it with a friend. In the absence of a playing companion, the table below presents an example of a game in which enough guesses have been made and scored to enable one to determine the secret code.

Example 2

The following guesses and scoring have occurred. Readers should try to determine the secret code themselves before reading our solution.

	Positions	Scoring
Guess 1	W G Bu R	●○
Guess 2	Bk Y R Bu	●○
Guess 3	R RG Y	○○○
Guess 4	Y RR W	●●

The critical color is red. Guess 3 with two reds and three scoring keys indicates that there must be at least one red peg in the secret code, but since the keys are white, the red(s) must be in positions 3 or 4. However, guess 4 has only black pegs, and red pegs are in positions 2 and 3. Then there cannot be two (or more) red pegs in the secret code; more specifically, we know from guess 3 that there cannot be a red peg in position 2 and so the red peg in this position in guess 4 is unscored (or its score would be a white peg). Thus the secret code must have exactly one red peg, and from guess 4 that red peg must be in position 3.

Guess 3 now implies that the green and yellow pegs must have earned two of the three white keys (since one of the reds received no key). If yellow is in the secret code, then guess 4 tells us that a yellow must be in position 1. So the secret code is of the form Y_R_. Since the two correct colors in guesses 2 and 4 are now known to be red and yellow, the other colors in these two guesses, black, blue, and white, cannot be in the secret code. Moreover, yellow cannot also occur in position 2 or 4, because it is used there in guesses 2 and 3, respectively, without earning a black key (the black key in guess 2 came from the red in position 3). Then the only possibility for the Secret Code is YGRG. ∎

More information about strategies for playing Mastermind can be found in *The Official Mastermind Handbook* by Ault.

EXERCISES

1. Determine the secret code for
G	Y	R	Bu	○○○○
G	Bu	R	Y	○○●●

2. Determine a possible secret code for (there are two possibilities: one of the two has no repeated colors)
Bk	Bu	Y	W	○○
R	W	Bu	G	○○●
Bu	G	R	G	○○
Y	R	W	Bu	○○●

3. Determine the secret code for
Bu	Bu	Bk	W	○●●
Y	W	Bk	W	●●
W	Bu	R	G	○●
Y	Bu	Bk	R	●●

4. Determine the two possible secret codes for
R	G	Y	W	○○
Y	Bk	W	Bu	○●
G	R	R	G	○●
Bk	Bu	G	R	○●
W	R	Bk	R	○

5. Find a fourth guess whose scoring will allow you to determine the secret code for

G	Y	Bk	R	●●
Y	Bu	G	W	○●
Bu	W	Y	Y	○

6. Find a fourth guess whose scoring will allow you to determine the secret code for

G	R	Bu	W	●
G	Y	Bu	G	●
Y	Bk	Bk	Y	○

7. A seventh color, orange (O) has been added in this game. Determine the secret code for

O	R	Bu	G	○○
Y	O	W	Y	○
R	Bk	Bu	G	○○○
O	R	R	Y	○●
Bk	Bu	O	R	○○

8. A seventh color, orange (O), has been added. Determine the two possible secret codes for

O	R	G	Bu	○○
Y	Y	W	O	○●
R	Bu	Bu	Bk	○●
Bk	O	Y	R	○●
Bu	Bk	O	Y	○○●

9. A seventh color, orange (O), has been added. Determine the seven possible secret codes for

O	Y	W	Bk	○
W	Bk	O	R	○●
G	O	R	R	●
Y	R	Bu	O	○○

10. There are now five positions and two extra colors: orange (O), and pink (P). Determine the secret code for

Bk	G	G	O	O	○●●
W	O	Bk	G	R	○●●
O	Bk	P	Y	Bu	○●
G	P	Y	R	W	○○

11. There are now five positions and three extra colors: orange (O), pink (P), and violet (V). Determine the secret code for

R	V	Bk	G	O	○○
V	R	V	R	Y	○●
Bk	W	W	Bu	Bk	●
P	O	P	O	G	○○
G	Y	Bu	W	P	○○●

W P V Bu Y ○○●
Y Bu R V Bk ○○○

12. Find the probability that your initial guess in a Mastermind game is correct.

13. (a) What collections of four or less black and white keys can never occur as Mastermind scores?
    (b) How many different collections of black and white keys can occur as Mastermind scores?

14. We call two guesses in Mastermind *similar* if one can be obtained from the other by a permutation of positions and/or a permutation of colors. How many different nonsimilar guesses are there?

15. Suppose your first guess uses four different colors and the score is four white keys. How many different secret codes are possible? (Note that all four-color guesses are similar; see Exercise 14 above.)

16. Suppose your first guess uses three different colors (one color is repeated) and its score is one black and three whites. How many different secret codes are possible? (Note that all three-color guesses are similar; see Exercise 14 above.)

17. Suppose your first guess uses at least two colors and its score is no keys. What is the minimum number of secret codes that are eliminated by this guess?

18. Consider the simplified Mastermind game in which there are four pegs, each of a different color and the secret code consists of some arrangements of these four colors. Develop a complete strategy for playing this game so that you can determine the secret code after at most three guesses (by the fourth guess, you get a score of four black keys).

19. Consider the simplified Mastermind game in which there are three positions and three colors of pegs. Find an optimal first guess for this game. A guess is *optimal* if for some number k, all possible scores of the guess leave at most k possible secret codes, and some possible score of any other guess leaves at least k possible secret codes.

20. Consider the simplified Mastermind game in which there are four positions but only two colors of pegs. Find an *optimal* first guess for this game (see Exercise 19 above).

21. Show that no matter what the scores of these four guesses, any secret code can be correctly guessed (using the scores of these first four guesses) in at most two more guesses.
    R   R   W   W
    R   Bk R   Bk
    Bu Bu G   G
    Bu Y   Bu Y

22. Write a computer program to make up secret codes and score guesses.

23. Write a computer program to determine an optimal first guess for the standard Mastermind game (see Exercise 19 above).

24. Write a computer program to play Mastermind.

Bibliography

GRAPH THEORY AND ENUMERATION COMBINED

G. Berman and K. Fryer, *Introduction to Combinatorics*, Academic Press, New York, 1969.

K. Bogart, *Introductory Combinatorics*, Pitnam, Marshfield, Mass., 1983.

R. Brualdi, *Introductory Combinatorics*, North Holland, New York, 1977.

D. Cohen, *Basic Techniques of Combinatorial Theory*, John Wiley & Sons, New York, 1978.

L. Lovasz, *Combinatorial Problems and Exercises*, North Holland, New York, 1979.

C. L. Liu, *Introduction to Combinatorial Mathematics*, McGraw-Hill, New York, 1968.

F. Roberts, *Applied Combinatorics*, Prentice-Hall, Englewood Cliffs, N.J., 1984.

GRAPH THEORY

M. Behzad, G. Chartrand and L. Lesniak-Foster, *Graphs and Digraphs*, Wadsworth, Belmont, Calif., 1979.

C. Berge, *Graphs and Hypergraphs*, North Holland, New York, 1973.

N. Biggs, E. Lloyd and R. Wilson, *Graph Theory 1736–1936*, Oxford University Press, New York, 1976.

J. Bondy and U. Murty, *Graph Theory and Applications*, American Elsevier, New York, 1976.

M. Capobianco and J. Molluzzo, *Examples and Counterexamples in Graph Theory*, North Holland, New York, 1978.

G. Chartrand, *Graphs as Mathematical Models*, Prindle, Weber and Schmidt, Boston, 1977.

N. Chrisofides, *Graph Theory: An Algorithmic Approach*, Academic Press, New York, 1975.

N. Deo, *Graph Theory with Applications to Engineering and Computer Science*, Prentice-Hall, Englewood Cliffs, New Jersey, 1974.

J. Graver and M. Watkins, *Combinatorics with Emphasis on the Theory of Graphs*, Springer-Verlag, New York, 1977.

F. Harary, *Graph Theory*, Addison-Wesley, Reading, Mass., 1969.

O. Ore, *Graphs and Their Uses*, Math. Assoc. of America, Washington, 1963.

F. Roberts, *Discrete Mathematical Models*, Prentice-Hall, Englewood Cliffs, New Jersey, 1976.

R. Trudeau, *Dots and Lines*, Kent State Press, Kent, Ohio, 1976.

R. Wilson, *Introduction to Graph Theory*, Academic Press, New York, 1972.

ENUMERATION

C. Berge, *Principles of Combinatorics*, Academic Press, New York, 1972.

M. Eisen, *Elementary Combinatorial Analysis*, Gordon Breach, New York, 1969.

M. Hall, *Combinatorial Theory*, Ginn Blaisdell, Boston, 1967.

A. Nijenhuis and H. Wilf, *Combinatorial Algorithms*, Academic Press, New York, 1974.

J. Riordan, *An Introduction to Combinatorial Analysis*, John Wiley & Sons, New York, 1958.

H. Ryser, *Combinatorial Mathematics*, Math. Assoc. of America, Washington, 1963.

A. Street and W. Wallis, *Combinatorial Theory*, Charles Babbage, London, 1975.

N. Vilenkin, *Combinatorics*, Academic Press, New York, 1971.

W. Whitworth, *Choice and Chance*, Hafner Press, New York, 1901.

Solutions to Selected Exercises

CHAPTER 1

Section 1.1:   1. (b) 2, (c) yes—one possibility is C-I90-E-I91-D-I82-B-I77-A-I85-F; 2. (b) *ADCB, BADC, CBAD, DCBA, DBAC*; 3. (a) a 5-vertex circuit; 5. (a) vertex = Board of Directors, edge = joins boards with common member; 6. (b) two days, Mary or Wendy; 8. (a) 2; 9. (b) 6; 13. (a) *bg, bf, fc, fd, ce*, (c) 3; 17. (b) block: 8, corner: 4; 20. (a) *A-c, B-a, C-b, D-d*, (b) none; 22. (a) *a-b, c-d, e-f*, (c) none; 26. (a) 20, (b) 10; 29. (a) yes, (b) yes, (c) no; 34. yes.

Section 1.2:   3. no; 5. (a) no, (b) yes: *a*-5, *b*-6, *c*-2, *d*-1, *e*-4, *f*-3, (c) no (see complement), (d) no; 7. first, second, and third graphs mutually isomorphic; 11. both 3; 13. (a) possible, (b) not possible.

Section 1.3:   1. (b) 9, (c) 8 or 10 or 20 or 40; 2. (a) yes, (c) no; 4. $n - 1$.

Section 1.4:   3. (a) nonplanar ($K_{3,3}$), (b) nonplanar ($K_{3,3}$); 10. (a) 1, (b) 1; 11. (a) (i) yes, (ii) yes; 13. $R = E - V + C + 1$.

Suppl. II: 1. 21; 3. 12; 9. 11; 10. (a) yes, (b) no; 12. (a) possible, (b) not possible; 30. (a) a 5-vertex circuit; 32. (a) possible, (b) not possible.

CHAPTER 2

Section 2.1:   2. (a) n odd, (b) only K_2; 4. twice; 5. impossible; 14. no; 16. (b) K_4.

Section 2.2:   2. (b) Ham. path *a-d-g-h-e-b-c-f-i*; 9. (c) (ii) $I = \{a,e,f,g,k\}$; 16. (a) $\frac{1}{2}(n - 1)!$

Section 2.3:   1. (a) 3, (b) 4, (c) 2, (d) 4; 3. (a) b, f, g; 8. no; 10. 3; 12. vertex = banquet, edge = joins banquets with common room, color = day of week: can associated graph be 7-colored; 13. (b) associated graph needs four colors.
Section 2.4:   1. (b) $K_{3,3}$ is one possibility.

CHAPTER 3

Section 3.1:   10. $(m^{h+1} - 1)/(m - 1)$; 12. $n - t$; 17. (a) internal vertices = arithmetic operations, (b) 7; 21. 8; 23. check if $X = A$, then if $X = B$, then $X = C$, etc.
Section 3.3:   2. 20: 1-4-2-6-3-5 or 1-6-3-2-4-5; 7. (a) 1-2-4-3.
Section 3.4:   1. (a) spanning tree is simply a path; 6. (a) $a, b, d, e, h, l, i, f, c, g, j, k$; 23. $n - 1$.
Section 3.5:   5. (a) 1, 8, 2, 7, 6, 3, 4, 5; 14. (b) $O(n \log_2 n)$.

CHAPTER 4

Section 4.1:   1. 14; 2. (a) 19 $(a, f, g, l, q, v, w, x, y)$, (b) 21; 3. (a) 31, (b) 32; 5. (a) 5, (b) 6.
Section 4.2:   1. (a) 39 (N, b), (b, c), (c, d), $\{(d, e)$ or $(g, h)\}$, (d, h), (e, f), (e, g), (f, i), $\{(g, j)$ or $(h, k)\}$, (j, k), (j, m), (m, R); 2. (a) total cost is 59; 4. total cost is 63; 5. total length is 133.
Section 4.3:   2. (a) maximal flow is 21, (b) maximal flow is 24; 3. maximal flow is 50; 4. (a) maximal flow is 19; 6. yes; 8. (a) not possible, maximal flow is 50, (b) possible; 9. five messengers possible; 11. (a) 3; 12. maximal flow is 9; 14. 3.
Section 4.4:   6. no; 8. 3.

CHAPTER 5

Section 5.1:   1. (a) 12, 56, (b) 3, 15; 3. 1920; 4. 1296; 9. (a) 12, (b) 14, (c) 196, (e) 122; 15. (a) 90,000, (b) 45,000, (c) $4 \cdot 8 \cdot 9^3 + 9^4 = 29{,}889$; 19. (a) $10 \cdot 9 \cdot 8^2 = 5760$, (b) $\frac{1}{8}$; 23. 96; 20. 125.

Section 5.2:   1.   52!;   4.   9!,   4!5!;   7. $\binom{10}{5}$, $\binom{8}{3}$;   11.   (a) $\binom{4}{2} \left[\binom{6}{4} + \binom{6}{5} + \binom{6}{6} \right] + \binom{4}{3}\binom{6}{6}$; 14. $(13!)^4(4!)^{13}/52!$; 16. (a) $\frac{1}{3}$; 23. (a) $\frac{1}{15}$; 29. (a) (i) 152; 42. (a) $2^{11} = 2048$, (b) 2048; 46. $1 - (P(365,25)/365^{25})$; 53. (a) $13 \binom{4}{2}\binom{12}{4} \, 4^4 / \binom{52}{6}$.

Section 5.3:   1. 60; 3. (a) 3^6, (b) $90/3^6$; 4. $3(6!/3!3!) + 3!(6!/3!2!1!) + 6!/2^3$; 6. $\binom{10 + 4 - 1}{10}$; 9. (a) $\binom{10 + 3 - 1}{10}$, (b) $\binom{15 + 3 - 1}{15} - \binom{4 + 3 - 1}{4}$, (c) 10; 10.

402;　　17.　　$\binom{8 + 25 - 1}{8}$,　　$\left(1 / \binom{8 + 25 - 1}{8}\right) \sum_{k=0}^{4} \binom{8 - 2k + 12 - 1}{8 - 2k}$.

$\binom{2k + 13 - 1}{2k}$; 22. (a) $\binom{15}{4}\binom{10}{5}$

Section 5.4:　1.　(a) $\binom{40 + 4 - 1}{40}$,　(b) 1,　(c) $\binom{36 + 4 - 1}{36}$; 3.　(a)

$\binom{13}{4}\binom{13}{3}^3 / \binom{52}{13}$,　(c) $4\binom{48}{9} / \binom{52}{13}$; 7. $5!\binom{6}{3}$; 9. (b) $\binom{23 + 5 - 1}{23}$;

10. $\binom{15 + 3 - 1}{15}$; 11. $\binom{95 + 5 - 1}{95}$; 14. (a) $n!/n^n$, (b) $n\binom{n}{2}(n-1)!/n^n$;

17. (a) $\sum_{k=0}^{7}(36 - 5k)$; 23. (a) $\binom{30 + 5 - 1}{30}$, (c) $\binom{40 + 5 - 1}{40} - 5\binom{19 + 5 - 1}{19}$;

35. (a) $31!/11!$.

CHAPTER 6

Section 6.1:　2. (a) $(1 + x + x^2 + x^3 + x^4)^5$, (c) $(x^3 + x^5 + x^7)(x^2 + x^4 + x^6 + x^8)(x^2 + x^3 + \cdots + x^8)^2$; 4. (b) $(x^3 + x^4 + \cdots + x^8)^4$, (d) $(1 + x + \cdots + x^5)(1 + x + x^2 + \cdots)^2$; 5. $(1 + x + x^2 + \cdots)^2(1 + x)^2$, coefficient of x^5; 9. $(1 + x + x^2 + \cdots)^n$; 13. $(x + x^2 + \cdots + x^6)^n$; 16. $(1 + x + \cdots + x^9)^6$; 17. $(1 + x^5 + x^{10} + \cdots)^8$; 19. $(1 + x + x^2 + \cdots)^2(x + x^2 + \cdots)^4$, coefficient of x^{15}.

Section 6.2:　2. $\binom{(r - 40) + 8 - 1}{(r - 40)}$;　4. $\binom{9 + 6 - 1}{9} - \binom{4 + 6 - 1}{4}$;

5. $\binom{10 + 4 - 1}{10} - \binom{4 + 4 - 1}{4}$ 4; 7. (b) $\binom{11 + 4 - 1}{11} - 3\binom{10 + 4 - 1}{10}$, (c)

$\sum_{k=0}^{5}\binom{5}{k}\binom{(12 - k) + 5 - 1}{(12 - k)}$; 8. (a) $(1 - x^{28})/(1 - x^4)$; 14. (a) $\binom{(r - 8) + 8 - 1}{(r - 8)}$,

(b) $\binom{r/2 + 8 - 1}{r/2}$; 18. $\binom{15 + 10 - 1}{15} - 10\binom{9 + 10 - 1}{9} + \binom{10}{2}\binom{3 + 10 - 1}{3}$;

22. $\frac{1}{2}\left[\binom{15 + 6 - 1}{15} - \binom{9 + 6 - 1}{9}6 + \binom{3 + 6 - 1}{3}\binom{6}{2}\right]$.

Section 6.3:　3. $(1 + x + x^2 + x^3)(1 + x^2 + x^4 + x^6)(1 + x^3 + x^6 + x^9) \cdots$;

5. $\dfrac{1}{(1 - x)(1 - x^5)(1 - x^{10})(1 - x^{25})}$; 7. (a) $(1 + x + \cdots)(1 + x^2 + x^4 + \cdots)$
$(1 + x^3 + x^6 + \cdots)$.

Section 6.4:　1. $(1 + x + x^2/2! + \cdots + x^5/5!)^5$; 3. $(1 + x)^5(1 + x + x^2/2! + \cdots)^{21}$; 5. Coefficient of $x^{13}/13!$ in $(1 + x + x^2/2! + x^3/3! + x^4/4!)^{13}$; 6. $4^8 - 3^8 - 8 \cdot 3^7$; 8. $\frac{1}{2}(4^r - 3^r - 2^r + 1)$; 11. $\frac{1}{4}4^r$; 16. $e^{(t-1)\mu}$; 20. $a_r = \sum_{k=0}^{r}\binom{k + n - 1}{k}P(n,k)$.

Section 6.5: 1. (a) $x/(1 - x)^2$, (e) $24x^4/(1 - x)^5$; 2. (a) $\binom{n + 1}{2}$, (e) $24 \binom{n + 1}{5}$; 5. (a) $(4x^2 - 3x + 1)/(1 - x)^3$.

CHAPTER 7

Section 7.1: 1. $a_n = 5a_{n-1}$, $a_1 = 5$; 3. $a_n = a_{n-1} + 2a_{n-2}$; 5. $a_n = 3a_{n-1} + 2a_{n-5} + a_{n-10} + a_{n-25}$; 6. (a) $a_n = a_{n-1} + a_{n-2}$; 7. $a_n = a_{n-1} + a_{n-2}$; 12. $a_n = a_{n-1} + 2(n - 1)$; 14. $a_n = 1.06(a_{n-1} + 50)$; 16. $a_n = a_{n-1} + \binom{n - 1}{1}$; 20. $a_n = (2n - 1)a_{n-1}$; 22. $a_{n,k} = a_{n,k-1} + a_{n-1,k}$; 26. Let $a_{n,k,i}$ ($i = 0$ or 1) be such sequences starting with an i: $a_{n,k,0} = a_{n-1,k,1}$, $a_{n,k,1} = a_{n-1,k,0} + a_{n-1,k-1,1}$; 29. (a) $a_n = 2a_{n-1} + 4^{n-1}$. 36. $a_n = \sum_{k=0}^{n-1} a_k a_{n-k-1}$.

Section 7.2: 1. (b) $a_n = 2n$; 2. $a_n = 2a_{n/2} + 1$, $a_n = n - 1$; 5. $a_n = 2a_{n/2} + n - 1$, $a_n = O(n \log_2 n)$.

Section 7.4: 2. $a_n = 2a_{n-1}$, $a_n = 3 \cdot 2^{n-1}$; 3(a) $a_n = (\frac{2}{5})4^n + (\frac{3}{5})(-1)^n$; 5. $a_n = 2a_{n-1} + a_{n-2}$; $a_n = \frac{1}{4}(2 - \sqrt{2})(1 - \sqrt{2})^n + \frac{1}{4}(2 + \sqrt{2})(1 + \sqrt{2})^n$; 7. $p_n - p_{n-1} = 2(p_{n-1} - p_{n-2})$, $p_n = 3 \cdot 2^n - 2$.

Section 7.5: 2. $a_n = a_{n-1} + 2$, $a_n = 2n$; 5. $a_n = a_{n-1} + 1 + \sum_{k=1}^{n-3} (k(n - 2 - k) + 1)$, $a_n = \binom{n}{4} + \binom{n}{2} - n + 1$; 8. (a) $a_n = 1.08a_{n-1} - 10n$, $a_n = (1000 + 10.8./0064)(1.08)^n - (10/.8)n - 10.8/.0064$; 13. $a_n = 2a_{n-1} - a_{n-2} + 10.2^n$, $a_n = 960n - 20 + 40 \cdot 2^n$.

Section 7.6: 7. $a_n = \sum_{k=2}^{n-1} a_{n+1-k}a_k$, $a_n = \frac{1}{n - 1}\binom{2n - 4}{n - 2}$; 10. (a) $a_n = \sum_{k=0}^{n-1} \binom{n - 1}{k} a_{n-k-1}$ 11. $F_n(x) = F_{n-1}(x) + xF_n(x)$, $F_n = (1 - x)^{-n}$.

CHAPTER 8

Section 8.1: 2. $10^9 - 10!$; 4. $\left[2^8 - \binom{8}{0} - \binom{8}{1}\right]/28$; 9. 100; 11. $10! - 2 \cdot 9! + 8!$; 13. (a) 35, (b) 35; 15. $3^{20} - 3 \cdot 2^{20} + 3$; 21. $\binom{3}{1}6!/2! - (2 \cdot 6!/3! + 6!/2!2!) + 6!/4!$; 23. $(15!/(3!)^5 - 3 \cdot 5! \cdot 10!/(2!)^5 + \binom{3}{2}5!^3 - 5!^3)/(15!/3!^5)$; 26. 100.

Section 8.2: 1. $10^n - 3 \cdot 9^n + 3 \cdot 8^n - 7^n$; 5. (a) $\left(\binom{52}{13} - 4\binom{39}{13} + \binom{4}{2}\binom{26}{13} - \binom{4}{3}\binom{13}{13}\right)/\binom{52}{13}$, (c) $\left(\binom{52}{13} - 4\binom{48}{13} + \binom{4}{2}\binom{44}{13} - \binom{4}{3}\binom{40}{13} + \binom{4}{4}\binom{36}{13}\right)/\binom{52}{13}$; 7. $9!/(3!)^3 - 3 \cdot 7!/(3!)^2 + 3 \cdot 5!/3! - 3!$; 10. (a) $\binom{30 + 4 - 1}{30} - 4\binom{19 + 4 - 1}{19} + \binom{4}{2}\binom{8 + 4 - 1}{8}$; 16. $\approx 26!e^{-1}$; 17. $\approx (10!)^2 e^{-1}$;

21. (a) $\sum_{k=0}^{n-1} \binom{n-1}{k}(n-k)!(-1)^k$; 30. $5^5 - 8 \cdot 5^4 + \binom{8}{2}5^3 - \left(4 \cdot 5^3 + \left(\binom{8}{3} - 4\right)5^2\right) + \left(25 \cdot 5^2 + \left(\binom{8}{4} - 25\right)5\right) - \left(4 \cdot 5^2 + \left(\binom{8}{5} - 4\right)5\right) + 21 \cdot 5$.

Section 8.3: 2. (a) $1 + 8x + 21x^2 + 20x^3 + 6x^4$, (c) $1 + 7x + 12x^2 + 4x^3$; 5. (a) 6/20; 6. Rook Polyn. $= (1 + 5x + 6x^2 + x^3)(1 + 2x)$, Ans. 24.

CHAPTER 9

Section 9.1: 1. only b is equiv. rel.; 2. all are groups except a; 4. (a) $(abcd)$, (c) $(ad)(bc)$, (e) $(ab)(cd)$; 10. (a) π_1, (c) π_2.

Section 9.2: 1. (b) 24, 3. $\frac{1}{3}(3^{15} + 2 \cdot 3^5)$; 7. n even, $\frac{1}{2}(5^n + 5^{n/2})$, n odd, $\frac{1}{2}(5^n + 5 \cdot 5^{(n-1)/2})$; 10. n even, $\frac{13}{3} \cdot 2^{n-1}$, n odd, $\frac{1}{2}(3 \cdot 2^{n-1} + 3 \cdot 2^{(n-1)/2})$.

Section 9.3: 1. 55; 3. (a) 130, (b) 92; 5. (b) $\frac{1}{8}(m^9 + 2m^3 + m^5 + 4m^6)$; 6. (a) $\frac{1}{8}(x_1^4 + 2x_4 + 3x_2^2 + 2x_1^2 x_2)$, (b) 21, (c) 954; 8. (a) (i) 2, (ii) 9; 11. (a) 136.

Section 9.4: 1. $1b^5 + 1b^4w + 2b^3w^2 + 2b^2w^3 + 1bw^4 + 1w^5$; 2. (b) $1b^9 + 1b^8w + 4b^7w^2 + 10b^6w^3 + 14b^5w^4 + 14b^4w^5 + 10b^3w^6 + 4b^2w^7 + 1bw^8 + 1w^9$; 4. $\frac{1}{4}((b + w)^{16} + 2(b^4 + w^4)^4 + (b^2 + w^2)^8)$; 5. (b) $\frac{1}{4}((b + w)^6 + 2(b^2 + w^2)^3 + (b + w)^2(b^2 + w^2)^2)$; 6. (a) $1b^5 + 2b^4w + 3b^3w^2 + 3b^2w^3 + 2bw^4 + 1w^5$; 8. $\frac{1}{24}((b + w)^{12} + 6(b^4 + w^4)^3 + 3(b^2 + w^2)^6 + 6(b + w)^2(b^2 + w^2)^5 + 8(b^3 + w^3)^4)$.

CHAPTER 10

Section 10.1: 1. (a) $s \Rightarrow 1s \Rightarrow 101s \Rightarrow 1011s \Rightarrow 101101s \Rightarrow 101101$; 2. (a) $s \Rightarrow tt \Rightarrow 0t \Rightarrow 001t \Rightarrow 0010$; 4. $s \to 1t0$, $t \to 1$, $t \to 0$, $t \to \emptyset$; 6. $s \to s_1$, $s_i \to 0s_{i+1}$, $s_i \to 1s_{i+1}$, $i = 1, 2, 3, 4, 5$, $s_6 \to \emptyset$.

Section 10.2: 1. (a) $(t_0,-) \xrightarrow{1} (s_1,1) \xrightarrow{1} (s_1,1) \xrightarrow{0} (t_1,1) \xrightarrow{1} (s_2,2) \xrightarrow{1} (s_2,2) \xrightarrow{0} (t_2,2)$; 2. (a) $(s_0,-) \xrightarrow{1} (s_0,Y) \xrightarrow{1} (s_1,Y) \xrightarrow{1} (s_0,Y) \xrightarrow{0} (s_1,Y) \xrightarrow{1} (s_0,Y)$.

Section 10.3: 7. Similar to Example 1 in this section, (1) is now $x_{i,1} \vee x_{i,2}, \bar{x}_{i,1} \vee \bar{x}_{i,2}$. 9. In Example 2, let k run from 1 to $n + 1$ and set $x_{1,n+1}$ True.

CHAPTER 11

Section 11.1: 2. three solutions exist; 3. seven solutions exist; 5. (a) true, (b) false; 6. (a) (a,b), (b,e), (e,g), (g,a), (c,d), (d,h), (h,f), (f,c). (b) no factor.
Section 11.2: 1. (a) a,c or b,d, (b) f; 2. first player wins by reducing pile size to a multiple of 6; 8. (a) either $g(a) = g(c) = 0$, $g(b) = g(d) = 1$ or else $g(a) = g(c) = 1$, $g(b) = g(d) = 0$.
Section 11.3: 1. (a) 7, remove 3 from fourth pile, (b) 0; 2. (a) 0, (b) 2; 4. (a) remove 3 from second or third pile; 9. (a) remove 3 in a line (along an edge of the triangular array).

APPENDIX

Appendix A.1: (a) $\{12,27\}$, (b) $\{2,3,6,7,9,12,15,17,18,21,22,24,27\}$; 2. (a) $A \cap B \cap C$, (b) $\overline{A \cup B \cup C}$; 5. (b) 4; 8. (a) $\overline{A} \cup B$; 17. (b) $(S \cup H \cup C) - (S \cap H \cap C)$.
Appendix A.3: 1. $\frac{1}{2}$; 3. (a) $\frac{1}{6}$, (b) $\frac{1}{2}$; 7. (a) $\frac{1}{9}$; 10. $1 - (\frac{5}{6})^5$; 11. $\frac{3}{10}$; 14. $\frac{1}{3}$, 15. (a) an elementary event is a sequence of flips ending in two heads, for example, THTTHTHH—such elementary events are not equally likely.
Appendix A.4: 1. $n + 1$; 4. 52; 6. 6.
Appendix A.5: 1. $R \, Bu \, G \, Y$; 3. $Y \, Bu \, Bu \, W$; 5. fourth guess must be either $Bu \, Bk \, Bk \, Bu$ or $Bk \, Bu \, Bu \, Bk$; 7. $G \, Y \, R \, Bk$; 12. 1/1296; 15. 9; 17. 1040.

Index